Communications
in Computer and Information Science 1769

Rationale

The CCIS series is devoted to the publication of proceedings of computer science conferences. Its aim is to efficiently disseminate original research results in informatics in printed and electronic form. While the focus is on publication of peer-reviewed full papers presenting mature work, inclusion of reviewed short papers reporting on work in progress is welcome, too. Besides globally relevant meetings with internationally representative program committees guaranteeing a strict peer-reviewing and paper selection process, conferences run by societies or of high regional or national relevance are also considered for publication.

Topics

The topical scope of CCIS spans the entire spectrum of informatics ranging from foundational topics in the theory of computing to information and communications science and technology and a broad variety of interdisciplinary application fields.

Information for Volume Editors and Authors

Publication in CCIS is free of charge. No royalties are paid, however, we offer registered conference participants temporary free access to the online version of the conference proceedings on SpringerLink (http://link.springer.com) by means of an http referrer from the conference website and/or a number of complimentary printed copies, as specified in the official acceptance email of the event.

CCIS proceedings can be published in time for distribution at conferences or as post-proceedings, and delivered in the form of printed books and/or electronically as USBs and/or e-content licenses for accessing proceedings at SpringerLink. Furthermore, CCIS proceedings are included in the CCIS electronic book series hosted in the SpringerLink digital library at http://link.springer.com/bookseries/7899. Conferences publishing in CCIS are allowed to use Online Conference Service (OCS) for managing the whole proceedings lifecycle (from submission and reviewing to preparing for publication) free of charge.

Publication process

The language of publication is exclusively English. Authors publishing in CCIS have to sign the Springer CCIS copyright transfer form, however, they are free to use their material published in CCIS for substantially changed, more elaborate subsequent publications elsewhere. For the preparation of the camera-ready papers/files, authors have to strictly adhere to the Springer CCIS Authors' Instructions and are strongly encouraged to use the CCIS LaTeX style files or templates.

Abstracting/Indexing

CCIS is abstracted/indexed in DBLP, Google Scholar, EI-Compendex, Mathematical Reviews, SCImago, Scopus. CCIS volumes are also submitted for the inclusion in ISI Proceedings.

How to start

To start the evaluation of your proposal for inclusion in the CCIS series, please send an e-mail to ccis@springer.com.

Mohammed Salem · Juan Julián Merelo ·
Patrick Siarry · Rochdi Bachir Bouiadjra ·
Mohamed Debakla · Fatima Debbat
Editors

Artificial Intelligence: Theories and Applications

First International Conference, ICAITA 2022
Mascara, Algeria, November 7–8, 2022
Revised Selected Papers

Editors
Mohammed Salem (iD)
University of Mascara
Mascara, Algeria

Juan Julián Merelo (iD)
University of Granada
Granada, Spain

Patrick Siarry
Université Paris-Est Créteil
Créteil, France

Rochdi Bachir Bouiadjra (iD)
University of Mascara
Mascara, Algeria

Mohamed Debakla
University of Mascara
Mascara, Algeria

Fatima Debbat
University of Mascara
Mascara, Algeria

ISSN 1865-0929   ISSN 1865-0937 (electronic)
Communications in Computer and Information Science
ISBN 978-3-031-28539-4   ISBN 978-3-031-28540-0 (eBook)
https://doi.org/10.1007/978-3-031-28540-0

This Springer imprint is published by the registered company Springer Nature Switzerland AG
The registered company address is: Gewerbestrasse 11, 6330 Cham, Switzerland

Preface

This book contains the proceedings of the first International Conference on Artificial Intelligence: Theory and Applications (ICAITA 22), which was hosted by the University of Mascara, Algeria, from November 7 to November 8th, 2022. It was the first in a series of conferences intended to bring together researchers from all over the world, focusing on innovations in the theory as well as implementation in the area of machine learning and artificial intelligence, which is why the topics covered in the conference include artificial intelligence, natural language processing, knowledge representation and reasoning, intelligent control methods, philosophical, ethical and theoretical foundations of artificial intelligence, pattern recognition and artificial/computer vision, and different approaches to machine learning, including new algorithms.

The launching of the conference reflects the increasing interest in artificial intelligence after the successful creation of new conversational tools, as well as the use of artificial intelligence in generational art. The wide availability of open-source software for the implementation of algorithms has been a watershed event in the whole industry, bringing in a large number of new developers, mainly from outside the world of academia. Learning from them, as well as transferring new results, is one of the main intentions in this new conference series.

Out of 66 papers sent for peer review, just 23 were accepted after a single-blind review by, on average, 3 reviewers. This reflects the quality of this conference, and the commitment of the organizers to create an outstanding meeting point for researchers and practitioners from academia and industry.

We would like to give thanks to everyone who has contributed to ICAITA 22, including the local organization and chairs, the keynote speakers, the authors of accepted papers, as well as all the conference delegates. We really appreciate the efforts of the Keynote speakers: Juan Julián Merelo, University of Granada, Spain, Leila Ben Ayed, University of la Manouba, Tunisia and Mostefa Merah, University of Mostaganem, Algeria.

We expect this to be the first of a successful series of conferences.

January 2023

Juan Julián Merelo
Mohammed Salem

Organization

Honorary Chairs

Samir Bentata   University of Mascara, Algeria
Ahmed Yahiaoui   University of Mascara, Algeria

General Chair

Rochdi Bachir Bouiadjra   University of Mascara, Algeria

Program Committee Chairs

Mohammed Salem   University of Mascara, Algeria
Juan Julián Merelo   University of Granada, Spain
Patrick Siarry   Paris-Est Créteil University, France
Mohamed Debakla   University of Mascara, Algeria
Fatima Debbat   University of Mascara, Algeria
Boudjelal Meftah   University of Mascara, Algeria

Steering Committee

Mohammed Salem   University of Mascara, Algeria
Juan Julián Merelo   University of Granada, Spain
Mohamed Debakla   University of Mascara, Algeria
Rochdi BACHIR BOUIADJRA   University of Mascara, Algeria
Mohamed Rebbah   University of Mascara, Algeria

Organizing Committee Chair

Mohamed Rebbah   University of Mascara, Algeria

Program Committee

Abdeslem Dennai	University of Bechar, Algeria
Sabrina Abid	University of Mascara, Algeria
Adham Alsharakwi	The University of Jordan, Jordan
Ahmed Louazani	University of Relizane, Algeria
Allali Mahamed Abdelmadjid	University of Chlef, Algeria
Antonio Mora	University of Granada, Spain
Ouarda Assas	University of Batna 2, Algeria
Soheyb Ayad	University of Biskra, Algeria
Sami Belkacem	National Higher School of Artificial Intelligence, Algeria
Nadjia Benblidia	Saad Dahlab University-Blida1, Algeria
Khaled Benmeriem	University of Mascara, Algeria
Asmaa Bengueddach	Oran 1 University, Algeria
Mohammed Benhammouda	Djillali Liabes University, Algeria
Nawal Benmoussat	USTO-MB, Algeria
Benyahia Kadda	University of Saida, Algeria
Hadj Ahmed Bouarara	University of Saida, Algeria
Chérifa Boudia	University of Mascara, Algeria
Fatma Boufera	University of Mascara, Algeria
Abdelouahid Bouhouche	Université 20 Août 1955 Skikda, Algeria
Brahami Menaouer	National Polytechnic School of Oran, Algeria
Abderrezak Brahmi	University of Abdelhamid Ibn Badis, Mostaganem, Algeria
Djamel Eddine Chaouch	University of Mascara, Algeria
Claude Tadonki	Mines Paris-Tech, France
Abdelkader Fekir	University of Mascara, Algeria
Djellali Hayet	Badji Mokhtar University, Algeria
Fahsi Mahmoud	Djillali Liabes University, Algeria
Hacene Belhadef	University of Constantine 2-Abdelhamid Mehri, Algeria
Hachem Slimani	University of Bejaia, Algeria
Hafed Zarzour	University of Souk Ahras, Algeria
Hajer Bouzaouache	Ecole Nationale d'Ingénieurs, Tunisia
Hanane Zermane	University of Batna 2, Algeria
Hassen Fourati	Grenoble Alpes University, France
Hedi Tmar	University of Sfax, Tunisia
Djalal Hedjazi	University of Batna 2, Algeria
Amina Houari	University of Mascara, Algeria
Hugues Marie Kamdjou	University of Dschang, Cameroon
Ishfaq Ahmad	University of Texas at Arlington, USA

Ivana Roncevic	Prince Sultan University, Saudi Arabia
Javed Ali Khan	University of Science and Technology Bannu, Pakistan
Jeba Sonia J	SRMIST Kattankulathur, India
Juan Julián Merelo	University of Granada, Spain
Khalifa Djemal	Université d'Évry Val d'Essonne, France
Laid Kahloul	University of Biskra, Algeria
Kamal Kant Hiran	Sir Padampat Singhania University, India
Okba Kazar	University of Biskra, Algeria
Ahlem Kenniche	Université de Mostaganem, Algeria
Khadidja Yahyaoui	University of Mascara, Algeria
Miloud Khaldi	Higher School of Computer Science, Sidi Bel-Abbès, Algeria
Mohamed Fayçal Khelfi	ESGEE Oran, Algeria
Abdelkader Khobzaoui	Djillali Liabes University, Algeria
Ahmed Khorsi	Imam Mohammad Ibn Saud Islamic University, Saudi Arabia
Zineddine Koualma	Université de Guelma, Algeria
Slimane Larabi	USTHB, Algeria
Leila Ben Ayed	University of la Manouba, Tunisia
Lotfi Boudjenah	University of Oran1, Algeria
Loveleen Gaur	Amity University, Noida, India
Mahi Faiza	University of Mascara, Algeria
Mariem Haoues	CCSE-PSAU, Saudi Arabia
Masheal Alghamdi	King Abdulaziz City for Science and Technology, Saudi Arabia
Mostefa Merah	University of Mostaganem, Algeria
Md. Sakir Hossain	American International University, Bangladesh
Mohamed B. Debbat	University of Mascara, Algeria
Mohd Yusuf	Universiti Teknologi Petronas, Malaysia
Djelloul Mokadem	University Moulay Tahar of Saida, Algeria
Monika Bansal	Punjabi University, Patiala, India
Mongi Besbes	University of Carthage, Tunisia
Mourad Loukam	Hassiba Benbouali University of Chlef, Algeria
Muhammad Talha Gul	Sharif College of Engineering and Technology, Pakistan
Mustafa Jarrar	Birzeit University, Palestine
Nawres Khlifa	University of Tunis El Manar, Tunisia
Ouajdi Korbaa	University of Sousse, Tunisia
Rashid Mehmood	King Abdulaziz University, Saudi Arabia
Mohammed Rebbah	University of Mascara, Algeria
Reshma V. K.	Hindusthan College of Engineering and Technology, India

Ruchi Doshi	Universidad Azteca, Mexico
S B Goyal	City University, Malaysia
Sahraoui Mustapha	University of Mascara, Algeria
Sailesh SuryanarayanIyer	Rai University, India
Samir Ladaci	National Polytechnic School of Algiers, Algeria
Seifedine Kadry	Noroff University College, Norway
Soraya Setti Ahmed	University of Mascara, Algeria
Seyedali Mirjalili	Torrens University Australia, Australia
Sofiane Boukli Hacene	Djillali Liabès University, Algeria
Surbhi Gupta	Punjab Agricultural University, India
Hamza Teggar	University of Mascara, Algeria
Najia Trache	Université Oran1 Ahmed Ben Bella, Algeria
Valentina Emilia Balas	Aurel Vlaicu University of Arad, Romania
Vicente Feliu Batlle	Universidad de Castilla-La Mancha, Spain
Wai Lok Woo	Northumbria University, UK
Xiao-Zhi Gao	University of Eastern Finland, Finland
Khadidja Yachba	University of Relizane, Algeria
Yaroub Elloumi	University of Monastir, Tunisia
Youcef Fekir	University of Mascara, Algeria
Mohammed Zagane	University of Mascara, Algeria
Sofiane Zaidi	University of Oum El Bouaghi, Algeria
Nacereddine Zarour	University of Constantine 2, Algeria
Meriem Amina Zingla	University of Mascara, Algeria
Mounir Zrigui	University of Monastir, Tunisia

Additional Reviewers

Mahmoudi Laouni
Cherouati Brahim
Samir Setaouti
Khaldi Brahim
Baligh Babaali

Contents

Evolutionary Algorithms Applications

Artificial Intelligence in Big Data and Natural Language Processing

Artificial Vision

Expanding Convolutional Neural Network Kernel for Facial Expression Recognition

Mohamed Amine Mahmoudi[1]([✉])(ⓘ), Fatma Boufera[1], Aladine Chetouani[2], and Hedi Tabia[3]

[1] Mustapha Stambouli University of Mascara, Mascara, Algeria
mohamed.mahmoudi@univ-mascara.dz
[2] Laboratoire PRISME, Université d'Orléans, Orléans, France
[3] IBISC laboratory, University of Paris-Saclay, Paris, France

Abstract. Facial Expression Recognition (FER) is increasingly gaining importance in various emerging affective computing applications. In this article, we propose a Facial Expression Recognition (FER) method, based on kernel enhanced Convolutional Neural Network (CNN) model. Our method improves the performance of a CNN without increasing its depth nor its width. It consists of expanding the linear kernel function, used at different levels of a CNN. The expansion is performed by combining multiple polynomial kernels with different degrees. By doing so, we allow the network to automatically learn the suitable kernel for the specific target task. The network can either uses one specific kernel or a combination of multiple kernels. In the latter case we will have a kernel in the form of a Taylor series kernel. This kernel function is more sensitive to subtle details than the linear one and is able to better fit the input data. The sensitivity to subtle visual details is a key factor for a better facial expression recognition. Furthermore, this method uses the same number of parameters as a convolution layer or a dense layer. The experiments conducted on FER datasets show that the use of our method allows the network to outperform ordinary CNNs.

Keywords: Emotion recognition · Facial expression recognition · Kernel function · Convolutional neural network

1 Introduction

The recognition of facial expressions is useful in various fields such as human machine interaction, games, alert systems and monitoring of patients, especially those who find it difficult to speak like disabled persons and autistic, to detect feelings of pain for example. The automatic recognition of facial expressions is not a recent field of research, in fact, the first research in this field dates back to a little over twenty years. However, new advances have been made in the last five years in the field of object recognition in general, have further revived research in the field of facial expressions recognition.

The classic facial expression recognition approach has been widely used over the past two decades with different methods at all levels giving varying results.

© The Author(s), under exclusive license to Springer Nature Switzerland AG 2023
M. Salem et al. (Eds.): ICAITA 2022, CCIS 1769, pp. 3–17, 2023.
https://doi.org/10.1007/978-3-031-28540-0_1

However, in the last five years a new approach, Deep Learning, has been used with impressive results. Deep Learning is a set of end-to-end methods, enabling a machine to be supplied with raw data and automatically discovering the representations necessary for detection or classification, with several levels of representation, obtained by composing simple modules. Convolutional Neural Networks (CNNs) are of the widely used models in deep learning and in particular computer vision. Their success was so big that researchers started to apply it in some difficult fields to test its capacity. Fine-grained recognition is an example of these topics. The goal in fine-grained recognition is recognize some classes that were seen as a single class in ordinary computer vision tasks. These classes have some small subtle details that differentiate them. Facial expression recognition is considered one of most challenging fine-grained recognition problems. Indeed, the difference in facial expression categories relies on small subtle areas in the facial images like the mouth, eyebrows and the noise.

To overcome this issue, facial expression recognition systems must be able to recognize these subtle differences efficiently. One of the solutions that computer vision engineers are using to solve this problem is using a large CNN network or replacing convolution with a more sophisticated function. In the former case, either the number of layers is increased (depth) or the length of the output is increased after each layer (width). This solution is working efficiently for a variety of problems and we obtained good results by doing so. However, we cannot keep using this method eternally. One of the drawbacks of this method is the network size (number of weights) which needs a relatively large computer memory to be stored and loaded. The second issue is the complexity which also increases accordingly to the network size and therefore needs highly sophisticated processing units to be run. The second solution is about modifying the underlying function of the network at all levels. This method is gaining interest from computer vision practitioners since it consumes less memory than the first one. even though they are harder to train.

In this article, we propose to enhance CNN performance without increasing the size of the model. Our method consists of replacing the linear kernel function, that is used at different levels of a CNN. The expansion is performed by combining different degrees of polynomial kernel functions. This method allows the network to automatically learn the suitable kernel for a specific task. The network can either uses one specific kernel or a combination of multiple kernels. In the latter case, we will have a Taylor series kernel. This type of kernel function has the advantage of having the ability to detect subtle details more than the linear kernel. The advantage is a key factor for better facial expression recognition. Furthermore, this technique does not use additional parameters or increase the size of the network in the convolution layer or a dense layer.

The remainder of this article is organized as follows: Sect. 2 reviews similar works that have been proposed for CNN improvement using kernel functions. Section 3 introduces the proposed expansion method for convolution and fully connected layers. Section 4 presents our experiments setting, the datasets we used and their related results. Finally, Sect. 5 concludes the article.

2   Related Work

Machine learning based on kernel functions (kernel Support Vector Machines (SVMs) [1], kernel Fisher Discriminant (KFD) [15], and kernel Principal Component Analysis (KPCA) [17]) was used in several fields. The intuition behind these techniques is that they use more complex functions than the linear ones and therefore be more accurate and precise. Deep learning model and more precisely CNNs have achieved very good by using only linear function as their core functions in every level. This big potential encouraged some researcher to include higher degree function in different levels of a CNN to enhance its performance. Recently, Zoumpourlis et al. [21] developed a second-order convolution method by exploring quadratic forms through the Volterra kernels. Constituting a richer function space, this method is used as approximation of the response profile of visual cells. However, in this method the number of parameters increases the training complexity exponentially. Kervolution was proposed by. [19] to overcome this problem in the convolution level of a CNN. It increased the model performance and allows it to capture more subtle details than the ordinary convolution operation.

In our previous studies, we have proposed many contributions that focus on the underlying function of CNN at different levels. We first adapted a improved bilinear pooling method. [11] to the field of facial expression recognition. This method consists of projecting the data before the classification phase into a higher, generating by that more data that helps to have a more accurate classification. The problem with this method is that it cannot be applied earlier in the network architecture. To resolve that problem, we proposed to incorporate a more sophisticated pooling. [10] layer that works similarly to an ordinary pooling layer yet using a more complex function. This layer has the ability to detect small visual details and keeps track of them while reducing the input size of a feature map. Therefore, instead of applying a max or average pooling and eliminate arbitrary useful data, our method learns the appropriate data to keep and the useless data that can be abandoned. After that, we have extended our method to fully connected layer where we have also replaced these layers with higher degree functions instead of linear kernel. These layer are called Kernelized Dense Layers (KDL) [9,12,13] and helps the network to learn to discriminate data in an accurate manner, since the learning will be back propagated to the earlier layers. Finally, we have also conducted some works [8,14] that studies the differences between our methods and the ordinary kernel methods proposed before (like Kernel SVM and Kervolution).

3   Method

A CNN is ruled by a linear kernel function throughout its layers. Whether it is convolution, pooling or fully connected layers. Each layer will receive an input data $x \in \mathbf{R}^n$ in the form of a vector. The later will be fed to the following layer L that apply a vector of weights related to it $W \in \mathbf{R}^n$. The linear function in these layers L will apply the weight vector $W \in \mathbf{R}^n$ to its input vector $x \in \mathbf{R}^n$ using the inner product $\langle .,. \rangle$ as follows:

$$K_{linear}(x, W) = \langle x, W \rangle, \tag{1}$$

The efficiency of linear kernel is particularly observable when the original data is linearly separable. In other words, the data has a high dimensional representation. Here the later can be separated by a linear boundary. However, having a high dimensional representation does not necessarily mean that the data can be linearly separable [16]. One of the most used high dimensional data is images, yet these are composed of pixels that do not provide much information. Moreover, in the case of CNN, we only process one small portion of the image at a time which reduces their dimension drastically. In such a case, a linear kernel function will not be able to sense the subtle visual details.

As described before, recent studies are emerging in which researchers propose new function that may replace the underlying linear functions of a CNN. In general, they focus on using more complex kernel functions with superior degree than the linear kernel. The idea behind these studies is to use these higher degree kernel functions to project the input data to a higher space (called Reproducing Kernel Hilbert Space or RKHS) where these data become more discriminative. In other words, instead of applying the linear classifier on the data, the later is transformed to higher dimensional data by projecting it into a superior feature space. After that, a linear kernel function is then applied to these high degree data which became linearly separable. This new linear classifier in the feature space is equivalent to a higher degree classifier in the first feature space. In such case, we will be able to detect more subtle features than using the ordinary linear classifiers. To map the data to these high degree spaces, we use to different methods according to [3]. We can use the kernel trick in which there is no need for explicit calculation for the high-dimensional vectors in the RKHS. This is called the explicit method and is used in kernel SVMs [2,18]. The following equation show the calculation operated in this method. Let $\varphi(.) : \mathbf{R}^n \longmapsto \mathbb{H}$ represent this RKHS embedding.

$$K(x, W)_{\mathbb{H}} = \langle \varphi(x), \varphi(W) \rangle_{\mathbb{H}}, \tag{2}$$

where $\langle .,. \rangle_{\mathbb{H}}$ denotes the inner product in the Hilbert space $\mathbb{H}$.

Equation 2 differs from Eq. 1 in the use of the inner product $\varphi(x)$ and $\varphi(W)$ instead of the dot product between x and W. This is because the Hilbert space $\mathbb{H}$ can be infinite-dimensional. This method has two main drawbacks according to [3,6]. First of all, the larger is the size of training data the larger will be the needed storage space and the longer time it would take to converge. Deep learning for instance, relies on large datasets. Therefore, such a technique is visibly inefficient. Moreover, using Stochastic Gradient Descent (SGD) for the training worsen the later.

The other way of projecting data into a superior feature space is to explicitly calculate the product of features. The disadvantage of this method is the size of the feature map that will be generated. It is impractical for real world scenarios.

However, even though these methods have some drawbacks concerning the size and the time of the training the resulting model will more efficient than a regular linear model without additional parameters or processing time. In other words, once a model is trained it works with the same storage and processing capacities yet it will be more discriminative.

Our proposition is to take advantage of these kernel properties and enhance the linear underlying function of a CNN ate the level of convolution layers and fully connected layers. This can be done using the two method of projection into a superior feature space namely, explicitly and implicitly. First of all, we implicitly calculate the feature projection using the kernel trick, by using a higher kernel degree function instead of the linear kernel function, as shown in Eq. 2. Second, we project the input data into a higher feature space a second time yet explicitly by computing several kernel functions from the first step and combining them in single output. By doing so, we will have an even higher projection into a more superior feature space. The combination can be done in many ways in this step yet we have chosen to use addition for that purpose. This two step expansion is done on two levels: convolution and fully connected layers. The following sections will describe the processing of each one of them. These obtained layers are similar to the ordinary convolution and fully connected layers in the way that they can be used in any level. Furthermore, they can be used in full extension fashion (stacked above of each other without including linear layers) or a mixed manner where we can use both types of layers to gain the advantages of each type of layers.

3.1   Convolution Layer Expansion

CNNs are named after the convolution function because it is its main function. Convolution layers take into consideration the fact that an image is made up from sub-features like edges, dots…etc. They analyze these sub-features solely first then combine the resulting analysis to create a general analysis about the whole image. This technique allowed CNN to reach results in computer vision that were unreachable before. The underlying analysis performed in the convolution layer is mainly a linear function as shown in Eq. 1. With this linear functions, convolution may not be able to linearly separate input features [6]. Hence our proposition of replacing convolution with a superior kernel function into ways explicit and implicit.

We first use polynomial kernel function with the kernel trick implicitly project the input data into a superior feature space (Eq. 3) as follows:

$$K_{Polynomial}(x, W) = \langle \varphi(x), \varphi(W) \rangle = \sum_{i=2}^{p} (x^T W)^i, \tag{3}$$

where $\varphi(.) : \mathbf{R}^n \longmapsto \mathbf{R}^d (d \gg n)$ is a non-linear mapping function.

This allows us to detect some features that only became visible in the newly generated feature space in which the computing is much complex than complexity of Eq. 1. However, using the kernel trick, we can avoid all these computation of the superior feature space $\varphi(x)$. In this step we use some polynomial kernels which degrees are $p \geq 2$. A single feature map will result of each polynomial kernel. In order to perform the next projection step, these feature maps should have the same height and the same width. However, for the size of the depth (number of channels in the output), no constraint is specified.

After the first step of projection, we will have a set of feature maps that are ready to be projected another time using the chosen combination. In our case, we have chosen addition. Since we have employed polynomial kernel function with ascending degrees in addition to convolution, which a special case of polynomial with degree one. The addition of these kernels will give us a special case kernel which is in the form of Taylor series kernel. Equation 4 describes a Taylor series kernel of order p as follows:

$$K_{Taylor}(x, W) = \langle \varphi(x), \varphi(W) \rangle = \sum_{i=1}^{p} \langle x, W \rangle^{i}, \qquad (4)$$

where for $p = 1$, K_{Taylor} is equivalent to a convolution kernel.

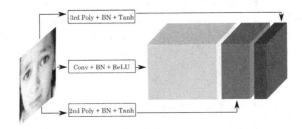

Fig. 1. The proposed expansion method consists of applying higher degree (≥ 2) polynomial kernels to the input, in addition to convolution. The result of these kernels is concatenated over the channel axis.

3.2   Dense Layer Expansion

Fully connected layers are an essential component of CNNs. These layers take as input the feature maps resulting from the successive convolution and pooling layers (commonly referred to as the feature vector f_v of the input image) in order to drive the final classification decision. Similarly to convolution layers, fully connected layers are driven by a linear kernel function (Eq. 1). As described above, this function, despite of being simple and computationally inexpensive, fails to learn fully linearly separable features [6]. Therefore, we propose to expand its linear kernel to a higher degree kernel function as well. This also will be achieved in both explicit and implicit ways.

The first step of fully connected layers is based on the Kernelized Dense Layer (KDL), proposed in [9]. These layers are similar to neuron layers yet the neuron they constitute is different from an ordinary artificial neuron. KDL are composed of kernelized neuron (KN) in which the kernel is a high degree kernel function instead of a linear kernel function. In other words, KN takes the input vector and a vector of weight and applies its underlying higher degree kernel function instead of a regular linear kernel and eventually adds bias ($b \geq 0$) vector.

Here, we also use some polynomial kernel functions (Eq. 3) that will project the input features to a superior dimensional feature space. Yet, in contrary to convolution expansion, the output has no size constraints. After that, project explicitly the resulting vectors. As for convolution expansion we compute simultaneously the linear dot product, which corresponds to the ordinary fully connected layers, in addition to layers of higher degree polynomial kernels. The result of these computations is concatenated to form a single vector. This vector is then in the form of Taylor series kernel. This resulted vector will constitute the input for the next fully connected layer. An illustration of this process is shown in Fig. 2.

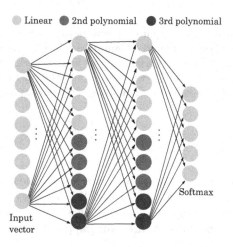

Fig. 2. Expansion of dense layer proceeds in two steps. First, it computes multiple polynomial KDL in addition to the linear layer. Then, it concatenates the resulting vectors in a single expanded dense layer vector. The latter will be finally fed to the next expanded dense layer or dense layer. Best viewed in color.

As illustrated in Fig. 2, we also propose to combine the two expansion methods described above. This combination will result in a fully expanded model in the form of a Taylor series kernel.

4 Experiments

In this section, we shed light on the experimental settings, that have been used to evaluate our expansion method. First, we briefly describe the FER datasets that have been used for the experiments. Then we detail the architecture of the models we have used and their training process. Recall from previous section that our expansion method operates mainly on two type of layers which are convolution and dense layers. Therefore, we study the impact of each expansion method, solely and jointly as shown in Fig. 3, on the network accuracy. This results are compared to the ordinary convolution and dense layers.

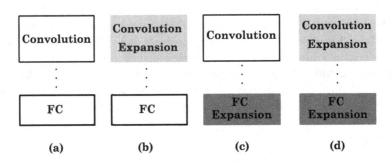

Fig. 3. In our experiments, we compare the results to an ordinary CNN (a) with our proposed method over three different configurations. In the first case (b), we replace convolution by the convolution expansion allover the network followed by fully connected layers. In the second case (c), we replace only the fully connected layers with our expansion method for these layers. Finally, we test our expansion methods allover the network (d).

4.1   Datasets

We have tested our proposed method with three of the most used FER datasets, namely RAF-DB [7], ExpW [20], and FER2013 [4]. They contain respectively 29,672 images, 91,793 images, and 35,000 images. They are all subdivided into seven output classes. These classes are fear, happiness, anger, disgust, surprise, sadness and neutral.

4.2   Models Architecture and Training Process

The assumption made in this article is that linear kernel, used at different levels on a CNN, can achieve better performance if they are expanded using our proposed method. Moreover, this improvement is reached without the need of additional weight parameters. To verify this assumption we compare the performance of an ordinary CNN to a network with the same architecture, yet using our proposed expansion method. For this purpose, we used two pre-trained networks namely, VGG-16 and VGG-19 in addition to a CNN built from scratch. For the pre-trained CNNs we took only the convolution part and added two dense layers of 256 unit each and a final softmax layer with 7 output units. This models will be referred to as VGG-16-bese and VGG-19-bese. Whereas, we will refer to the model built from scratch as Model-1. Model-1 is a five block convolution architecture. These blocks are composed of a convolution layer, on top of which we add a batch normalization layer on top of which there is a RELU layer. After each one of these blocks we add a Max pooling layer followed by a dropout layer. Finally, three fully connected layers are added on top of the last convolution block with respectively 256, 256 and 7 units (Fig. 4). Furthermore, we build expanded models, with the same architecture as the three models described above, following the configuration shown in Fig. 3.

Our training process is the following. We have started with a learning rate of 0.001 using Adam optimizer and gradually decreased the learning rate by 0.5 if the validation accuracy does not improve over after five epochs until reaching a learning rate of 5e−5. We have also used data augmentation with 20° of random rotations, 0.2 of shear intensity, a range of 0.2 of random zoom, and a random horizontal flip. To avoid over fitting, we also used early stopping after 10 epochs if the validation accuracy does not improve by 0.01. We have initialized our layers with the *He* normal distribution [5] and a weight decay of 0.0001. Finally, the input image were cropped around the face region then resized to 100 × 100 pixels.

4.3   Ablation Study

This section explores the impact of the use of the proposed expansion method on the overall accuracy of VGG-16, VGG-19 and Model-1. The obtained results using these models are reported in Tables 1, 2 and 4, as base models. As shown in Tables 1, 2, and 4 with VGG-16-base we have obtained 69.38% on FER2013, 85.42% on RAF-DB, and 77.75% on ExpW. In the case of VGG-19-base, we obtained 69.52%, 85.99%, and 77.92% on FE2013, RAF-DB and ExpW, respectively. Finally, for Model-1 the obtained result are 70.13% for FER2013, 87.05% for RAF-DB, and 75.91% for ExpW.

After that, we evaluated the performance of these network architectures, with our proposed method, on different levels and with different kernel function degrees (Fig. 3). These expanded models of VGG-16, VGG-19 and Model-1, were, as mentioned above, built and trained from scratch. As shown in Fig. 3, we performed CNN layers expansion following three main configurations. First, we used the expansion on the convolution level. After that, we expanded the fully connected layers. Finally, we tested full expansion by combining the two previous expansion methods. We have expanded all the layers following three configuration which are: i) a Taylor series kernel with polynomial of first and second degree with proportions of 80% and 20% respectively; ii) a Taylor series kernel with polynomial of first, second and third degree with proportions of 70% and 20% and 10% respectively; and iii) the linear kernel only. In the first case, for instance, if the depth of the output is 128 channels then 102 channels will be resulting from convolution, 26 resulting from second degree polynomial. In the second case following the same output size then 90 channels will be resulting from convolution, 26 resulting from second degree polynomial and 12 from third degree polynomial. We also used the same proportion in the dense layers expansion and the full network expansion. The sections bellow discuss the obtained results at each level solely, then the results of their combinations.

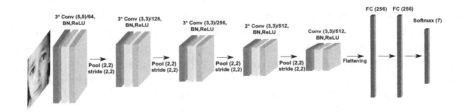

Fig. 4. Model-1 architecture.

Convolution Layer Expansion. Table 1 shows the impact of the convolution expansion on the three models namely, VGG-16 model, VGG-19 model and model-1. First of all, the accuracy rate obtained using the convolution expansion of second degree is 70.07% for FER2013, 86.13% for RAF-DB, and 78.17% for ExpW. This is an enhancement of 0.7% over the full linear model. Whereas, with the convolution expansion of third degree we reached 70.31% for FER2013, 86.24% for RAF-DB, and 78.61% for ExpW. This is an enhancement of 1% more than its full linear counterpart. On the other hand, the results reached with VGG-19 like model are 70.82% for FER2013, 86.44% for RAF-DB, and 78.33% for ExpW, with the convolution expansion of second degree. This is an enhancement of 0.8% more than the linear kernel model. Whereas, with the convolution expansion of third degree, the obtained results are 71.08% for FER2013, 87.04% for RAF-DB, and 79.16% for ExpW, which represent an enhancement of 1.5% compared to its linear counterpart. Finally, Model-1 reached 70.84% for FER2013, 87.69% for RAF-DB, and 76.52% for ExpW, with the convolution expansion of second degree. This is an improvement of 0.6% more than the linear kernel model. Whereas, with the convolution expansion of third degree, the obtained results are 71% for FER2013, 87.84% for RAF-DB, and 79.71% for ExpW. Therefore, this an improvement of 0.8% compared to its linear counterpart.

Dense Layer Expansion. First of all, as shown in Table 2, the results obtained with VGG-16 like model are 70.98% for FER2013, 86.89% for RAF-DB, and 78.85% for ExpW, with the dense layer expansion of second degree. This is an enhancement of 1.3% more than its full linear counterpart. Whereas, with the dense layer expansion of third degree, the obtained results are 71.58% for FER2013, 87.04% for RAF-DB, and 79.16% for ExpW. This is an improvement of 2% over the full linear model. On the other hand, the results obtained with VGG-19 like model are 71.46% for FER2013, 87.18% for RAF-DB, and 79.21% for ExpW, with the dense layer expansion of second degree. Similarly to VGG-16, the use of this is an enhancement of 1.9% more than the linear kernel model. Whereas, with the dense layer expansion of third degree, the obtained results are 71.95% for FER2013, 87.29% for RAF-DB, and 79.39% for ExpW. This is an improvement of 2.5% over its linear counterpart. Finally, the results obtained with Model-1 are 71.41% for FER2013, 88.26% for RAF-DB, and 76.11% for

Table 1. Results of convolution layer expansion method.

Level	Model	Kernel	FER2013	RAF	ExpW
VGG-16-base			69.38%	85.42%	77.75%
VGG-19-base			69.52%	85.99%	77.92%
Model-1			70.13%	87.05%	75.91%
Conv-Expansion	VGG-16	Conv-2nd	70.07%	86.13%	78.17%
		Conv-3rd	**70.31%**	**86.24%**	**78.61%**
	VGG-19	Conv-2nd	70.82%	86.44%	78.33%
		Conv-3rd	**71.08%**	**86.57%**	**78.71%**
	Model-1	Conv-2nd	70.84%	87.69%	76.52 %
		Conv-3rd	71%	87.84%	76.71 %

ExpW, with the dense layer expansion of second degree. This is an improvement of 1.3% over its linear counterpart. With the dense layer expansion of third degree, Model-1 reached 71.86%, 88.59% and 76.63% on FE2013, RAF-DB and ExpW, respectively. This is an improvement up to 1.8% over its linear counterpart.

Table 2. Results of dense layer method.

Level	Model	Kernel	FER2013	RAF	ExpW
VGG-16-base			69.38%	85.42%	77.75%
VGG-19-base			69.52%	85.99%	77.92%
Model-1			70.13%	87.05%	75.91%
Dense-Expansion	VGG-16	Dense-2nd	70.98%	86.89%	78.85%
		Dense-3rd	**71.58%**	**87.04%**	**79.16%**
	VGG-19	Dense-2nd	71.46%	87.18%	79.21%
		Dense-3rd	**71.95%**	**87.29%**	**79.39%**
	Model-1	Conv-2nd	71.41%	88.26%	76.11 %
		Conv-3rd	71.86%	88.59%	76.63%

To further evaluate the efficiency of dense layer expansion, we tested these layers solely in an MLP fashion. The goal, here, is not achieve state-of-the-art or competitive results. It is rather to demonstrate the improvement that an expanded dense layer can bring when used, solely, in an MLP fashion. We built an MLP with two hidden layers of 256 units each and a softmax output layer of 7 units. We followed the same configuration as the previous dense layer expansion, that is: i) a full linear MLP; ii) an MLP with a Taylor series kernel up to the second degree; and ii) an MLP with a Taylor series kernel up to the third degree. Given the small size of these MLP networks, We resized the

FER dateset images to 48 × 48 pixels. The results of these configuration are reported in 3. As one can see in Table 3, the linear MLP achieved 43.18%, 51.29%, 39.65% on FE2013, RAF-DB and ExpW, respectively. Whereas, the MLP with Taylor series expanded dense layers up to the second degree reached 45.32% on FER2013, 51.84% on RAF and 40.11% on ExpW. This is represents an improvement up to 2.12% in accuracy rate compared to its linear counterpart. Finally, the MLP with Taylor series expanded dense layers up to the third degree reached 45.80%, 52.26%, 40.81% on FE2013, RAF-DB and ExpW, respectively. Once again, the MLP with Taylor series expanded dense layers up to the third degree outperformed the other expansion kernels and enhanced the accuracy for about 2.62%. This short experiment show the inherent descriminative power of neuron with a kernel of high degree. In other words, it show the enhancement that such a neuron can bring at any level in any type of neural network.

Table 3. Results of dense layer as MLP.

Kernel	FER2013	RAF	ExpW
Linear-MLP	43.18%	51.29%	39.65%
Dense-2nd-MLP	45.32%	51.84%	40.11%
Dense-3nd-MLP	45.80%	52.26%	40.81%

Full Expansion. After testing expansion on the two main levels of the CNN, namely: convolution layer and fully connected layer, we have also tested the efficiency of this expansion method jointly on these layer types. As shown in Table 4, the results obtained with VGG-16 like model are 70.22% for FER2013, 86.17% for RAF-DB, and 78.32% for ExpW, with the full expansion of second degree. This is an improvement of the accuracy of the model up to 0.8% more than its full linear counterpart. Whereas, with the full expansion of third degree, the obtained results are 70.28% for FER2013, 86.20% for RAF-DB, and 78.41% for ExpW. This is an improvement of 0.9% over the full linear model. On the other hand, the results obtained with VGG-19 like model are 70.91% for FER2013, 86.48% for RAF-DB, and 78.52% for ExpW, with the full expansion of second degree. Similarly to VGG-16, the use of this kernel also enhances the accuracy for VGG-19 of 1.5% more than the linear kernel model. Whereas, with the full expansion of third degree, the obtained results are 70.97% for FER2013, 86.53% for RAF-DB, and 78.63% for ExpW. This is an improvement up to 1.6% over its linear counterpart. Unfortunately, the expansion to the third degree Taylor series kernel does not enhance much the overall accuracy of the network compared with the second degree expansion.

Table 4. Results of full expansion method.

Level	Model	Kernel	FER2013	RAF	ExpW
VGG-16-base			69.38%	85.42%	77.75%
VGG-19-base			69.52%	85.99%	77.92%
Model-1			70.13%	87.05%	75.91%
Full-Expansion	VGG-16	Full-2nd	70.22%	86.17%	78.32 %
		Full-3rd	70.28%	86.20%	78.41%
	VGG-19	Full-2nd	70.91%	86.48%	78.52%
		Full-3rd	70.97%	86.53%	78.63%
	Model-1	Conv-2nd	70.95%	87.78%	76.68%
		Conv-3rd	71.12%	87.93%	76.84%

The obtained results with our expansion method show that the use of higher order kernel along with the linear kernel is beneficial to the overall accuracy of the network. However, the impact of our method changes according to the level where it is applied. For instance, the use of our expansion method at the convolution level increases the accuracy according to the kernel degree. The higher is the kernel degree, the better is the accuracy. Similarly to the convolution layers, the fully connected layers also increase the accuracy rates according to the kernel degree. However, the use of our expansion method on the fully connected layers performs better than the convolution expansion. Finally, even though the full network expansion increases the overall accuracy over the linear model, it slightly outperforms the convolution expansion method. Furthermore, the increase of the accuracy with respect to the kernel degree is not truly perceptible. Also, the use of the full expansion seems to be prone to over-fitting. Taking into consideration the computation cost of a full expansion, the latter seems to be less useful than the precedent uses of our expansion method.

5   Conclusion

In this article, we proposed to improve CNN performance without increasing the number of learnable parameters. Our method consists of expanding the linear kernel function, used at different levels of a CNN. The expansion is performed by combining multiple polynomial kernels with different degrees. By doing so, the network automatically learns the suitable kernel that optimizes the target objective. In our settings, a network can either use a single kernel or a combination of multiple ones which make a Taylor series kernel. We demonstrated that the used kernel function is more sensitive to subtle details than the linear one. This is important for fine-grained classification in particular it increases both the representation and the classification power of the CNN for facial expression recognition. The experiments conducted on FER datasets showed that the use of our method allows the network to outperform conventional CNNs. The obtained results showed that the use of higher order kernel along with the linear one is beneficial to the overall accuracy of the network. However, we noticed that the position of the plugged kernel impacts on the accuracy of the full network. For instance, the kernel expansion at the convolutional level increases the accuracy according to the kernel degree. We also observed that, the fully connected layers react similarly as the convolutional layers, to the kernel degree. However, the kernel expansion used on the fully connected layers performs better than any convolutional layer in the network. We finally observed that, even though the full network expansion increases the overall accuracy over the linear model, it slightly outperforms the convolution. Furthermore, in the full network expansion, the increase of the accuracy with respect to the kernel degree is not truly perceptible. Also, the use of the full expansion seems to be prone to over-fitting. Taking into consideration the computation cost of a full expansion, one may prefer to use either the convolutional layer expansion or the fully connected layer expansion taken separately.

References

1. Burges, C.J., Scholkopf, B., Smola, A.J.: Advances in Kernel Methods: Support Vector Learning. MIT Press, Cambridge (1999)
2. Cortes, C., Vapnik, V.: Support-vector networks. Mach. learn. **20**(3), 273–297 (1995)
3. Cui, Y., Zhou, F., Wang, J., Liu, X., Lin, Y., Belongie, S.: Kernel pooling for convolutional neural networks. In: Proceedings of the IEEE Conference on Computer Vision and Pattern Recognition, pp. 2921–2930 (2017)
4. Goodfellow, I.J., et al.: Challenges in representation learning: a report on three machine learning contests. In: Lee, M., Hirose, A., Hou, Z.-G., Kil, R.M. (eds.) ICONIP 2013. LNCS, vol. 8228, pp. 117–124. Springer, Heidelberg (2013). https://doi.org/10.1007/978-3-642-42051-1_16
5. He, K., Zhang, X., Ren, S., Sun, J.: Delving deep into rectifiers: surpassing human-level performance on imagenet classification. In: Proceedings of the IEEE International Conference on Computer Vision (ICCV) (2015)
6. Jayasumana, S., Ramalingam, S., Kumar, S.: Kernelized classification in deep networks. arXiv preprint arXiv:2012.09607 (2020)
7. Li, S., Deng, W., Du, J.: Reliable crowdsourcing and deep locality-preserving learning for expression recognition in the wild. In: 2017 IEEE Conference on Computer Vision and Pattern Recognition (CVPR), pp. 2584–2593. IEEE (2017)
8. Mahmoudi, M.A., Chetouani, A., Boufera, F., Tabia, H.: Deep kernelized network for fine-grained recognition. In: Mantoro, T., Lee, M., Ayu, M.A., Wong, K.W., Hidayanto, A.N. (eds.) ICONIP 2021. LNCS, vol. 13110, pp. 100–111. Springer, Cham (2021). https://doi.org/10.1007/978-3-030-92238-2_9
9. Mahmoudi, M.A., Chetouani, A., Boufera, F., Tabia, H.: Kernelized dense layers for facial expression recognition. In: 2020 IEEE International Conference on Image Processing (ICIP), pp. 2226–2230 (2020)
10. Mahmoudi, M.A., Chetouani, A., Boufera, F., Tabia, H.: Learnable pooling weights for facial expression recognition. Pattern Recogn. Lett. **138** (2020)
11. Mahmoudi, M.A., Chetouani, A., Boufera, F., Tabia, H.: Improved bilinear model for facial expression recognition. Pattern Recogn. Artif. Intell. **1322**, 47 (2021)

12. Mahmoudi, M.A., Chetouani, A., Boufera, F., Tabia, H.: Taylor series kernelized layer for fine-grained recognition. In: 2021 IEEE International Conference on Image Processing (ICIP), pp. 1914–1918. IEEE (2021)
13. Mahmoudi, M.A., Chetouani, A., Boufera, F., Tabia, H.: Kernel-based convolution expansion for facial expression recognition. Pattern Recogn. Lett. **160**, 128–134 (2022)
14. Mahmoudi, M.A.: Deep learning for emotion recognition. Ph.D. thesis (2022)
15. Mika, S., Ratsch, G., Weston, J., Scholkopf, B., Mullers, K.R.: Fisher discriminant analysis with kernels. In: Neural networks for signal processing IX: Proceedings of the 1999 IEEE Signal Processing Society Workshop (cat. no. 98th8468), pp. 41–48. IEEE (1999)
16. Robert, C.: Machine learning, a probabilistic perspective (2014)
17. Schölkopf, B., Smola, A., Müller, K.R.: Nonlinear component analysis as a kernel eigenvalue problem. Neural Comput. **10**(5), 1299–1319 (1998)
18. Schölkopf, B., Smola, A.J., Bach, F., et al.: Learning with Kernels: Support Vector Machines, Regularization, Optimization, and Beyond. MIT Press, Cambridge (2002)
19. Wang, C., Yang, J., Xie, L., Yuan, J.: Kervolutional neural networks. In: Proceedings of the IEEE Conference on Computer Vision and Pattern Recognition, pp. 31–40 (2019)
20. Zhang, Z., Luo, P., Loy, C.C., Tang, X.: From facial expression recognition to interpersonal relation prediction. Int. J. Comput. Vis. **126**(5), 550–569 (2017). https://doi.org/10.1007/s11263-017-1055-1
21. Zoumpourlis, G., Doumanoglou, A., Vretos, N., Daras, P.: Non-linear convolution filters for cnn-based learning. In: Proceedings of the IEEE International Conference on Computer Vision, pp. 4761–4769 (2017)

Robust Method for Breast Cancer Classification Based on Feature Selection Using RGWO Algorithm

Ali Mezaghrani[1]([✉]), Mohamed Debakla[1], and Khalifa Djemal[2]

[1] Faculty of Science Exact, University of Mustapha Stambouli Mascara, Mascara, Algeria
{Ali.mezaghrani,Debakla_med}@univ-mascara.dz
[2] IBISC Laboratory, Evry Val d'Essone University, Evry, France

Abstract. Breast cancer is a leading cause of mortality in women all over the world. According to the worldwide cancer statistics, early detection and treatment are keys components for improving the recovery rate of breast cancer and lowering the death rate. Machine learning solutions have been proved to be particularly very successful in exploring the origins of such severe diseases, which requires processing vast amounts of data.

In the present study, robust grey wolf optimisation-Random Forest (RGWO-RF) approach was proposed. Our proposed approach based on two steps feature selection process and classification. Modified Grey Wolf Optimizer is used to locate and determine the most significant features. Then, utilizing the prior optimum selections of features, by using Random Forest (RF) classifier to classify breast cancer disease. The reason for using RF it's robustness and highest accuracy.

We apply the proposed approach on Wisconsin Diagnostic Breast Cancer (WDBC) database. The experimental result improve that the hybridation between RGWO for feature selection and RF classifier increase the accuracy rate of classification and demonstrating it's robustness in identifying the breast cancer.

Keywords: Breast cancer · Grey wolf optimizer · Feature selection · Random Forest

1 Introduction

Breast cancer disease has been considered as one of the deadly disease in the world [1]. To increase the odds of survival and save more women's lives, early detection of breast cancer is critical and very important factor in the diagnostic. Breast cancer identification requires precise categorization of the tumor as benign or malignant [2]. Several methods proposed by researchers to enhance the classification capability of their system of breast cancer diagnosis. But still, there is a huge opportunity to create a breast cancer categorization system that is more efficient.

In this study we try to develop an approach which effectively classifies the breast cancer tumor using RGWO for feature selection and RF for disease classification. The suggested strategy aims to extract the most significant and optimum subset of features

© The Author(s), under exclusive license to Springer Nature Switzerland AG 2023
M. Salem et al. (Eds.): ICAITA 2022, CCIS 1769, pp. 18–27, 2023.
https://doi.org/10.1007/978-3-031-28540-0_2

from the dataset which helps to make an efficient and effective classification of breast cancer. In this study, the RGWO-RF method is suggested to identify the ideal set of features that would improve the RF classifier classification performance. The RF classifier will be trained on the optimized subset of features identified by RGWO.

The WDBC breast cancer dataset, which has a total of 569 instances and 33 characteristics. The WDBC dataset accessible through the UCI Machine Learning Repository [3], was used to test the suggested methodology.

The experimental result shown that the proposed system increases the accuracy rate when we use RGWO for feature selection. The suggested method outperformed recent studies, obtaining an accuracy of 98.60%, Precision (98.1%), F1-Score (98.1%), Sensitivity (98.1%) and a Specificity value of 98.9%.

The remainder of the essay is structured as follows: Sect. 2 discusses relevant research and several cutting-edge methods for diagnosing breast cancer. The description of the GWO algorithm and the Mathematical model of GWO. In Sect. 4, We provide a thorough explanation of the strategy we suggest. Experimentation and debate are covered in Sect. 5. Finally, in Sect. 6, we conclude our paper with a summary and outlooks for future work.

2 Related Works

The current section provides a summary of the techniques and algorithms used in the suggested study. Based on feature selection and machine learning methodologies, several strategies have been established to identify breast cancer.

Based on tumor traits, the study in [4] sought to make a diagnosis of breast cancer. K-means and K-SVM were combined in order to extract meaningful information from WDBC dataset. Hidden patterns of benign and malignant tumors are found using the K-means method. The outcomes showed that the suggested approach was effective in diagnosing breast cancer while also reducing training time.

Using various data mining approaches, the study in [5] aimed to evaluate the likelihood of developing breast cancer as well as the likelihood of the disease returning. The Wisconsin dataset of UCI machine learning was used to collect cancer patient data. The results show that the Naive Bayes and the decision tree algorithm are more accurate and deliver superior outcomes.

Breast cancer detection was investigated using the SVM approach in [6]. The accuracy, ROC, measurement, and computational time of training were employed as benchmarks in this study. The results supported the SVM algorithm's better performance over other classification methods.

Dora et al. [7] suggested a new technique for calculating the ideal weight coefficients in order to train samples termed GNRBA (Gauss–Newton representation-based approach). The purpose of this method is to reduce computing complexity while also reducing reaction time. GNRBA beats the previous methods in both the UCI cancer datasets.

Shahnaz et al. [8] performed and examined many statistical and deep learning data studies on a dataset of breast cancer cases in order to improve the classification accuracy through feature selection.

Li et al. [9] introduced a medical diagnostic system that used the grey wolf optimizer with the kernel extreme learning machine to determine the ideal feature subset for medical data. With the use of the wrapper approach and a novel fitness function,

Liu et al. [10], presented a novel breast cancer intelligent detection method, Implementing a feature selection process using Information gain directed simulated annealing genetic algorithm wrapper (IGSAGAW). In this procedure, they rank the features using the IG method, and then they use the cost-sensitive support vector machine (CSSVM) learning algorithm to extract the top m optimum features. The efficacy of the suggested approach is tested on Wisconsin Original Breast Cancer (WBC) and Wisconsin Diagnostic Breast Cancer (WDBC) breast cancer data sets, and the outcomes show that the suggested hybrid algorithm works better than existing techniques.

A chaotic crow search algorithm technique, which was a meta-heuristic optimizer, was recommended by Sayed et al. [11] to address the issue of poor convergence rate.

A technique for determining the ideal qualities for a decision tree's input using a bee colony algorithm was described by Rao et al. [12], along with a way for generating decisions using an artificial bee colony algorithm.

In [13], To handle feature selection for classification issues based on wrapper approaches employing KNN classifier, M. Abdel-Basset and D. El-Shahat developed a novel Grey Wolf Optimizer algorithm coupled with a Two-phase Mutation. To demonstrate the effectiveness and performance of the suggested method, statistical studies was performed.

From the mini MIAS dataset, 80 digital mammograms of normal breasts, 40 benign cases, and 40 malignant cases were selected in [14]. In this study, comparison of classification process performance of support vector machines (SVM), artificial neural networks (ANN), then a hybrid SVM-ANN model was used to create a computer-aided detection (CAD) system, and the later model demonstrated a respectable accuracy of 98%.

In [15], In order to accurately identify benign and malignant tumors, the best group of traits must be chosen. an enhanced GWO has been suggested with SVM applying on WDBC dataset, Experimental result show that the new method improve accuracy by 98.24%.

3   Grey Wolf Optimizer

The Grey Wolf Optimizer, developed by Seyedali Mirjalili et al. in 2014 [16], is a population-based meta-heuristics algorithm that mimics the natural leadership structure and hunting behavior of grey wolves. A social hierarchy is observed within the grey wolf group. According to their functions within the pack, the wolves in the pack are assigned positions in the hierarchy. The wolf pack is typically divided into four different groups: alpha (α), beta (β), delta (δ), and omega (ω), depending on how each wolf participates to the hunting process. Figure 1 shows the social hierarchy of the grey wolves, Alpha (α) wolf is the pack commander and should be obeyed by the other wolves in the pack. The second place of hierarchy is occupied by beta wolves (β), Beta wolf assist the alpha in making decisions and are seen to be the best candidate to be the alpha wolf. Delta (δ) wolf represent the third rank in the pack, must subordinate to the alpha and beta, but they rule the omega (ω). Omega (ω) wolves are the least significant members of the pack and are only permitted to eat last, occupy the lowest rung of the hierarchy. They are viewed as the scapegoats in the pack.

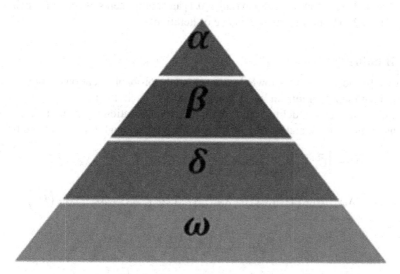

Fig. 1. Grey wolf social structure.

3.1 Mathematical Model of the GWO

In a hunting formation, the pack leaders serve as the spearheads. Once a broad location is established for the prey, they send the omega wolves to encircle it, getting closer as the precise position is sought. The wolves strike after completely encircling their victim.

The procedure might be broken down into three separate parts to represent this as a mathematical model.

3.1.1 Encircling

In the GWO algorithm, hunting is guided by alpha, beta, and delta wolves, and Omega wolves follow them.

The mathematical simulation of grey wolves encircling is shown in Eqs. (1) and (2):

$$\vec{X}(t+1) = \vec{X}_P(t) + \vec{A} \cdot \vec{D} \tag{1}$$

$$\vec{D} = \left| \vec{C} \cdot \vec{X}_P(t) - \vec{X}(t) \right| \tag{2}$$

where t represent the current iteration, A and C are coefficient vectors, $\vec{X}_P$ is the position vector of the prey, and $\vec{X}$ indicates the position vector of a grey wolf. Following are the calculations for the vectors A and C:

$$\vec{A} = 2\vec{a} \cdot \vec{r_1} - \vec{a} \tag{3}$$

$$\vec{C} = 2 \cdot \vec{r_2} \qquad (4)$$

where r1 and r2 are random vectors in range [0,1] and components of vector a are linearly reduced from 2 to 0 throughout the course of iterations.

3.1.2 Hunting

Because α, β, and δ are more knowledgeable about the probable locations of prey, omega wolves adjust their positions in line with α, β, and δ in each iteration.

The mathematical model to adjust a search agent's location in accordance with the positions of alpha, beta, and delta search agents is represented by the equations below:

$$\vec{D_\alpha} = \left| \vec{C_1}.\vec{X_\alpha} - \vec{X} \right|, \ \vec{D_\beta} = \left| \vec{C_2}.\vec{X_\beta} - \vec{X} \right|, \ \vec{D_\delta} = \left| \vec{C_3}.\vec{X_\delta} - \vec{X} \right| \qquad (5)$$

$$\vec{X_1} = \vec{X_\alpha} - A_1 \cdot \left(\vec{D_\alpha} \right), \ \vec{X_2} = \vec{X_\beta} - A_2 \cdot \left(\vec{D_\beta} \right), \ \vec{X_3} = \vec{X_\delta} - A_3 \cdot \left(\vec{D_\delta} \right) \qquad (6)$$

$$\vec{X}(t+1) = \frac{\vec{X_1} + \vec{X_2} + \vec{X_3}}{3} \qquad (7)$$

3.1.3 Attaking

The coefficient vector A plays a crucial role in the GWO Algorithm, is a random value in the range [−a,a] where a decreases from 2 to 0 throughout the duration of iterations. When random values of $\vec{A}$ are in [−1,1], the next position of a search agent can be in any position between its current position and the position of the prey. If |A| < 1 compels the wolves to attack towards the prey (**Exploitation**), |A| > 1 compels the grey wolves to diverge from the prey to find a best prey (**Exploration**), there is an other component favors the exploration is $\vec{C}$. In contrast to A, the vector C does not decrease linearly and has random values between [0, 2]. However, C may also be thought of as natural impediments that prevent approaching to the prey [16].

4   Proposed Approach

The GWO algorithm has become very popular among other swarm intelligence techniques, Due to its many benefits, including its ease of use, scalability, and most importantly, its capacity to deliver faster convergence by maintaining the proper balance between exploration and exploitation during the search. We are aware that the placements of the alpha, beta, and delta search agents influence where the prey will be found in the best possible way. These top three search agents are in charge of pointing all other search agents in the right direction for the best prey. In order for the other search agents to be effectively led to approach the prey, it is crucial to ensure that these three search agents are the fittest in each iteration. From Eqs. (5) and (6) it can be seen that each search agent position is updated in relation to the locations of the alpha, beta, and delta search agents. The Wisconsin Diagnostic Breast Cancer (WDBC) dataset, which comprises 569 instances and 32 characteristics, was used in the current study to evaluate the effectiveness of the suggested methodology. The WDBC dataset was retrieved from the UCI Machine Learning Repository. A digital fine needle aspirate (FNA) scan is used to determine the breast mass attribute. The characteristics depict several parameters that might be helpful in determining whether a tumor is benign or malignant. The RGWO-RF approach is suggested to choose the best subset of characteristics that would produce the highest level of classification accuracy to obtain better results the following points was implemented:

- We use a metaheuristic algorithm to eliminate redundant features and irrelevant features to increase classification accuracy and minimise time consuming in classification process, RGWO algorithm was implemented for variable selection on WDBC dataset, then we use RF classifier to classifier breast cancer based on the subset of optimal features gained by RGWO algorithm. The features with the highest accuracy of classification and the fewest number of selected characteristics is the optimal and the best. The fitness function which is also employed in RGWO to assess the selected features and utilized to optimize classification accuracy, is indicated as Eq. (8):

$$Fitness = w * accuracy + (1 - w) * 1/(len(features)) \tag{8}$$

- In relation to the locations of alpha, beta, and delta wolves, the basic GWO updates the position of search agents wolves (Omega wolves). The three best locations of grey wolves are averaged for the position update of search agent (Eq. (7)). This approach results in early convergence and poor solutions. The update method of the positions should not be considered the same in Eq. (7). In this work, To enhance base GWO performance, We use weighted position update concept which proposed by S. Kumar, M. Singh [15], Then we implement a RGWO algorithm in combination with RF classifier and modify the update technique of the position. The mathematical model of weighted position update technique is represented in Eqs. (9) and (10).

$$W1 = A1 * C1 \quad W2 = A2 * C2 \quad W3 = A3 * C3 \tag{9}$$

$$X(t + 1) = (W1 * X1 + W2 * X2 + W3 * X3)/(W1 + W2 + W3) \tag{10}$$

4.1 Methodology

As is depicted in Fig. 2, in this work, we implement two scenario, For the diagnosis of breast cancer, the first one with three major phases was proposed. We start with data preprocessing which is a common phase for two scenario, which mean data cleaning and filtering were performed to avoid the establishment of ineffective rules and patterns. The breast cancer dataset was preprocessed in this article, and outliers were removed using the outer line approach, then classification using RF classifier of breast cancer on all features of WDBC dataset, the third phase is the assessment of classification accuracy. **The second scenario** consisting four main phases, data preprocessing. Then Using RGWO for feature selection, feature selection was used to identify the key characteristics of a given outcome. After that, RF was used for classification process and in the end, the evaluation of classification accuracy.

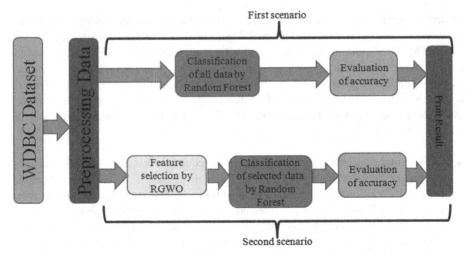

Fig. 2. Suggested methodology for accurately classifying breast cancer tumors.

5   Experimental Result

By employing the RGWO to reduce the dimensions of features, the primary goal of the current study was to increase diagnostic accuracy while enhancing classification performance. In the experiments, the number of iterations for RGWO algorithm has been fixed at 20 iteration with 10 search agents. We used RF classifier in order to classify the data between malignant and benign tumors, RF is regarded as a very reliable and precise technique. In order to choose the appropriate subset of features, a hybrid strategy using the RF classifier and the RGWO produced the best results. In the proposed approach, the results presented in Table 1 show that when we use dimensionality reduction with RGWO-RF algorithm The Sensitivity, Specificity, Precision, F1-Score, and Accuracy was increased.

5.1   Comparison of the Suggested RGWO-RF Method with the Base GWO-RF

Table 2 compare the RGWO-RF method performance with that of the standard GWO-RF technique. Use of the weighted position update technique in basic GWO is being evaluated through this comparison. The findings have improved in terms of accuracy, F1-score, and sensitivity, as can be seen from the table, which employed the RGWO in the suggested technique.

5.2   Comparing the Suggested Methodology to the Current Feature Selection Methods

The suggested approach was compared with current methods for feature selection-based breast cancer detection approaches in Table 3. It is evident that the recommended RGWO-RF technique outperforms all of the approaches that were considered.

Table 1. Classification results of the suggested RGWO-RF approach using different performance measures.

Performance measures	Classification results (%)	
	Without feature selection	Feature selection with RGWO-RF
Sensitivity	96,3	98,1
Specificity	97,8	98,9
Precision	96,3	98,1
F1-score	96,3	98,1
Accuracy	97,2	98,6

Table 2. Comparison of proposed RGWO-RF approach with the base GWO-RF approaches.

Performance measures	Classification results (%)	
	Proposed RGWO-RF	Base GWO-RF
Sensitivity	98,1	96,3
Specificity	98,9	98,9
Precision	98,1	98,1
F1-score	98,1	97,2
Accuracy	98,6	97,9

Table 3. Evaluation of the proposed RGWO-RF methodology by comparing results with existing feature selection methods.

Approaches	Authors	Years	Number of features	Accuracy %
FS-KNN	Sayed et al. [11]	2019	14	90,28
FS-GBDT	Rao et al. [12]	2019	14	92.80
FS-KNN	Abdel-Basset et al. [13]	2020	16	94,82
FS+EGWO-SVM	S. Kumar and M. Singh [15]	2021	6	98,24
Proposed approach	Proposed	2022	12	98,60

6 Conclusion

The presence of a large number of variables is not always correlated with improved classification performance, as some of them may be redundant, irrelevant, or a source of noise. As a result, a Feature Selection phase is frequently applied to high-dimensional datasets.

In this paper, we proposed a Robust GWO in conjunction with RF classifier. The later has been used to get the best parameters for our new approach. We have shown that this step increases the accuracy of the GWO and hence reduces the mortality rate. PYTHON and WDBC datasets were used to get experimental findings. We discovered that the outcomes of our proposed technique outperform other efforts in term of accuracy measurement, Specificity and Precision. In the near future, we plan to adopt this approach in the diagnosis of other disease with other dataset like heart diseases dataset and diabetes dataset.

References

1. Bray, F., Ferlay, J., Soerjomataram, I., Siegel, R.L., Torre, L.A., Jemal, A.: Global cancer statistics 2018: GLOBOCAN estimates of incidence and mortality worldwide for 36 cancers in 185 countries. CA Cancer J. Clin. **68**(6), 394–424 (2018)
2. Ades, F., et al.: Luminal breast cancer: Molecular characterization, clinical management, and future perspectives. J. Clin. Oncol. **32**, 2794–2803 (2014)
3. Dua, D., Gra®, C.: UCI Machine Learning Repository. University of California, School of Information and Computer Science, Irvine, CA (2019). http://archive.ics.uci.edu/ml
4. Zheng, B., Yoon, S.W., Lam, S.S.: Breast cancer diagnosis based on feature extraction using a hybrid of K-means and support vector machine algorithms. Expert Syst. Appl. **41**(4), 1476–1482 (2014)
5. Pritom, A.I., Munshi, M.A.R., Sabab, S.A., Shihab, S.: Predicting breast cancer recurrence using effective classification and feature selection technique. In: 2016 19th International Conference on Computer and Information Technology (ICCIT), pp. 310–314. IEEE, New York (2016)
6. Huang, M.W., Chen, C.W., Lin, W.C., Ke, S.W., Tsai, C.F.: SVM and SVM ensembles in breast cancer prediction. PLoS One **12**(1), e0161501 (2017)
7. Dora, L., Agarwal, S., Panda, R., Abraham, A.: Optimal breast cancer classification using Gauss-Newton representation based algorithm. Expert Syst. Appl. **85**, 134–145 (2017)
8. Shahnaz, C., Hossain, J., Fattah, S.A., Ghosh, S.: Efficient approaches for accuracy improvement of breast cancer classification using Wisconsin database. In: IEEE Region 10 Humanitarian Technology Conference (R10-HTC) (2017)
9. Li, Q., et al.: An enhanced grey wolf optimization based feature selection wrapped kernel extreme learning machine for medical diagnosis. Comput. Math. Methods Med. **2017**, 1–15 (2017). https://doi.org/10.1155/2017/9512741
10. Liu, N., Qi, E., Xu, M., Liu, G.: A novel intelligent classification model for breast cancer diagnosis. Inf. Process. Manage. **56**, 609–623 (2019)
11. Sayed, G.I., Hassanien, A.E., Azar, A.T.: Feature selection via a novel chaotic crow search algorithm. Neural Comput. Appl. **31**(1), 171–188 (2017). https://doi.org/10.1007/s00521-017-2988-6
12. Rao, H., Shi, X., Rodrigue, A., Feng, J., Xia, Y.: Feature selection based on artificial bee colony and gradient boosting decision tree. Appl. Soft Comput. **74**, 634–642 (2019)
13. Abdel-Basset, M., El-Shahat, D., El-Henawy, I., Mirjalili, S.: A new fusion of grey wolf optimizer algorithm with a two-phase mutation for feature selection. Expert Syst. Appl. **139**, 112824 (2020)
14. Lim, T.S., Tay, K.G., Huong, A., Lim, X.Y.: Breast cancer diagnosis system using hybrid support vector machine-artificial neural network. Int. J. Electr. Comput. Eng. **11**(4), 3059 (2021). https://doi.org/10.11591/ijece.v11i4.pp3059-3069

15. Kumar, S., Singh, M.: Breast cancer detection based on feature selection using enhanced grey wolf optimizer and support vector machine algorithms. Vietnam J. Comput. Sci. **8**(2), 177–197 (2021)
16. Mirjalili, S., Mirjalili, S.M., Lewis, A.: Grey wolf optimizer. Adv. Eng. Softw. **69**, 46–61 (2014)

Convolution Neural Network Based Approach for Glaucoma Disease Detection

Sahraoui Mustapha$^{(\boxtimes)}$ ⓘ, Salem Mohammed ⓘ, and Smail Oussama

University of Mascara, Mascara, Algeria
{sahraoui.musta,salem}@univ-mascara.dz, abdou94220@live.fr

Abstract. Glaucoma is among the discreet and incurable eye diseases; it causes a slow and progressive decay of the retina of human eyes. The absence of clear symptoms during the early stages makes it hard to detect. This paper is dedicated to design a convolutional neural network (CNN) based approach to detect and diagnose glaucoma based on the processed funds images. It is divided into two phases: Learning and Classification. Results are promising to detect such a disease from retina images even when compared to other architectures.

Keywords: Artificial intelligence · Glaucoma · Convolutional neural networks (CNN) · Optic nerve head (ONH)

1 Introduction

Within the field of ophthalmology, glaucoma could be a family of eye illnesses that influence the optic nerve. Optic nerve harm is caused by raised IOP in grown-ups and people over 60 a long time ancient [1, 3]. Numerous diverse methods have been created to distinguish diverse infections of retina as glaucoma. Almazroa et al. [8] have been able to prepare a machine to investigations utilizing the freely accessible RIGA (Retinal-fundus Pictures for Glaucoma Butt-centricysis) information set with impressive precision. Hatanaka et al. [9] were able to consequently degree the CDR by performing online profile examination of retinal pictures. Artifacts other than OC and OD can be utilized to identify glaucoma from a retinal picture. Within the audit conducted by Haleem et al. [10], different anatomical features to help early discovery of glaucoma have been talked about Optic nerve corruption happens within the region of the Optic circle (OD) known as the optic nerve head (ONH). Ophthalmology is the medical specialty that makes greater use of artificial intelligence. While the use of classification and segmentation algorithms based on machine learning and computer vision techniques demonstrated effective and correct diagnosis of this eye disease [17].

Convolutional neural networks (CNNs) with more hidden layers have a more complicated network structure and stronger feature learning and feature expression capabilities than conventional machine learning techniques as a result of the advent of the massive data era.

The remainder of this paper is divided into five sections. Brief description of glaucoma disease is presented in the second section. The principal ideas of concepts of CNN

M. Salem et al. (Eds.): ICAITA 2022, CCIS 1769, pp. 28–39, 2023.
https://doi.org/10.1007/978-3-031-28540-0_3

are presented in the third section, where CNN based glaucoma approach is explain in section four. The principle of image classification is given in the Sect. 5. In Sect. 6 the implementation of the CNN-based glaucoma detection approach is validated by Python simulations. The last section is intended for a comparative analysis between the proposed approach and the VGGNet-16 architecture. We have obtained reliable and precise results for several cases.

2 Glaucoma Disease

In the field of ophthalmology, the optic nerve is impacted by the family of eye conditions known as glaucoma. Intraocular disorders (eye pressure, IOP) are linked to glaucoma but can also be caused by high blood pressure, migraines, ethnicity and family history. Optic nerve damage is caused by elevated IOP [1, 3] in adults and humans over 60 year. Optic nerve degradation occurs in the region of the Optic disc (OD) known as the optic nerve head (ONH). All forms of glaucoma are incurable and their damage is mostly irreparable.

Abating the rate of movement of this malady is among the as it were alternatives accessible to patients. Subsequently, the viability of all medicines depends on early conclusion of the malady. The nonattendance of self-evident indications causes be that as it may, early conclusion is uncommon. The investigation of fundus pictures permitted us to analyze glaucoma, in specific by measuring the sizes of the OC and OD (a discouragement within the OD). In [5] states that the CDR for the ordinary eye is 0.65 (see Fig. 1). The alter in CDR is relative to the nearness of glaucoma. The utilize of calculations makes a difference us to have a conclusion with certainty of glaucoma, it is basic that it must be able to confine, identify and section both OC and OD from the retinal picture. Examinations of restorative pictures and classification by visual computer frameworks have demonstrated to be especially successful.

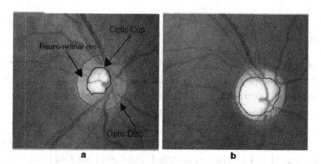

Fig. 1. Digital fundus images cropped around optic disc. **a** Main structures of a healthy optic disc, **b** glaucoma optic disc.

Datasets are collections of pertinent photos with accurately identified regions of interest in the context of image analysis. There can be anywhere from a few hundred and several thousand photos. It has been demonstrated that when dataset size increases, analysis quality improves noticeably [2, 19, 18].

3   Convolution Neural Networks

Convolutional neural networks are a particular type of constructed neural system that, in at least one of their layers, uses the numerical operation known as convolution rather than conventional framework multiplication [6, 17]. An input layer, a layer that is covered, and a layer that is yielding make up CNN. Since the actuation work and final convolution conceal the inputs and outputs of the center layers in any feed-forward neural arrangement, these layers are referred to as covered up layers. In a convolutional neural network, the hidden layers contain convolution layers. This frequently includes a layer that only executes a tiny portion of the convolution section using the layer's input lattice. Typically, this item is a Frobenius internal item, and ReLU is frequently used as its actuation mechanism. A highlight outline is produced by the convolution musical dramation as it moves through the input network for the layer, and this highlight outline then serves as input for the layer below. Other layers, including normalizing layers, pooling layers, and layers that are entirely connected, come after this [6] (see Fig. 2).

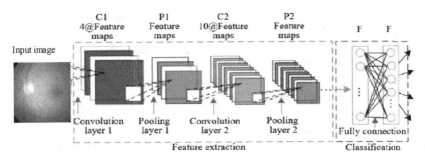

Fig. 2. Typical convolutional neural network (CNN) structure

3.1   Convolutional Layers

The convolutional layers in a CNN convolve the input and then pass the result to the following layer [6]. Data processing for each convolutional neuron's specific receptive field only. Convolution lowers the amount of open parameters, enabling a deeper network [7]. Traditional neural networks' vanishing gradients and ballooning gradients issues are avoided by using regularized weights across fewer parameters [9, 10, 17]. The input is convolved by the convolutional layers, which then send the result to a subsequent layer. Additionally, as spatial links between isolated highlights are taken into account during convolution and/or pooling, convolutional neural nets are ideal for information having a grid-like architecture (such as pictures).

3.2   Pooling Layers

The layer of pooling comes following. Convolutional systems may too have standard convolutional layers and nearby or worldwide pooling layers. By combining the yields of neuron clusters at one layer into a single neuron at the taking after layer, pooling layers lower the dimensionality of the information. Down sampling is what we call this. In a CNN engineering, it is typical to contributed a pooling layer each so frequently between succeeding convolutional layers (each of which is regularly taken after by an actuation work, such as a ReLU layer) [9]. Little clusters are combined utilizing nearby pooling, which as often as possible employments tiling sizes of 2 × 2. They include map's neurons in add up to are influenced by worldwide pooling [11]. Max and normal are the two most broadly utilized sorts of pooling. Whereas normal pooling utilizes the normal esteem, max pooling utilizes the greatest esteem of each nearby cluster of neurons within the include map [9]. The pooling layer typically resizes the input spatially and independently on each depth, or slice (see Fig. 3). A layer with filters of size 2 × 2, applied with a stride of 2, subsamples each depth slice in the input by 2 along both width and height, rejecting 75% of the activations, and is a very typical example of max pooling:

$$f_{X,Y}(S) = max^1_{a,b=0} S_{2X+a,2Y+b} \tag{1}$$

The depth of the output volume determines the amount of neurons in a layer that connect to the same region of the input volume, whereas Stride controls how depth columns around the width and height are assigned.

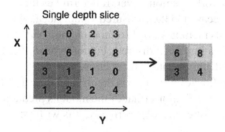

Fig. 3. With a 2 × 2 filter and a stride of 2, maximum pooling

3.3 Receptive Field

In neural systems, each neuron gets input from a few number of areas within the past layer. In a convolutional layer, each neuron gets input from because it were a kept run of the past layer called the neuron's open field. Regularly the run may be a square (e.g. 5 by 5 neurons). While, in a totally related layer, the responsive field is the full past layer. When utilizing broadened layers, the number of pixels inside the open field remains steady, but the field is more meagerly populated as its estimations create when combining the affect of some layers [10].

3.4 Weights

Each neuron in a neural network calculates a yield value by performing a specific operation on the input values obtained from the open field in the preceding layer. A bias and a vector of weights determine the work that is related to the input values. Iteratively changing these inclinations and weights is what learning entails. The weights and inclinations vectors, also known as channels, speak to particular highlights of the input. The ability of several neurons to share one channel is a distinguishing feature of CNNs. This lessens the memory imprint because instead of each open field having its own predisposition and vector weighting, a single predisposition and a single vector of weights are used over all open areas that share that channel [12].

3.5 ReLU Layer

The term "rectified linear unit" (abbreviated "ReLU") applies the non-saturating activation function [13] as shown in (2):

$$f(x) = \max(0, x) \tag{2}$$

By setting them to zero, it effectively eliminates negative values from an activation map [13].

3.6 Loss Layer

The loss layer is the network's bottom layer. It determines the difference in inaccuracy between the network forecast and the measured value. Because it can only have a value of 0 or 1, which represents membership (1) or non-membership (0) in a class, the random variable used in a classification task is discrete. For this reason, the cross-entropy function is the most prevalent and appropriate loss function.

This comes from the field of information theory, and measures the overall difference between two probability distributions (that of the model's prediction, that of reality) for a random variable or a set of events [4]. Formally, it is written:

$$loss(x, class) = -\sum_{class=1}^{c} Y_{x,\text{class}} \log(P_{x,class}) \tag{3}$$

With Y the estimated probability of x belonging to class i, P the real probability of x belonging to class i, knowing that there are C classes [13].

4   CNN Based Glaucoma Detection Approach

In this work, we propose a CNN approach to diagnose glaucoma from fundus images (from the dataset) and classify accurately to declare whether the patient has the disease or not. CNN Machine Learning framework for image classification is one of the objectives of this work to detect glaucoma disease.

This approach is divided into two parts; the first consists in obtaining a model of all the images of the dataset after machine learning of the CNN (see Fig. 4). The second concerns the image of the patient's fundus. After the pre-processing and classification via CNN, the cost Binary_crossentropy function is operational in order to have a decision on the existence of glaucoma or not (see Fig. 5).

4.1 Learning Phase

The first step of CNN glaucoma detection approach is given in the following flow-chart (see Fig. 4). In this learning phase, the first step is considered to import dataset, where is trained by different step of pre-processing. The next step, the CNN training is lunch to obtain a trained model to be saved.

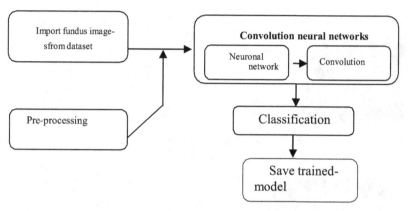

Fig. 4. Flowchart of the learning phase

4.2 Classification Phase

The second part of the CNN-based glaucoma detection approach is based on the patient's fundus image. After having the model resulting from the training of the CNN (see Fig. 4), the decision is represented by the output of the binary function. The latter is interpreted in two cases, either the patient has glaucoma disease or not as in Fig. 5.

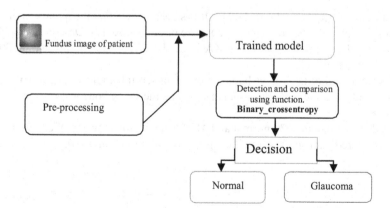

Fig. 5. Flowchart represents the algorithmic steps for classification face

5 Experimentation

In this work we propose a CNN approach to diagnose glaucoma from fundus images (of dataset) and classify accurately in order to declare whether the patient has the disease or not. This requires pre-processing on medical images (modification RGB grayscale and each image in dataset is encoded by it histogram vector (see Figs. 6 and 7).

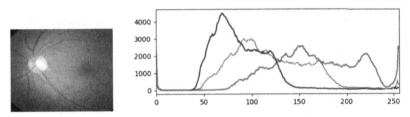

Fig. 6. RGB histogram of fundus image

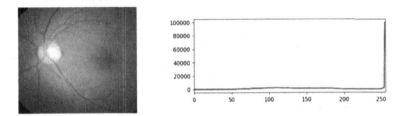

Fig. 7. Grayscale histogram of fundus image

5.1 Datasets Settings

Learn effectively requires a large dataset. It has been shown that the quality of analysis increases substantially with the increase in size of the dataset [7]. Dataset to use in this experiment is from online web source KAGGLE-Data. This dataset presenting 30,000 images of all eye diseases.

After filtering this dataset, we based on two data-sets that indicate glaucoma disease: The first one for people with glaucoma disease and the second one is for normal people (see Fig. 8).

This dataset contains 670 normal and 413 Glaucomatous images. We have selected these two filtered data-sets (240 sick images and 240 normal images).

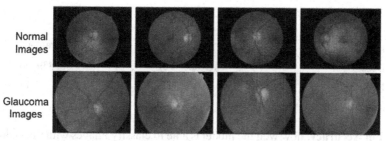

Fig. 8. Examples of new publicly available dataset. Normal and glaucoma fundus images from the new publicly available dataset (KAGGLE-Data).

5.2 Simulation Results

After the launch of the CNN training with 100 iteration an estimate of the number of hearable parameters of our model in order to better visualize the complexity of the latter. We can see a total approaching is important as in following Table 1, Table 2:

Table 1. Parameters learning

Total parameters	Trainable parameters	Non-trainableparameters
63,996,993	63,996,097	896

Table 2. Parameter result of CNN model training

Layer (type)	Output shape	Param #
conv2d (Conv2D)	(None, 198, 198, 64)	1792
conv2d_1 (Conv2D)	(None, 196, 196, 64)	36928
max_pooling2d (MaxPooling2D)	(None, 98, 98, 64)	0
batch_normalization	(None, 98, 98, 64)	256
conv2d_2 (Conv2D)	(None, 96, 96, 128)	73856
conv2d_3 (Conv2D)	(None, 94, 94, 128)	147584
max_pooling2d_1(MaxPooling)	(None, 47, 47, 128)	0
batch_normalization	(None, 47, 47, 128)	512
conv2d_4 (Conv2D)	(None, 45, 45, 256)	295168
max_pooling2d_2(MaxPooling)	(None, 22, 22, 256)	0
batch_normalization_2 (HN)	(None, 22, 22, 256)	1024
flatten (Flatten)	(None, 123904)	0

(*continued*)

Table 2. (*continued*)

Layer (type)	Output shape	Param #
dense (Dense)	(None,512)	63439360
dense_1 (Dense)	(None, 1)	513

Validation Accuracy: how well the model is able to classify images with the validation dataset. (A validation dataset is a sample of data retained while training a model that is used to give an estimate of the model's skill when training the model).

So while the percentage of this Validation accuracy parameter is more than 95% then the classification gives good result.

Training Loss: is a parameter that measures how well a deep learning model matches the training set. In other words, it assesses the model's inaccuracy on the training set. It should be noted that the training set is a portion of a dataset utilized to train the model at first [12]. The training graphs in Fig. 9 (training loss-validation loss-training accuracy-validation accuracy) show the consistency of learning that supports the diagnosis of the glaucoma illness.

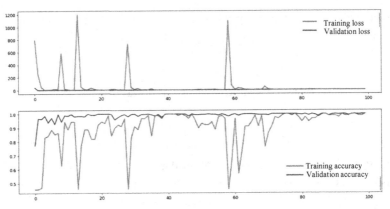

Fig.9. Training graph (training loss-validation loss-training accuracy-validation accuracy)

In this paper, CNN architecture is implemented after choosing the dataset and importing it to perform the pre-processing (cited in Sect. 4) of the fundus images of the class which represents normal images and the images indicating glaucoma, the implementation of CNN with the adjustment of its parameters is launch. The learning of the CNN allows us to have a model of a digital image of the dataset, which will be compared with the image of the fundus of the patient's eye, in order to obtain a precise decision.

The test result presented in this paper is an important significant Parameter Represented by the cost function. **Binary_crossentropy**. The binary classification challenge makes use of the later function. These are assignments that offer a single, two-option response to a query (yes or no, A or B, 0 or 1, left or right).

By utilizing the convolution layer, pool layer, entire connection layer, and other fundamental components of a convolution neural network, you may enable the network structure to learn and extract the necessary features for application. A very complicated modeling approach is not necessary because of this characteristic, which offers numerous conveniences for many investigations. Deep learning is also currently producing very significant findings and advancements in image classification, object detection, attitude estimation, picture segmentation, and other areas.

5.3 Comparative Analysis

To prove the validity of the approach proposed in this paper, a comparative study is realized. One of the deep learning architectures VGGNet-16, is chosen for this comparison. The VGGNet-16 architecture consists of two components and has 16 layers. Three fully linked layers are placed before a stack of convolution layers, which is subsequently followed by another stack of convolution layers [20, 21].

The statistical measure accuracy (ACC) is performed for achieving the numerical results. The performance of approach proposed for detection of is analyzed with the ACC parameter that is mathematically described in the following equation:

$$Accuracy(ACC) = (TP + TN)/(TP + FN + TN + FP)) \qquad (4)$$

If the number of correctly predicted positive cases is TP, the number of correctly forecasted negative cases is TN, when the case is negative; the number of cases that were mistakenly predicted is FN, and when the case is positive, the number of cases that were incorrectly anticipated is FP. The Validation accuracy parameter of our suggested CNN approach performs better than that of VGGNet-16 under the identical experimental simulation conditions, demonstrating the effectiveness of the proposed CNN approach in this work. The outcomes are listed in following Table 3.

Table 3. Comparison of accuracy of the proposed CNN model with VGGNet-16

Detection of Glaucoma eye disease	
Methodologies	Accuracy parameter
CNN approach	96%
VGGNet-16	87%

This outcome is also due to the two major drawbacks of the VGGNet-16 architecture. The first disadvantage is that training is excruciatingly slow, and the second is that the network architecture weights themselves are quite large (in terms of disk/bandwidth). VGG16 is larger than 533 MB due to its depth and number of fully connected nodes. However, because it is simple to implement, it is an excellent learning tool.

6 Conclusion

In this paper, we have presented CNN based glaucoma detection approach. Moreover, the proposed architecture is flexible, and that makes it possible to integrate other dataset. The results obtained by this approach improve a CNN machine learning for the classification image. In this work we have fond through several experiments that the limitation of the CNN approach is that increasing the growth rate translates into many other parameters, which can make the training harder and the tests slower. In the future, research could focus on simplifying the network structure while maintaining its capacity.

Finally, this work creates numerous research opportunities. More research is required to refine some of the concepts that we have introduced.

References

1. Hollows, F.C., Graham, P.A.: Intra-ocular pressure, glaucoma, and glaucoma suspects in a denned population. Br. J. Ophthalmol. **50**(5954089), 570–586 (1966)
2. Ræfl, R.E.W., Gonzalez, C.: Digital Image Processing (2008)
3. Leibowitz, H.M., et al.: The Framingham eye study monograph: an ophthalmological and epidemiological study of cataract, glaucoma, diabetic retinopathy, macular degeneration, and visual acuity. J. Surv. Ophthalmol. **24**, 335–610 (1980)
4. Spalton, D.J., Hitchings, R.A., Hunter, P.: Atlas of Clinical Ophthalmology with CD-ROM. Elsevier, Amsterdam (2013)
5. Zhang, F., et al.: An online retinal fundus image database for glaucoma analysis and research. In: Proc. Int. Conf. IEEE Eng. Med. Biol., pp. 3065–3068 (2010)
6. Ciresan, D.C., Meier, U., Masci, J., Gambardella, L.M., Schmidhuber, J.: Flexible, high performance convolutional neural networks for image classification. In: Proceedings of the Twenty-Second International Joint Conference on Artificial Intelligence, vol. 2, pp. 1237–1242 (2013)
7. Aghdam, H.H., Heravi, E.J.: Guide to Convolutional Neural Networks. Springer International Publishing, Cham (2017). https://doi.org/10.1007/978-3-319-57550-6
8. Balas, V.E., Kumar, R., Srivastava, R.: Recent Trends and Advances in Artificial Intelligence and Internet of Things. Springer Nature (2019)
9. Fuente-Arriaga, E., Felipe-Riverón, M., Garduño-Calderón, E.: Application of vascular bundle displacement in the optic disc for glaucoma detection using fundus images. J. Comput. Biol. Med. **47**, 222–235 (2014)
10. Ciresan, D., Meier, U., Masci, J.: Flexible, high performance convolutional neural networks for image classification. Proceedings of the 22nd International Joint Conference on Artificial Intelligence, Barcelona, Catalonia, Spain, July 16–22 (2011)
11. Azulay, A., Weiss, Y.: Why do deep convolutional networks generalize so poorly to small image transformations. J. Mach. Learn. Res. **20**(184) (2019)
12. LeCun, Y.: Convolutional Neural Networks (2013)
13. Krizhevsky, A., Sutskever, I., Hinton, G.E.: ImageNet classification with deep convolutional neural networks. Commun. ACM **60**(6), 84–90 (2017). https://doi.org/10.1145/3065386
14. https://www.journaldunet.fr/web-tech/guide-de-l-intelligenceartificielle/1501309apprentissage-non-supervise/. 10 Apr 2022
15. https://intelligence-artificielle.com/classification-d-image-guide-complet. 5 Apr 2022
16. Pooja, K., et al.: A survey on image classification approaches and techniques. Int. J. Adv. Res. Comput. Commun. Eng. **2** (2013)

17. Chen, X., Xu, Y., Wong, D.W.K., Wong, T.Y., Liu, J.: Glaucoma detection based on deep convolutional neural network. In: Annual International Conference of the IEEE Engineering in Medicine and Biology Society (EMBC), pp. 715–718 (2015)

18. World Health Organization. Bulletin of the World Health Organization. http://www.who.int/bulletin/. Accessed 5 May 2016

19. Bock, R., Meier, J., Nyúl, L.G., Hornegger, J., Michelson, G.: Glaucoma risk index: automated glaucoma detection from color fundus images. Med. Image Anal. **14**(3), 471–481 (2010). https://doi.org/10.1016/j.media.2009.12.006

20. Khambra, G., Shukla, P.: Novel machine learning applications on fly ash based concrete: an overview. Mater. Today Proc. **7**, 2214–7853 (2021)

21. Shubham, J., et al.: Glaucoma detection using image processing and supervised learning for classification. J. Healthc. Eng. **7**, 1–12 (2022)

Deep Multi-Scale Hashing for Image Retrieval (DMSH)

Adil Redaoui[1]📷, Kamel Belloulata[1(✉)]📷, and Amina Belalia[2]📷

[1] Telecommunications Department, RCAM Laboratory, Sidi Bel Abbès, Algeria
{adil.redaoui,kamel.belloulata}@univ-sba.dz
[2] High School of Computer Sciences of Sidi Bel Abbes, Sidi Bel Abbès, Algeria
a.belalia@esi-sba.dz

Abstract. In recent years, with the great success of deep learning, deep networks-based hashing has become a leading approach for image retrieval. Most existing deep hashing methods extract the semantic representations only from the last layer, resulting in structure information being ignored which contains additional semantic details that are useful for hash learning. To enhance the image retrieval accuracy by exploring the semantic information and the structure information (local information), We propose a new method of deep hashing called Deep Multi-Scale Hashing (DMSH). This is achieved, firstly, by extracting multiscale features from multiple convolutional layers. Secondly, the features extracted from the convolutional layers are fused to generate more robust representations for efficient image retrieval. The experiments on the CIFAR10 and NUS-WIDE datasets show the superiority of our method.

Keywords: Deep learning · Deep supervised hashing · Image retrieval · Feature pyramid · Multi-scale feature

1 Introduction

The rapid advances in the internet and communication have resulted in images' massive overloading to the internet [22,40,41], making it extremely difficult to retrieve large-scale data accurately and effectively, representing a practical research problem. The hash-based image retrieval technique [36] has attracted increasing attention to guarantee retrieval quality and computation efficiency. The hashing methods perform to map images into compact binary codes to take advantage of binary codes' superior computational and storage capabilities.

Existing hashing techniques can be divided into groups that are data-dependent and Data-independent. Data-independent approaches, representing locally-sensitive hashing (LSH) [11], are used in random projections as the hash functions to learn binary codes. However, these methods do have some limitations. The data-independent methods do not use auxiliary information, which makes them suffer in terms of poor image retrieval accuracy. Furthermore, They require long codes that cost plenty of storage [1,11,15]. To address this problem, data-driven methods exploit training information to create hash functions, obtaining shorter binary codes with better performance. Moreover, they can be categorized-into unsupervised hashing methods [12,13,25,29] and supervised hashing methods [7,9,20,23,24,33,34].

On the other hand, deep hashing techniques [4,10,21,42] have been proposed after the great success of deep neural networks on many computer vision tasks. Compared with traditional hashing approaches, deep hashing techniques exhibit advan- tages in extracting high-level semantic features and producing binary codes into an end-to-end framework. However, many recent works on deep hashing methods [14,17,31] used the penultimate layer features in fully connected layers as the global image descriptor. However, they are undesirable image features because the high-level features exhibit global details but lack local characteristics.

To address the above problems, in this paper, we propose a new deep hashing method called Deep Multi-Scale Hashing (DMSH), which use FPN in deep hashing—using FPN. Extracting the multi-level semantic and visual information from the input image is facilitated. To be more specific, An FPN builds a pathway of bottom-up and the top-down features with links between the features the network produces at different scales. The hash codes are then learned from these feature scales and fuse to obtain the final ones. Moreover, the network employs various hashing results depending on different scale features. Consequently, the network would improve retrieval recall while preventing precision degradation. The main contributions of the work, in brief, are as follows:

1. A Deep Multi-Scale Hashing (DMSH) learns hash codes from various feature scales and fuses them to generate the final binary codes. As a result, the network will be able to retrieve better.
2. A new deep hashing method is proposed that conducts joint optimization of feature representation learning and binary codes learning in a deep, unified framework.
3. Experimentation based on two large-scale data-sets show that DMSH provides state-of-the-art performance in real applications.

2   Proposed Method

2.1   Problem Definition

Let $X = \{x_i\}_{i=1}^{N} \in \mathbb{R}^{d \times N}$ denote the training dataset with N image samples, where $Y = \{y_i\}_{i=1}^{N} \in \mathbb{R}^{K \times N}$ represents the ground truth labels of the x_i, K is the number of classes. Let's say that the pairwise labels for training images are represented by the matrix $S = \{s_{ij}\}$, where $s_{ij} = 0$ represents no semantic similarity between samples x_i and x_j and $s_{ij} = 1$ represents semantic similarity. The objective of deep hashing methods with pairwise labels is to Learning a nonlinear hash function $f : x \mapsto B \in \{-1,1\}^{L}$, which can the ability to convert each input data xi into binary codes $b_i \in \{-1,1\}^{L}$, L is the Hash code length.

2.2 Model Architecture

Lin [26] suggested that a feature pyramid network was applied to improve object detection in RetinaNet [27]. FPN aids the network in learning more effectively and detecting objects at various scales present in the image. Some previous methods worked by providing input to the network, such as an image pyramid. While doing this does enhance the feature extraction procedure, it also lengthens processing time and is less effective.

FPN solved this issue by creating the bottom-up and top-down connection of entities and merging them with network entities created at different levels via a lateral connection.

Figure 1 displays the architecture of our proposed method (DMSH). We used VGG-19 as the backbone network. The FPN we used is similar to the original FPN [26], with the difference that, in light of the authors' expertise, we employed concatenation layers rather than adding layers within the feature pyramid. FPN extracts four final features, each presenting the input image's features at various scales. The features obtained are fused using a convolutional (Conv1 × 1) layer to get a merged feature and then apply dropout layers, followed by the first hash layers.

At the end of the design, we assembled the five hash layers. We then connected the latter to the final hash layer, and finally, we connected the final hash layer to the classification layer (the number of neurons in the classification layer is equal to the number of categories of the dataset). With this procedure, the network uses various hash results based on different features scale. The network will be able to retrieve images more effectively.

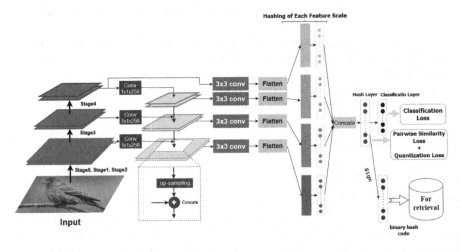

Fig. 1. The architecture of our proposed method: Deep Multi-Scale Hashing (DMSH).

Table 1. Details of the feature extraction network. Note that we use the features of the layers marked by "#". For simplicity, we omit ReLU and batch normalization layers.

Conv block	Layers	Kernel size	Feature size
1	Conv2D	$64 \times 3 \times 3$	224×224
	Conv2D#	$64 \times 3 \times 3$	
	MaxPooling	2×2	
2	Conv2D	$128 \times 3 \times 3$	112×112
	Conv2D#	$128 \times 3 \times 3$	
	MaxPooling	2×2	
3	Conv2D	$256 \times 3 \times 3$	56×56
	Conv2D	$256 \times 3 \times 3$	
	Conv2D	$256 \times 3 \times 3$	
	Conv2D#	$256 \times 3 \times 3$	
	MaxPooling	2×2	
4	Conv2D	$256 \times 3 \times 3$	28×28
	Conv2D	$256 \times 3 \times 3$	
	Conv2D	$256 \times 3 \times 3$	
	Conv2D#	$256 \times 3 \times 3$	
	MaxPooling	2×2	
5	Conv2D	$256 \times 3 \times 3$	14×14
	Conv2D	$256 \times 3 \times 3$	
	Conv2D	$256 \times 3 \times 3$	
	Conv2D#	$256 \times 3 \times 3$	
	MaxPooling	2×2	

2.3 Objective Function

Pairwise Similarity Loss. We carry out our deep hashing method by maintaining the greatest similarity between each pair of images in the Hamming space. The inner product is used to calculate the pairwise similarity. For two binary codes, b_i and b_j, the inner product is written as follows: $dist_H(b_i, b_j) = \frac{1}{2}b_i^T b_j$

Given that all binary codes of points are $B = \{b_i\}_{i=1}^{N}$, we can write the likelihood of pairwise labels $S = \{s_{ij}\}$ as follows:

$$p(s_{ij}|B) = \begin{cases} \sigma(w_{ij}) & s_{ij} = 1 \\ 1 - \sigma(w_{ij}) & s_{ij} = 0 \end{cases} \tag{1}$$

where $\sigma(w_{ij}) = \frac{1}{1+e^{-w_{ij}}}$, and $w_{ij} = \frac{1}{2}b_i^T b_j$

According to the equation above, we may deduce that the greater the inner product $\langle b_i, b_j \rangle$ corresponds to a smaller equivalent $dist_H(b_i, b_j)$ and a larger $p(1|b_i, b_j)$.

This indicates that when $s_{ij} = 1$, the hash codes b_i and b_j are regarded as similar, and vice versa.

We may get the following optimization problem by considering the negative log-likelihood of the pairwise labels in S:

$$J_1 = -\log p\left(S|B\right) = -\sum_{s_{ij} \in S} \left(s_{ij}w_{ij} - \log(1 + e^{w_{ij}})\right) \tag{2}$$

The above optimization problem makes the distance between two similar points as small as possible, precisely what pairwise similarity-based hashing approaches aim to achieve.

Pairwise Quantization Loss. In real-world applications, discrete hash codes (binary codes) measure similarity. Optimizing discrete hash coding (binary) in CNN is difficult, however. Therefore, gradient disappearance through the back-propagation stage is avoided using the continuous hash coding version.

Discrete hash codes are utilized in real-world applications to determine similarity. However, discrete hash coding in CNN is challenging to optimize. Therefore, continuous hash coding prevents gradient disappearance during the back-propagation phase. Where the hash layer output is u_i and $b_i = \text{sgn}\left(u_i\right)$. Hence, quantization loss has been introduced to reduce the gap between discrete and continuous hash codes. An objective function is defined as

$$J_2 = \sum_{i=1}^{Q} \parallel b_i - u_i \parallel_2^2 \tag{3}$$

Q is the mini-batches.

Classification Loss. We apply the classification loss (the cross-entropy loss) to identify the classes to obtain robust multiscale features. The classification loss function can be expressed as follows:

$$J_3 = -\sum_{i=1}^{Q}\sum_{k=1}^{K} y_{i,k} \log(p_{i,k}), \tag{4}$$

where $y_{i,k}$ is the label, $p_{i,k}$ is the output of the $i - th$ training sample, which corresponds to the $k - th$ class.

In conclusion, quantization loss, pairwise similarity loss, and classification loss can be combined to produce the overall loss function:

$$J = J_1 + \beta J_2 + \gamma J_3 \tag{5}$$

3   Experiments

3.1   Datasets

The **CIFAR-10** [19] contains 60,000 images of 10 different categories, and each image is 32×32 in size. Following [2], we sampled 100 images per category as a query set (a total of 1000) and the remaining images as a database. Additionally, 500 images per category are selected (a total of 5000) from the retrieval database as a training set.

NUS-WIDE. [8] is a multi-label data set containing approximately 270,000 images collected from Flickr consisting of 81 ground-truth concepts. We randomly sampled 2100 images from the 21 most frequent classes as the query set, while the rest as a database, and selected 10,000 images from the retrieval database as a training set.

3.2 Experimental Settings

Our DMSH implementation utilizes PyTorch. We use a convolutional network called VGG-19, pre-trained on ImageNet [30]. In all training, we use the Adam [18] algorithm. For the hyperparameters of the objective function, the beta to 0.1 and the alpha is set to 0.01.

3.3 Evaluation Metrics

We use four evaluation metrics to measure the retrieval performance of different hashing methods: Mean Average Precision (MAP), Precision-Recall curves, precision curves w.r.t, and Precision curve within Hamming radius 2. Five unsupervised approaches are included in our comparison of the proposed DMSH, i.e., SGH [16], LSH [11], SH [38], ITQ [13], PCAH [37], and two supervised approaches, i.e., KSH [28], SDH [32], and eight deep hashing approaches, i.e., DPH [3], CNNH [39], DNNH [21], DCH [5], DHN [42], HashNet [6], LRH [2], DHDW [35].

3.4 Results

Tables 2 and 3 displays the MAP results for all methods on the CIFAR-10 and NUS-WIDE data sets with different code lengths. The results in the table demonstrate how well the proposed DMSH method performs compared to all other techniques. In particular, on the datasets CIFAR-10 and NUS-WIDE, DMSH delivers absolute gains in average mAP of 49.4% and 24.44%, respectively, above SDH, the top shallow hashing technique. However, we can observe that deep hashing performs better than classical hashing techniques, mainly because it can produce more reliable representations of features. For deep hashing methods, On CIFAR-10 and NUS-WIDE, respectively, the average mAP of our suggested DMSH approach increased by 11.5% and 8.3% per cent compared to the second-best hashing method LRH.

Table 2. Mean Average Precision (MAP) of hamming ranking for different number of bits on CIFAR-10.

	CIFAR-10 (MAP)			
Method	12 bits	24 bits	32 bits	48 bits
SH [38]	0.127	0.128	0.126	0.129
ITQ [13]	0.162	0.169	0.172	0.175
KSH [28]	0.303	0.337	0.346	0.356
SDH [32]	0.285	0.329	0.341	0.356
CNNH [39]	0.439	0.511	0.509	0.522
DNNH [21]	0.552	0.566	0.558	0.581
DHN [42]	0.555	0.594	0.603	0.621
HashNet [6]	0.609	0.644	0.632	0.646
DPH [3]	0.698	0.729	0.749	0.755
LRH [2]	0.684	0.700	0.727	0.730
DMSH	**0.800**	**0.823**	**0.838**	**0.840**

Table 3. Mean Average Precision (MAP) of hamming ranking for different number of bits on NUS-WIDE, The MAP values are calculated on the top 5,000 retrieval images.

	NUS-WIDE (MAP)			
Method	12 bits	24 bits	32 bits	48 bits
SH [38]	0.454	0.406	0.405	0.400
ITQ [13]	0.452	0.468	0.472	0.477
KSH [28]	0.556	0.572	0.581	0.588
SDH [32]	0.568	0.600	0.608	0.637
CNNH [39]	0.611	0.618	0.625	0.608
DNNH [21]	0.674	0.697	0.713	0.715
DHN [42]	0.708	0.735	0.748	0.758
HashNet [6]	0.643	0.694	0.737	0.750
DPH [3]	0.770	0.784	0.790	0.786
LRH [2]	0.726	0.775	0.774	0.780
DMSH	**0.826**	**0.850**	**0.853**	**0.859**

Precision curves represent the retrieval performance in Figs. 2a and 3a (P@H = 2). The proposed DMSH performs noticeably better than alternative approaches. According to the precision curves, the proposed DMSH approach still has the highest precision rate when the code length rises.

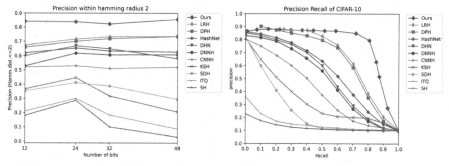

(a) Precision within hamming radius 2   (b) Precision Recall Curve on 48 bits

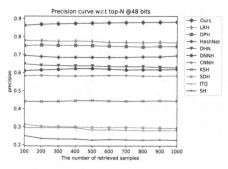

(c) Precision curve w.r.t top-N @48 bits

Fig. 2. The comparison results on the CIFAR-10 dataset under three evaluation metrics.

We further highlight the effectiveness of our method in Figs. 2b, 3b, 2c and 3c, where we compare our method's precision concerning top returned samples and Precision-Recall with other techniques. Figures 2c and 3c show that, for return sample counts between 100 and 1000, the suggested DMSH approach provides the highest precision with 48 bits. Based on Figs. 2a and 3b, it can be shown that our DMSH delivers significantly high precision at a low recall level, which is necessary for precision-first retrieval and is frequently utilized in real-world systems. In summary, our technique, DMSH, outperforms the compared techniques.

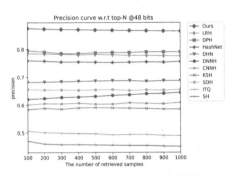

(a) Precision within hamming radius 2   (b) Precision Recall Curve on 48 bits

(c) Precision curve w.r.t top-N @48 bits

Fig. 3. The comparison results on the NUS-WIDE dataset under three evaluation metrics.

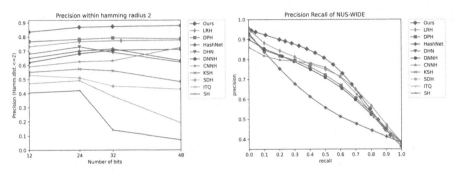

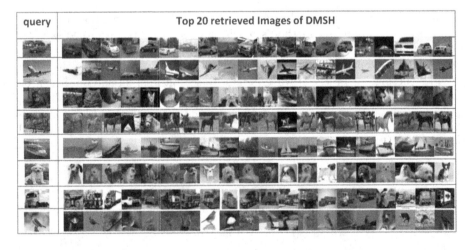

Fig. 4. Top 20 retrieved results from CIFAR-10 dataset by DMSH with 48-bit hash codes. The first column shows the query images, the retrieval results of DMSH are shown at other columns.

4 Conclusions

In this paper, we developed an end-to-end Deep Multi-Scale Hashing (DMSH) for large-scale image retrieval, Which generates robust hash codes by optimizing the semantic loss, similarity loss, and quantization loss. In addition, the network employs different hashing results depending on various scale features. As a result, the network would improve retrieval recall while preventing precision degradation. The experimental results on two image retrieval data sets demonstrated that our method outperforms other state-of-the-art hashing methods. The suggested model can provide robust representative features due to its scalable structure, which allows for use in many additional computer vision tasks.

References

1. Andoni, A., Indyk, P.: Near-optimal hashing algorithms for approximate nearest neighbor in high dimensions. Commun. ACM **51**(1), 117–122 (2008)
2. Bai, J., et al.: Loopy residual hashing: filling the quantization gap for image retrieval. IEEE Trans. Multimedia **22**(1), 215–228 (2019)
3. Bai, J., et al.: Deep progressive hashing for image retrieval. IEEE Trans. Multimedia **21**(12), 3178–3193 (2019)
4. Cakir, F., He, K., Bargal, S.A., Sclaroff, S.: Hashing with mutual information. IEEE Trans. Pattern Anal. Mach. Intell. **41**(10), 2424–2437 (2019)
5. Cao, Y., Long, M., Liu, B., Wang, J.: Deep cauchy hashing for hamming space retrieval. In: Proceedings of the IEEE Conference on Computer Vision and Pattern Recognition, pp. 1229–1237 (2018)
6. Cao, Z., Long, M., Wang, J., Yu, P.S.: Hashnet: deep learning to hash by continuation. In: Proceedings of the IEEE International Conference on Computer Vision, pp. 5608–5617 (2017)
7. Chen, Z., Zhou, J.: Collaborative multiview hashing. Pattern Recogn. **75**, 149–160 (2018)
8. Chua, T.S., Tang, J., Hong, R., Li, H., Luo, Z., Zheng, Y.: Nus-wide: a real-world web image database from national university of Singapore. In: Proceedings of the ACM International Conference on Image and Video Retrieval, pp. 1–9 (2009)
9. Cui, Y., Jiang, J., Lai, Z., Hu, Z., Wong, W.: Supervised discrete discriminant hashing for image retrieval. Pattern Recogn. **78**, 79–90 (2018)
10. Erin Liong, V., Lu, J., Wang, G., Moulin, P., Zhou, J.: Deep hashing for compact binary codes learning. In: Proceedings of the IEEE Conference on Computer Vision and Pattern Recognition, pp. 2475–2483 (2015)
11. Gionis, A., Indyk, P., Motwani, R., et al.: Similarity search in high dimensions via hashing. In: Vldb, vol. 99, pp. 518–529 (1999)
12. Gong, Y., Kumar, S., Rowley, H.A., Lazebnik, S.: Learning binary codes for high-dimensional data using bilinear projections. In: Proceedings of the IEEE Conference on Computer Vision and Pattern Recognition, pp. 484–491 (2013)
13. Gong, Y., Lazebnik, S., Gordo, A., Perronnin, F.: Iterative quantization: a procrustean approach to learning binary codes for large-scale image retrieval. IEEE Trans. Pattern Anal. Mach. Intell. **35**(12), 2916–2929 (2012)
14. Gordo, A., Almazán, J., Revaud, J., Larlus, D.: Deep image retrieval: learning global representations for image search. In: Leibe, B., Matas, J., Sebe, N., Welling, M. (eds.) ECCV 2016. LNCS, vol. 9910, pp. 241–257. Springer, Cham (2016). https://doi.org/10.1007/978-3-319-46466-4_15

15. Indyk, P., Motwani, R.: Approximate nearest neighbors: towards removing the curse of dimensionality. In: Proceedings of the 30th Annual ACM Symposium on Theory of Computing, pp. 604–613 (1998)
16. Jiang, Q.Y., Li, W.J.: Scalable graph hashing with feature transformation. In: 24th International Joint Conference on Artificial Intelligence (2015)
17. Jiang, Q.Y., Li, W.J.: Asymmetric deep supervised hashing. In: Proceedings of the AAAI Conference on Artificial Intelligence, vol. 32 (2018)
18. Kingma, D.P., Ba, J.: Adam: A method for stochastic optimization. arXiv preprint arXiv:1412.6980 (2014)
19. Krizhevsky, A., Hinton, G., et al.: Learning multiple layers of features from tiny images (2009)
20. Kulis, B., Darrell, T.: Learning to hash with binary reconstructive embeddings. In: Advances in Neural Information Processing Systems, vol. 22 (2009)
21. Lai, H., Pan, Y., Liu, Y., Yan, S.: Simultaneous feature learning and hash coding with deep neural networks. In: Proceedings of the IEEE Conference on Computer Vision and Pattern Recognition, pp. 3270–3278 (2015)
22. Li, S., Chen, Z., Li, X., Lu, J., Zhou, J.: Unsupervised variational video hashing with 1d-cnn-lstm networks. IEEE Trans. Multimedia **22**(6), 1542–1554 (2019)
23. Lin, G., Shen, C., Shi, Q., Van den Hengel, A., Suter, D.: Fast supervised hashing with decision trees for high-dimensional data. In: Proceedings of the IEEE Conference on Computer Vision and Pattern Recognition, pp. 1963–1970 (2014)
24. Lin, G., Shen, C., Suter, D., Van Den Hengel, A.: A general two-step approach to learning-based hashing. In: Proceedings of the IEEE International Conference on Computer Vision, pp. 2552–2559 (2013)
25. Lin, G., Shen, C., Wu, J.: Optimizing ranking measures for compact binary code learning. In: Fleet, D., Pajdla, T., Schiele, B., Tuytelaars, T. (eds.) ECCV 2014. LNCS, vol. 8691, pp. 613–627. Springer, Cham (2014). https://doi.org/10.1007/978-3-319-10578-9_40
26. Lin, T.Y., Dollár, P., Girshick, R., He, K., Hariharan, B., Belongie, S.: Feature pyramid networks for object detection. In: Proceedings of the IEEE Conference on Computer Vision and Pattern Recognition. pp. 2117–2125 (2017)
27. Lin, T.Y., Goyal, P., Girshick, R., He, K., Dollár, P.: Focal loss for dense object detection. In: Proceedings of the IEEE International Conference on Computer Vision, pp. 2980–2988 (2017)
28. Liu, W., Wang, J., Ji, R., Jiang, Y.G., Chang, S.F.: Supervised hashing with kernels. In: 2012 IEEE Conference on Computer Vision and Pattern Recognition, pp. 2074–2081. IEEE (2012)
29. Liu, W., Wang, J., Mu, Y., Kumar, S., Chang, S.F.: Compact hyperplane hashing with bilinear functions. arXiv preprint arXiv:1206.4618 (2012)
30. Russakovsky, O., et al.: Imagenet large scale visual recognition challenge. Int. J. Comput. Vis. **115**(3), 211–252 (2015)
31. Shen, F., Gao, X., Liu, L., Yang, Y., Shen, H.T.: Deep asymmetric pairwise hashing. In: Proceedings of the 25th ACM International Conference on Multimedia, pp. 1522–1530 (2017)
32. Shen, F., Shen, C., Liu, W., Tao Shen, H.: Supervised discrete hashing. In: Proceedings of the IEEE Conference on Computer Vision and Pattern Recognition, pp. 37–45 (2015)
33. Song, J., Gao, L., Liu, L., Zhu, X., Sebe, N.: Quantization-based hashing: a general framework for scalable image and video retrieval. Pattern Recogn. **75**, 175–187 (2018)

34. Strecha, C., Bronstein, A., Bronstein, M., Fua, P.: Ldahash: Improved matching with smaller descriptors. IEEE Trans. Pattern Anal. Mach. Intell. **34**(1), 66–78 (2011)
35. Sun, Y., Yu, S.: Deep supervised hashing with dynamic weighting scheme. In: 2020 5th IEEE International Conference on Big Data Analytics (ICBDA), pp. 57–62. IEEE (2020)
36. Wang, J., Zhang, T., Sebe, N., Shen, H.T., et al.: A survey on learning to hash. IEEE Trans. Pattern Anal. Mach. Intell. **40**(4), 769–790 (2017)
37. Wang, J., Kumar, S., Chang, S.F.: Semi-supervised hashing for large-scale search. IEEE Trans. Pattern Anal. Mach. Intell. **34**(12), 2393–2406 (2012)
38. Weiss, Y., Torralba, A., Fergus, R.: Spectral hashing. In: Advances in Neural Information Processing Systems, vol. 21 (2008)
39. Xia, R., Pan, Y., Lai, H., Liu, C., Yan, S.: Supervised hashing for image retrieval via image representation learning. In: 28th AAAI Conference on Artificial Intelligence (2014)
40. Yan, C., Li, Z., Zhang, Y., Liu, Y., Ji, X., Zhang, Y.: Depth image denoising using nuclear norm and learning graph model. ACM Trans. Multimedia Comput. Commun. Appl. (TOMM) **16**(4), 1–17 (2020)
41. Yan, C., Shao, B., Zhao, H., Ning, R., Zhang, Y., Xu, F.: 3d room layout estimation from a single RGB image. IEEE Trans. Multimedia **22**(11), 3014–3024 (2020)
42. Zhu, H., Long, M., Wang, J., Cao, Y.: Deep hashing network for efficient similarity retrieval. In: Proceedings of the AAAI Conference on Artificial Intelligence, vol. 30 (2016)

Genetic Programming for Screen Content Image Quality Assessment

Naima Merzougui[1,2](✉) ⓘ and Leila Djerou[1]

[1] LESIA Laboratory, University Mohamed Khider, Biskra, Algeria
`merzougui.naima@univ-ouargla.dz, l.djerou@univ-biskra.dz`
[2] Department of Computer Science and Information Technologies, University Kasdi Merbah, Ouargla, Algeria

Abstract. The study of Screen Content Image Quality Assessment (SCI-QA) is a new and interesting topic due to its excellent potential for instruction and optimization in various processing systems, it has been attractive recently. In this paper, we proposed a full reference quality assessment method for screen content images named Genetic Programming based Screen Content Image Quality (GP-SCIQ). The proposed method operates via a symbolic regression technique using Genetic Programming (GP). Hence, for predicting subject scores of images in datasets we combined the objective scores of a set of Image Quality Metrics (IQM). Two largest publicly available image databases (namely SICAD and SCID) are used for training and testing the predictive models, according the k-fold-cross-validation strategy. The performance of the proposed approach is evaluated, and several experiments are carried using four performance indices (SRCC, PCC, KROCC and RMSE). The results achieve superior performance to state-of-the-art methods in predicting the perceptual quality of SCIs.

Keywords: Full reference (FR) · Genetic programming (GP) · Image quality assessment (IQA) · Screen content image (SCI)

1 Introduction

With the rapid development of multimedia and social networking, numerous typical applications, such as online news and advertising, online education, electronic brochures, Facebook, Twitter, etc., involve computer generalized screen content images (SCIs) [1–3]. Typical SCIs contain a mixture of sources which include natural images, text, charts, maps, and logos, etc. Therefore, in our everyday lives, screen content images have become very popular.

The results in [4] show that SCIs exhibit several distinct characteristics from Natural Images (NIs). Such NIs have rich color variations, smooth edges and complex texture content with thick lines, while SCI images are mostly thinner lines, crisp edges, with limited color variations and simple shapes; Fig. 1 shows samples of both types of images. These differences render the methods developed specifically to assess the quality of natural scene images less reliable.

M. Salem et al. (Eds.): ICAITA 2022, CCIS 1769, pp. 52–64, 2023.
https://doi.org/10.1007/978-3-031-28540-0_5

Examples of Natural Image   Examples of Screen Content Image

Fig. 1. Examples of natural images versus screen content images

During the different stages of the distribution chain, digital images might introduce a variety of distortions, which degrade their visual quality. There are several methods of evaluating image quality that fall into two categories: subjective and objective. The subjective methods of image quality assessment (IQA) are carried out by human subjects, these methods are precise in estimating the visual quality of an image but they are expensive, time-consuming and nor can they be automated. On the other hand, the objective IQA methods are computer-based methods that can automatically predict the perceived image quality. Therefore, these methods purport to correlate well with perceived quality; they would be relatively quicker and cheaper than subjective assessment.

In general, objective IQA methods are classified into three categories according to the availability of a reference image[5]: no reference (NR), reduced-reference (RR), and full reference (FR) methods; where in the case of NR methods require no access to any information about the reference image, in the case of RR methods comparing a description of the image to be evaluated with just partial information about the original image and in the case of FR comparing the version of the degraded image with a fully reference version. In this paper, we primarily focus on FR approaches due to their universality and high performance.

In the processing of SCIs (e.g., generation, compression and transmission), all types of distortion (e.g., blurring, compression, change in color saturation, change in contrast) will surely lead to image quality degradation. Therefore, a key problem relevant to SCI's image quality assessment is: how to objectively evaluate the image quality and ensure high consistency of the measurement derived from the designed objective model with a judgment of the human visual system (HVS).

Given that the number of established IQA metrics has increased in recent years and each one has its own intended use or construction. It would be ideal if there is a universal metric that exploits the advantages of some metrics and reduces the influence of their limitations; which can be achieved by adopting the method of combining metrics.

Through inspiring studies [6–9] using the approach of combining metrics on natural images which have demonstrated its success in the tasks of evaluating natural image quality, this can be seen as a promising advantage of our proposed approach. Based on our best knowledge, there is no method of combining metrics in the case of screen content images.

The primary contributions of this study are:

1. Our metric is a universal scorer since they are not designed for specific distortion types; it fuses scores from a set of image quality metrics, where each one can be designed to predict the quality of some specific image distortion types. The proposed metric can perform well in situations where an individual metric cannot perform well. And it is linearly correlated with how individuals perceive various types of distortions.

2. It shows that a linear combination based solution, together with Genetic Programming (GP), is capable of finding a well-performing fusion of IQA measures.

3. No need for a metrics selection stage, which can degrade the overall performance; Using GP aims to find an explicit formulation of image quality metrics (IQM) combination, in a regression model form, which best fits a given dataset, both in terms of accuracy and simplicity.

The remaining of this paper is organized as follows. Section2 introduces the proposed IQA model for the SCIs in detail. Section3 presents the experimental results and discussions. Finally, Sect. 4 provides conclusions.

2 Related Work

To investigate the SCI-QA (Screen Content Image Quality Assessments), several algorithms have been proposed. Among them GFM(Gabor Feature-based Model)[10], GSS(Gradient Similarity for SCI) [11], SCDM(Structural Contrast Distortion Metric) [12], SCQI(Structural Contrast-Quality Index) [12], SQMS(Saliency-guided Quality Measure) [13], SVQI(Structural Variation based Quality Index) [14], ESIM(Edge SIMilarity) [15]. All these methods designed for SCIs have achieved encouraging results.

Gu et al. proposed a quality measure for SCIs (SQMS) guided by the identification of salient regions [13]. This method relies primarily on a simple convolution operator, first showing the degradation of the structure caused by different types of distortion, then detecting the significant region of the distortion that is generally more noticeable, and finally predicting the overall image quality.

Gu et al. evaluated the quality of SCIs by analyzing structural changes caused by compression, transmission, etc., and proposed a quality index based on structural variation (SVQI) [14]. They divide the structure into global structures and local structures that correspond to the basic perception and the detailed perception of human beings. Factors related to the basic perceptions (such as brightness, contrast, and complexity) are assigned to an overall structure, and factors related to perception of detail (such as edge information) are reduced to local structures. Finally, to predict the final quality score of SCIs, a measurement of variations in global and local structures is systematically combined.

Ni et al. [11] proposed an SC-IQA approach called GSS that operates by using similarity in gradient direction and similarity in gradient magnitude between the reference and distorted SCIs. Then a variance based pooling technique is used by combining the two similarity maps to obtain the overall quality estimate.

Ni et al. [10] also proposed an effective FR-IQA model for SCIs, called the Gabor feature model (GFM), which was developed based on Gabor's horizontal and vertical imaginary part. This approach is similar to HVS interpretation than other approaches, and the measurement complexity is therefore smaller.

S. Bae and M. Kim [12] proposed a new IQA method, called the Structural Contrast-Quality Index (SCQI), using a structural contrast index, which may well characterize local and global visual quality perceptions for different image characteristics with structural-distortion types. Furthermore, they developed a Structural Contrast Distortion Metric (SCDM), a tweaked SCQI with the desirable mathematical properties of valid distance metric ability and quasi-convexity.

Recently, Ni et al. [15] used three salient edge attributes (edge contrast, edge width, and edge direction information) extracted from the SCIs in their developed IQA model, called edge similarity measurement (ESIM). It is worth to mention that they also developed and presented a new SCI database, denoted as SCID.

3 Proposed Method

In this paper, we introduce a novel idea of combining image quality metrics (IQMs) by applying Genetic Programming (GP) [16]. We have selected seven IQMs: GFM [10], GSS [11], SCDM [12], SCQI [12], SQMS [13], SVQI [14], ESIM [15] based on their intended use or construction and on the public accessibility of the code of all of them. The GP may exploit huge search space that consists of all possible combinations of objective scores of these IQMs and find the best regression model, considering its performance on the training data as well as its simplicity, which describes the weighted sum of IQM objective scores for predicting the subject scores of images in datasets, without making any prior assumptions about the model structure.

Genetic Programming (GP) is an evolutionary algorithm that solves user-defined tasks through the evolution of computer programs, which are represented as trees and have evolved using genetic operators such as crossover and mutation. It has a wide range of applications such as symbolic regression, classification and program synthesis, of which Symbolic Regression (SR) is the most popular application. The SR aims to find the relationship between the input variables and the responses in the given data set. It discovers the relationship by combining various mathematical expressions without assuming a model beforehand unlike traditional regression tasks, this property makes it more flexible for practical applications, and also it has a strong expressive power because it tries to arrive at clear white box models to interpret.

For facilitating the implementation of Genetic Programming, in our study we adopt the widely used technology platform for symbolic regression through GP "GPTIPS", which is a free open-source MATLAB toolbox developed by Dominic Searson [17]. Since May 2015, an improved version of GPTIPS2 [18] has been available. The choice of this optimization algorithm is motivated by its ability to evolve the white box model structure and its coefficients automatically, using linear least squares regression.

For training data, we use the seven Full Reference IQA measures previously mentioned with publicly available source code, and the corresponding output values (target) are the subjective Mean Opinion Score (MOS). As these metrics are with full reference, they require the degraded image and its reference image. The model that performed best on the holdout validation data was chosen.

The new approach starts with an initial population of randomly-generated computer programs that are composed of functions and terminals. The functions used here are " +" and "−" for linear combination, and the terminals are seven FR-IQA metrics. Then, from generation to generation, after applying selection, crossover and mutation operators, better computer programs (represented by tree structures) emerge; with a maximum tree depth are 5 and 3 genes allowed in each individual. The GP was run for 100 generations, with a population of 100 individuals. Scattered crossover (85%), Gaussian mutation (5%) and Stochastic uniform selection (10%) rules were used.

The GP's parameters were determined experimentally, by observing convergence of the objectives functions over the generations, to evolve models that perform well but are not overly complex.

The two objectives are used here:

1. Maximise the goodness-of-fit, where we use the coefficient of determination (R^2) measure:

$$R^2 = \sum_1^m (objM - sub)^2 / \sum_1^m (objM - mean(objM))^2 \qquad (1)$$

   where:

   ObjM: vector of objective scores of the obtained model,
   Sub: MOS or differential MOS (DMOS),
   m: the total number of images.

   The coefficient R^2 (between the MOS and the predicted output *ObjM* for the obtained model) is calculated to determine the model performance in the population over the generations of a run.

2. Minimize the complexity of the model, where "Expressional Complexity" is defined as the sum of nodes of all sub-trees within a tree as defined in [19]; this sort of metric has the advantage of favouring fewer layers as well as providing more resolution at the low end of the complexity axis of the Pareto front so that more simple solutions may be included in the Pareto front.

So the goal is to obtain models that make a good tradeoff and perform well on both objectives. As the two objectives are often contradictory, we say that the optimization is multi-objectives.

The performance of the proposed model is evaluated on two benchmark image datasets; they are succinctly described as follows:

The Screen Image Quality Assessment Database (SIQAD) [1]. This database, which is the first of its kind, is made up of 20 references screen content images, as shown in Fig. 1, and 980 corresponding images distorted by seven categories of distortions: Gaussian Blur (GB), Gaussian Noise (GN), JPEG2000 Compression (J2C), JPEG Compression (JC), Contrast Change (CC), Motion Blur (MB), and Layer segmentation-backed Coding (LC), each at seven levels of distortions. In this database, SCIs are with different resolutions, and the subjective ratings (in terms of DMOS values) were collected using a single-stimulus scaling paradigm.

The Screen content image quality database (SCID) [15] is composed of 40 references SCIs and 1800 distorted SCIs that suffered nine types of distortions at five levels, these distortion contains GN, MB, GB, CC, JPEG, JPEG2000, color saturation change (CSC), High Efficiency Video Coding based the Screen Content Coding (HEVC-SCC) and Color Quantization with Dithering (CQD).The resolution is fixed at 1280×720. The subjective ratings (in terms of MOS values) were collected using a double-stimulus scaling paradigm.

Note that the original SCIs were collected in the two databases with various combinations of text, graphics and images through screen snapshots from web pages, Power-Point slides, PDF files, digital magazines, etc. [1, 15].

For all experiments, we apply the k-fold-cross-validation strategy (with k = 5) to select our training and testing sets. On the SIQAD datasets, 980 distorted SCIs are randomly divided into two subsets; the 80% of data are oriented for training where it contains 784 distorted SCI associated with 16 reference images, and the remaining 20% i.e. 196 distorted SCI associated with 4 reference images are used as test set. Pristine images are included in the training/testing set. This procedure of k-fold-cross-validation strategy can prevent the overfitting problem.

After 50 runs with random training/testing dataset splits, a fitted model can be written as:

$$GP - SCIQ = 0, 23 * GFM + 0, 43 * SVQI + 0, 37 * ESIM - 1, 9 \qquad (2)$$

where, in order to select the predictive model, from a Pareto-optimal model, a trade-off must be made between the complexity and the performance of the model. It can be seen that resulted in aggregate models used only three metrics among seven.

4 Experimental Evaluation

4.1 Evaluation of Prediction Performance

To check the efficiency of the proposed approach, we use a MOS scatter plot versus the results obtained; where the diagram should have a compact form without outliers and with a tendency to a monotonic behavior.

Figure 2 shows the scatter plots of the score given by the results of our new method versus the subjective MOS and the seven IQA metrics used. Additionally, a fit with a logistic function as suggested in [20] and [21] is shown for easier comparison. Each point on the plot represents one image in the benchmark, the horizontal axis corresponds to the score of a new approach for that image and the vertical axis corresponds to the subjective MOS for that image. It can be seen that the objective scores predicted by the proposed method correlate consistently with the subjective scores. We note that the result of our method is adequate to human perception; where the percentage of outliers is decreased and a tendency to a monotonic behavior is increased.

For evaluation and development of the developed model, according to the widely-used protocol [20, 21], we use the following four indices of prediction accuracy, monotonicity, and consistency to measure IQA model performance. The Root Mean Square Error (RMSE), Pearson linear Correlation Coefficient (PCC), Spearman Rank order Correlation Coefficient (SRCC) and Kendall Rank order Correlation Coefficient (KRCC). These performance indices are calculated after a nonlinear mapping between a vector of objective scores, and subjective scores MOS, using the following mapping function for the nonlinear regression [20]:

$$Q_i = \beta_1 \left(\frac{1}{2} - \frac{1}{1 + \exp(\beta_2(S_i - \beta_3))} \right) + \beta_4 S_i + \beta_5 \tag{3}$$

where $\beta = [\beta_1, \beta_2, ..., \beta_5]$ are parameters of the non-linear regression model, and Q_i is the mapped objective score of the ith distorted SCI computed from an IQA model, S_i is the corresponding subjective MOS or DMOS values.

The four correlation coefficients are calculated as Eq. (4), (5), (6) and (7):

$$PCC = \frac{\sum_{i=1}^{n}(Q_i - \overline{Q})(m_i - \overline{m})}{\sqrt{\sum_{i=1}^{n}(Q_i - \overline{Q})^2(m_i - \overline{m})^2}} \tag{4}$$

$$SRCC = 1 - \frac{6\sum_{i=1}^{N} d_i^2}{N(N^2 - 1)} \tag{5}$$

$$KRCC = \frac{2(N_c - N_d)}{N(N - 1)} \tag{6}$$

$$RMSE = \sqrt{\frac{1}{n}\sum_{i=1}^{n}(Q_i - \overline{Q})^2} \tag{7}$$

where n represents the number of distorted images, d_i is the difference between the ith image's ranks in subjective and objective levels. Letters $\overline{m}$ and $\overline{Q}$ express the mean values of m_i and Q_i. m_i and Q_i are the objective and subjective scores of ith SCI in the database which has n SCIs. N_c and N_d denote the number of concordant and discordant pairs found in the database, respectively.

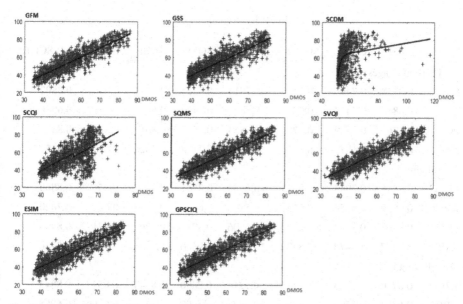

Fig. 2. Correlations of subjective assessment obtained and the seven state-of-the-art IQA metrics for SIQAD dataset (each data point represent one test image)

Table 1 shows the evaluation results, between our new method and the methods used for fusion in the two benchmarks. The top one model for each criterion is shown in boldface. The values of RMSE closer to 0 are considered better, in contrary to SRCC, KRCC, and PCC whose values should be close to 1. Results show that our method (GP-SCIQ) is clearly a top performing model relative to the individual metrics.

The table also contains direct and weighted averages of obtained values. For the weighted average, the number of images in the database is used as its weight. Results show the outstanding performance of our new model.

Table 1. Performance comparison of the proposed model with the seven metrics used in the fusion

	GFM	GSS	SCDM	SCQI	SQMS	SVQI	ESIM	GP-SCIQ
SIQAD (980 images)								
SRCC	0,8734	0,8436	0,6124	0,6112	0,8803	0,8835	0,8632	**0,8917**
KRCC	0,6875	0,6486	0,4565	0,4546	0,6935	0,6985	0,6742	**0,7094**
PCC	0,8831	0,8515	0,4517	0,6138	0,8872	0,891	0,8791	**0,9005**
RMSE	6,7158	7,5051	12,7703	11,3	6,6039	6,4961	6,8220	**6,2239**

(*continued*)

Table 1. (*continued*)

	GFM	GSS	SCDM	SCQI	SQMS	SVQI	ESIM	GP-SCIQ
SCID (1800 images)								
SRCC	**0,8758**	0,749	0,7715	0,7814	0,832	0,8385	0,8486	**0,8802**
KRCC	**0,6843**	0,5422	0,5737	0,5826	0,6429	0,6501	0,6517	**0,6946**
PCC	**0,876**	0,7609	0,7781	0,7854	0,8563	0,861	0,8642	**0,8875**
RMSE	**6,8296**	9,189	8,8956	8,7646	7,3134	7,2026	7,1259	**6,5252**
Overall direct								
SRCC	0,8746	0,7963	0,6919	0,6963	0,8561	0,861	0,8559	**0,8860**
KRCC	0,6859	0,5954	0,5151	0,5186	0,6682	0,6743	0,6630	**0,7020**
PCC	0,8795	0,8062	0,6149	0,6996	0,8717	0,876	0,8717	**0,8940**
RMSE	6,7727	8,3470	10,8329	10,0323	6,9586	6,8493	6,9740	**6,3746**
Overall weighted								
SRCC	0,8750	0,7823	0,7154	0,7214	0,8490	0,8544	0,8537	**0,8843**
KRCC	0,6854	0,5797	0,5324	0,5375	0,6607	0,6672	0,6596	**0,6998**
PCC	0,8785	0,7928	0,6630	0,7249	0,8672	0,8716	0,8695	**0,8921**
RMSE	6,7895	8,5954	10,2615	9,6584	7,0633	6,9535	7,0188	**6,4190**

4.2 Performance Comparisons with SCI Methods on SIQAD

Here, the proposed GP-SCIQ is compared, on SIQAD database, with state-of-the-art approaches (SFUW [3], PICNN [22], MDOGS [23], FQI [24], SIQADFII [25], Xia19 [26]) which are specifically developed IQA methods for SCIs.

Table 2 outlines the performance comparison of the proposed GP-SCIQ and seven state-of-the-art methods for SCI on SIQAD database. It is observed that the proposed GP-SCIQ outperforms all other methods, which can be demonstrated by the highest PLCC and SRCC values and the lowest RMSE value in Table 2. Although the SRCC of our GP-SCIQ is closed to the SRCC of PICNN, but is outperforms it in terms of PLCC and RMSE.

Table 2. Experimental results of proposed method and other existing methods on SIQAD database

Method	PLCC	SRCC	RMSE
SFUW (2017)	0.8910	0.8800	6.4990
PICNN (2018)	0.896	**0.897**	6.790
MDOGS (2018)	0.8839	0.8822	6.6951

(*continued*)

Table 2. (*continued*)

Method	PLCC	SRCC	RMSE
FQI (2019)	0.8018	0.7729	8.7312
SIQADFII(2019)	0.9000	0.8880	6.2422
Xia19 (2020)	0.8343	0.8213	7.8924
GPSCIQ	**0.9005**	**0.8917**	6.2239

4.3 Cross Database Test

Besides, to verify the robustness and generalization ability of our model, we trained the system based on SIQAD, and tested on the subset of SCID that includes 1200 distorted images with the 6 common distortion types of noises (GN, GB, MB, CC, JPEG, JPEG2000), (Where 7 types of noises in SIQAD are characterized that overlaps with 6/9 of the noise types in SCID). The cross-data performance (in terms of SPCC) of our proposed model is compared with the following state-of-the-art new models: QODCNN-FR [27], RIQA [28], SIQA-DFII [25] and MTDL-SCI [29], the values of SPCC are taken from the originally published papers. From Table 3, it can be observed that our model achieves an outstanding performance improvement which means the proposed model owns stronger generalization ability.

Table 3. Cross-database evaluation of our proposed model and other models in SCID

	GP-SCIQ	QODCNN-FR2019 [27]	RIQA 2020 [28]	SIQA-DFII 2020 [25]	MTDL-SCI 2021[29]
SPCC	**0,9126**	0,876	0.8771	0,8507	0.8186

4.4 Statistical Significance

To assess whether or not the numerical difference between IQA algorithm's performances were statistically significant, an F-test method [20] is carried out with a significance level of 0.05; it was employed to compare the prediction residuals of two algorithms, assuming that the residuals are Gaussian distributed.

The results of the statistical performance, on the both SIQAD and SCID databases, of the proposed approach (GP-SCIQ) compared to other algorithms are presented in Table 4.

Each entry in the table is a code word made up of two symbols; the first symbol designates the result on the SIQAD database and the second is the result on the SCID database. A value of "1" indicates that the row metric is significantly greater than the column metric, a value of "0" indicates that the row metric is significantly comparable to the column metric.

For both image bases, the results of our model are all "1". In summary, the significance tests confirm the excellent performance of the developed model against all the listed methods.

Table 4. Statistical significance tests on the database SIQAD and SCID ('--': statistically indistinguishable)

SIQAD/SCID	GFM	GSS	SCQI	SQMS	SVQI	ESIM	GP-SCIQ
GFM	--	1 1	1 1	0 1	0 1	0 0	0 0
GSS	0 0	--	0 1	0 0	0 0	0 0	0 0
SCQI	0 0	0 0	--	0 0	0 0	0 0	0 0
SQMS	0 0	1 1	1 1	--	0 0	1 0	0 0
SVQI	0 0	1 1	1 1	0 0	--	1 0	0 0
ESIM	0 0	0 0	1 1	0 0	0 0	--	0 0
GP-SCIQ	**1 1**	**1 1**	**1 1**	**1 1**	**1 1**	**1 1**	--

5   Conclusion

In this article, the evaluation of screen content images is examined using the symbolic regression by the method of genetic programming (GP); where the objective scores of seven image quality metrics are employed as predictors, and the appropriate metrics are determined to formulate a new metric for evaluating screen content image quality, this is by optimizing simultaneously two competing objectives of model goodness of fit to data and model complexity.

The performance of the trained model is evaluated on two public large-scale SCI databases (SIQAD and SCID). Experimental evaluation shows that the proposed approach generates the greatest PLCC, SRCC, KRCC and RMSE scores in both Direct Average and Weighted Average performance comparison, meaning that the proposed model outperforms all the state-of-the-art models under comparison by a significant margin. We also perform the tests across databases to show the generality and robustness of the proposed metric.

In future works, we would like to extend this approach to different image contents including both classical Natural Images and popular Screen Content images.

References

1. Yang, H., Fang, Y., Lin, W.: Perceptual quality assessment of screen content images. IEEE Trans. Image Process. **24**, 4408–4421 (2015). https://doi.org/10.1109/TIP.2015.2465145
2. Wang, S., Ma, L., Fang, Y., Lin, W., Ma, S., Gao, W.: Just noticeable difference estimation for screen content images. IEEE Trans. Image Process. **25**, 14 (2016)

3. Fang, Y., Yan, J., Liu, J., Wang, S., Li, Q., Guo, Z.: Objective quality assessment of screen content images by uncertainty weighting. IEEE Trans. Image Process. **26**, 2016–2027 (2017). https://doi.org/10.1109/TIP.2017.2669840
4. Gu, K., Zhai, G., Lin, W., Yang, X., Zhang, W.: Learning a blind quality evaluation engine of screen content images. Neurocomputing **196**, 140–149 (2016). https://doi.org/10.1016/j.neu com.2015.11.101
5. Chandler, D.M.: Seven challenges in image quality assessment: past, present, and future research. ISRN Sig. Process. **2013**, 1–53 (2013). https://doi.org/10.1155/2013/905685
6. Ieremeiev, O., Lukin, V., Ponomarenko, N., Egiazarian, K.: Combined no-reference IQA metric and its performance analysis. Electron. Imaging **31**(11), 260-1–260-7 (2019). https://doi.org/10.2352/ISSN.2470-1173.2019.11.IPAS-260
7. Ieremeiev, O., Lukin, V., Ponomarenko, N., Egiazarian, K.: Robust linearized combined metrics of image visual quality. Electron. Imaging **2018**, 260-1–260-6 (2018). https://doi.org/10.2352/ISSN.2470-1173.2018.13.IPAS-260
8. Merzougui, N., Djerou, L.: Multi-gene genetic programming based predictive models for full-reference image quality assessment. J. Imaging Sci. Technol. **65**, 60409-1–60409-13 (2021)
9. Oszust, M.: Decision fusion for image quality assessment using an optimization approach. IEEE Signal Process. Lett. **23**, 65–69 (2016). https://doi.org/10.1109/LSP.2015.2500819
10. Ni, Z., Zeng, H., Ma, L., Hou, J., Chen, J., Ma, K.-K.: A gabor feature-based quality assessment model for the screen content images. IEEE Trans. Image Process. **27**, 4516–4528 (2018). https://doi.org/10.1109/TIP.2018.2839890
11. Ni, Z., Ma, L., Zeng, H., Cai, C., Ma, K.-K.: Gradient direction for screen content image quality assessment. IEEE Signal Process. Lett. **23**, 5 (2016)
12. Bae, S.-H., Kim, M.: A novel image quality assessment with globally and locally consilient visual quality perception. IEEE Trans. Image Process. **25**, 2392–2406 (2016). https://doi.org/10.1109/TIP.2016.2545863
13. Gu, K., et al.: Saliency-guided quality assessment of screen content images. IEEE Trans. Multimedia **18**, 1098–1110 (2016). https://doi.org/10.1109/TMM.2016.2547343
14. Gu, K., Qiao, J., Min, X., Yue, G., Lin, W., Thalmann, D.: Evaluating quality of screen content images via structural variation analysis. IEEE Trans. Visual Comput. Graphics **24**, 2689–2701 (2018). https://doi.org/10.1109/TVCG.2017.2771284
15. Ni, Z., Ma, L., Zeng, H., Chen, J., Cai, C.: ESIM: edge similarity for screen content image quality assessment. iEEE Trans. Image Process. **26**, 14 (2017)
16. Gandomi, A.H., Alavi, A.H.: A new multi-gene genetic programming approach to nonlinear system modeling. Part I: materials and structural engineering problems. Neural Comput. Applic. **21**(1), 171–187 (2012). https://doi.org/10.1007/s00521-011-0734-z
17. Hii, C., Searson, D.P., Willis, M.J.: Evolving toxicity models using multigenesymbolic regression and multiple objectives. Int. J. Mach. Learn. Comput. **1**, 30–35 (2011). https://doi.org/10.7763/IJMLC.2011.V1.5
18. Searson, D.P.: GPTIPS 2: an open-source software platform for symbolic data mining. In: Gandomi, A.H., Alavi, A.H., Ryan, C. (eds.) Handbook of Genetic Programming Applications, pp. 551–573. Springer, Cham (2015). https://doi.org/10.1007/978-3-319-20883-1_22
19. Smits, G.F., Kotanchek, M.: Pareto-front exploitation in symbolic regression. In: O'Reilly, U.-M., Yu, T., Riolo, R., Worzel, B. (eds.) Genetic Programming Theory and Practice II, vol. 8, pp. 283–299. Springer-Verlag, New York (2005)
20. Sheikh, H.R., Sabir, M.F., Bovik, A.C.: A statistical evaluation of recent full reference image quality assessment algorithms. IEEE Trans. Image Process. **15**, 3440–3451 (2006). https://doi.org/10.1109/TIP.2006.881959
21. MRohaly: VQEG: Final Report from the video quality experts group on the Validation of objective models of video quality assessment. FR-TV Phase II, http://Www.Vqeg.Org/ (2000)

22. Chen, J., Shen, L., Zheng, L., Jiang, X.: Naturalization module in neural networks for screen content image quality assessment. IEEE Signal Process. Lett. **25**, 1685–1689 (2018). https://doi.org/10.1109/LSP.2018.2871250

23. Fu, Y., Zeng, H., Ma, L., Ni, Z., Zhu, J., Ma, K.-K.: Screen content image quality assessment using multi-scale difference of Gaussian. IEEE Trans. Circuits Syst. Video Technol. **28**, 2428–2432 (2018). https://doi.org/10.1109/TCSVT.2018.2854176

24. Rahul, K., Tiwari, A.K.: FQI: feature-based reduced-reference image quality assessment method for screen content images. IET Image Proc. **13**, 1170–1180 (2019). https://doi.org/10.1049/iet-ipr.2018.5496

25. Jiang, X., Shen, L., Ding, Q., Zheng, L., An, P.: Screen content image quality assessment based on convolutional neural networks. J. Vis. Commun. Image Represent. **67**, 102745 (2020). https://doi.org/10.1016/j.jvcir.2019.102745

26. Xia, Z., Gu, K., Wang, S., Liu, H., Kwong, S.: Toward accurate quality estimation of screen content pictures with very sparse reference information. IEEE Trans. Industr. Electron. **67**, 2251–2261 (2020). https://doi.org/10.1109/TIE.2019.2905831

27. Jiang, X., Shen, L., Feng, G., Yu, L., An, P.: Deep Optimization model for Screen Content Image Quality Assessment using Neural Networks. ArXiv:190300705 [Cs] (2019)

28. Jiang, X., Shen, L., Yu, L., Jiang, M., Feng, G.: No-reference screen content image quality assessment based on multi-region features. Neurocomputing **386**, 30–41 (2020). https://doi.org/10.1016/j.neucom.2019.12.027

29. Gao, R., Huang, Z., Liu, S.: Multi-task deep learning for no-reference screen content image quality assessment. In: Lokoč, J., et al. (eds.) MMM 2021. LNCS, vol. 12572, pp. 213–226. Springer, Cham (2021). https://doi.org/10.1007/978-3-030-67832-6_18

A Progressive Deep Transfer Learning for the Diagnosis of Alzheimer's Disease on Brain MRI Images

Norelhouda Laribi[1]([✉]), Djamel Gaceb[1], Akram Benmira[1], Sara Bakiri[1], Amira Tadrist[1], Abdellah Rezoug[1], Ayoub Titoun[2], and Fayçal Touazi[1]

[1] LIMOSE Laboratory, University M'Hamed Bougara of Boumerdes, Boumerdes, Algeria
{n.laribi,d.gaceb,a.rezoug,f.touazi}@univ-boumerdes.dz
[2] National School of Computer Science (ESI), Algiers, Algeria

Abstract. Alzheimer's Disease (AD) is a serious public health issue that affects elderly people. According to the World Health Organization (WHO), nearly 100,000 people suffer from it in Algeria. Therefore, early diagnosis is vital for the patient's treatment. In this paper, to improve the precision of early diagnosis, MRI images are employed to extract hidden features using a variety of transfer learning strategies on convolutional neural network models with different fine-tuning levels. The learned features are transferred from a large dataset of natural images to a small dataset of AD MRI images. For this purpose, progressive transfer learning is proposed, using the brain cancer MRI dataset as an intermediate to maintain both general features and domain-specific features relevant to the target domain. Accordingly, our tests showed that the performance results of the proposed approach produce a high accuracy of 99.22%, 98.90%, 98.28%, and 99.37% using respectively the VGG16, MobileNetV2, Xception and Inception models. These results demonstrate that, even with a highly diverse database like ImageNet, selecting the appropriate architecture and level of fine-tuning yields an improved adaptability and specialization of the pre-trained model.

Keywords: Convolutional neural network · Transfer learning · Fine-tuning · Alzheimer disease · Brain cancer · MRI images

1 Introduction

Alzheimer's disease (AD) is a progressive, neuro-degenerative brain disorder that destroys neurotransmitters, brain cells, and nerves, altering brain processes, memory, and behaviors, ultimately leading to dementia in elderly individuals. However, there are medical procedures that are only accessible with a doctor's prescription that can assist in slowing the condition's progression. Thus, early diagnosis of AD is essential for patient care and relevant research studies [1].

The researchers in this area tried to better understand the biological mechanisms of this disease, especially since the causes of the disease have not been identified yet. Pathophysiological: AD is characterized by the association of brain neuro-pathological lesions,

M. Salem et al. (Eds.): ICAITA 2022, CCIS 1769, pp. 65–78, 2023.
https://doi.org/10.1007/978-3-031-28540-0_6

which will progress over time to the entire brain cortex, explaining the progression of the disorders with the appearance of aphasia, dyspraxia, visuospatial disorders, and executive function disorders. Doctors diagnose patients concerning many requirements where imaging scanning is an essential part by determining the presence of certain symptoms using medical imaging analysis, including a thorough medical history, mental status testing, a physical and neurological exam, blood tests, and brain imaging exams, including computed tomography (CT), a positron emission tomography (PET) scan or scanning, and magnetic resonance imaging (MRI). MRI can detect brain abnormalities associated with mild cognitive impairment (MCI) and show a decrease in the size of different areas of the brain (mainly affecting the temporal and parietal lobes) [2].

With the advent of new deep learning (DL) models aimed at recognizing the salient low-level elements in an image, it is universally recognized that artificial intelligence may help in handling the growing complexity of Alzheimer's imaging data.

It is the most popular and effective method for determining a disease's classification from a large dataset [3]. Thus, several studies have used structural MRI-based biomarkers to classify AD, which discussed techniques of DL like CNN, Deep Belief Networks, Deep Boltzmann Machine, RNN [2, 4–8], etc. While several approaches have been proposed to different AD stages, using transfer learning with a relatively small dataset to exploit the usage of learned features by a pre-trained network over a given problem to solve different problems.

In this paper, we explore the use of progressive deep transfer learning to enhance AD diagnosis on small training dataset. The main contribution of the proposed method is how to investigate a progressive transfer that can transfer the learned features from a large dataset of natural image data to a small AD dataset of a specific domain. We find that applying the learned features from ImageNet dataset to the 15000 images of brain cancers and then applying the learned features to these latter ones based on Alzheimer's MRI images considerably improves the performance of the model.

The remainder of this paper is organized as follows. Section 2 introduces the concept of transfer learning and summarizes a part of the literature related to the subject. Section 3 describes the proposed method of progressive deep transfer learning. The experimental results are presented in Sect. 4.

2   Related Works

The current literature shows that approaches based on deep learning offer better performance compared to approaches based on traditional machine learning, especially in the domain of computer vision. Consequently, the application of deep learning in diagnosis using medical imaging has led to a real revolution for the treatment of different diseases in early phases. Alzheimer's disease (AD) is a remarkable challenge for healthcare in the 21st century. The application of deep learning to the early detection and automated classification of AD has recently gained considerable attention, as rapid progress in neuroimaging techniques has generated large-scale multimodal neuroimaging data. Several studies have shown how to analyze medical imaging; especially those related to AD, by using transfer learning (TL) [9]. Several reviews or comparative studies have been published to trace the evolution of scientific research on the application of DL in the diagnosis of AD.

A systematic review was published in 2019 in [10], retraces several works (between January 2013 and July 2018) on the application of deep learning approaches on neuroimaging data for diagnostic classification of AD. These works were reviewed, evaluated, and classified by algorithm and neuroimaging type, and the findings were summarized. Of 16 studies meeting full inclusion criteria, 4 used a combination of deep learning and traditional machine learning approaches, and 12 used only deep learning approaches. The combination of traditional machine learning for classification and stacked auto-encoder (SAE) for feature selection produced accuracies of up to 98.8% for AD classification. Deep learning approaches (CNN or RNN) that use neuroimaging data without pre-processing for feature selection have yielded accuracies of up to 96.0% for AD classification. Another study, published in late 2021, in [11] presents a systematic review of the current state of early AD detection by using deep learning models with transfer learning and neuroimaging biomarkers. Five databases were used and the results before screening reported 215 studies published between 2010 and 2020. After screening, 13 studies met the inclusion criteria. We noted that the maximum accuracy achieved by the works studied for AD classification is 98.20% by using the combination of 3D CNN and local transfer learning. The results show that transfer learning helps researchers in developing a more accurate system for the early diagnosis of AD. Other works that emphasize the importance of transfer learning by developing different strategies have been published. Hon et al. [12] as well as Khan et al. [13] investigated how the training size impacts the nature of transfer learning; for this, they performed three experiments with small datasets: VGG16 from scratch, VGG16 and Inception pre-trained models. Khan et al. [13] performed four experiments of fine-tuning convolutional layers on VGG19: from block 3 to block 5 (from layer 5 to 16); from block 4 to block 5 (from layer 9 to 16), block 5 (from layer 13 to 16) or fine-tuning all the layers. Oktavian et al. [14] proposed a transfer learning with the Residual Network 18 Layer (ResNet-18) by changing the network activation function to a mish activation function. Hosseini-Asl et al. [6] proposed an AD diagnostic framework that uses hierarchical feature extraction in hidden layers (3D-CNN) to extract features of brain MRI with a source-domain-trained 3D-CAE and performs task-specific classification with a deeply supervised target-domain-adaptable 3DCNN. Zoetmulder et al. [15] compared pre-trained CNNs on classification, segmentation, and self-supervised tasks using various source and target domains. Zhu et al. [16] compared the performance of a transfer learning approach to a transfer learning with co-train (TLCO), which is used for re-training the pre-trained memory prediction model by using a combination of the tuning and training sets with the site indicator. Helaly et al. [17] compared two methods: The first method uses simple 2D and 3D convolution architectures; the second method applies transfer learning to take advantage of the pre-trained models for medical image classification: the VGG19 model. Tan et al. [18] proposed a new method named: Instance Transfer Learning algorithm (ITL). ITL uses gravity transfer to transfer the source domain data closer to the target data set. This method is based on wrapper mode to get the best deviation between the transferred source domain samples and the target domain. Finally, the optimal transferred domain samples and the target domain training samples are combined for classification; they showed that transferred source domain samples by the ITL algorithm can enlarge the target domain training samples and assist in improving the classification accuracy significantly.

Previous studies that demonstrated the benefit of transfer learning for medical segmentation target tasks using various source tasks and domains focused primarily on VGG [12, 13, 17, 19, 20], ResNet [14], Alex Net [19–22], Inception v3 [19, 23], and GoogLeNet [11]. They showed that the number of fine-tuned layers depends on the size of the training set, [12, 13] used for this entropy to select the sample set. Although, fine-tuned of the entire network [19, 22] or the last layers [23] achieved the higher performance while in the other approaches, the pre-trained model is fine-tuned as feature extractor to improve the accuracy [12, 17].

Other studies, however, have shown that in order to transfer learning for AD diagnosis, it is necessary to investigate additional discriminating features representation to separate similar brain patterns [11] or to change the nature of the input images by adding brain radioactive tracers. After studying these different approaches to transfer learning, it was shown that fine-tuning the entire network was the method most frequently used to transfer the features of natural images to the task/domain (medical images). However, for this kind of image, the best transfer learning is accomplished with a comparable pre-training task and domain. The issue with target domains (downstream) is that we only have a small training dataset because we haven't collected relevant images pertaining to the target domain. Our goal is to develop a new progressive transfer learning approach that combines progressively general and specialized features from a large dataset of natural and medical images, then retrain the pre-trained model on the smaller, more specialized dataset.

3 Proposed Approach

In this paper, multiple CNN architectures are investigated for the classification of Alzheimer's MRI images using various transfer learning strategies. Due to the small dataset size of Alzheimer's MRI images, it has been investigated if learned features from natural images of ImageNet dataset (containing more than 14 millions of non-medical images) and particular médical images, can be transferred to this target dataset. A progressive transfer learning is then investigated in order to transfer the learned features to the downstream domain based on Alzheimer's MRI images. This involves first transferring the learnt features from natural images to the 15,000-image brain cancer MRI dataset. This kind of transfer may be useful when our dataset is too small to train a DL algorithm from scratch.

The VGG-16 model, pre-trained on ImageNet, is first refined and retrained to classify brain cancer images that have the same nature as the targeted images. After that the obtained model is refined one more time and retrained on AD dataset. The evaluation metrics used are: accuracy, precision, recall, AUC and the F1-score (or F-measure). The configuration of fine-tuning levels that give better results will be applied to the MobileNetV2, Inception and Xception architectures, to be able to compare the results obtained (Fig. 1).

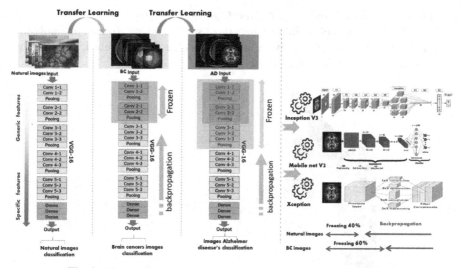

Fig. 1. Illustration of the proposed progressive deep transfer learning.

3.1 Approach 01: Transfer Learning on Pre-trained VGG-16 Model

Fine-tuning the pre-trained VGG-16 model from the source domain to the target domain, which is based on ImageNet and AD's dataset respectively. There are three scenarios being constructed:

1) **Without fine-tuning**

   At this stage, the CNN model will only be trained using the images from the Alzheimer's dataset, and the five frozen convolution blocks will be reused for feature extraction from pre-trained networks (Fig. 2).

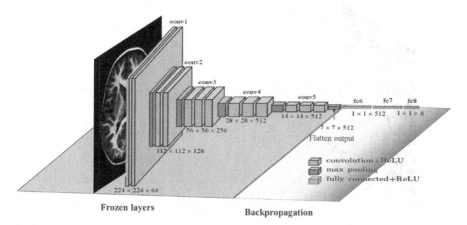

Fig. 2. Feature extraction using a pre-trained network (the entire convolution blocks is frozen while the classifier blocks is matched)

2) **Fine-tuning block 5 and keep the first four blocks frozen**
The convolution layers of the first four blocks were initially frozen, and only the fifth block of the convolution basis and the classifier FC are refined, enabling the model to learn more AD's features (Fig. 3).

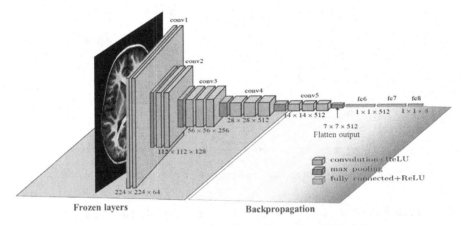

Fig. 3. Feature extraction with the fine-tuning the block 5.

3) **Fine-tuning of blocs 4 and 5**
The first three blocks are frozen and the fourth and fifth blocks are fine-tuned, the pre-trained network is refined on the Alzheimer's dataset to give the resulting model additional specificity (Fig. 4).

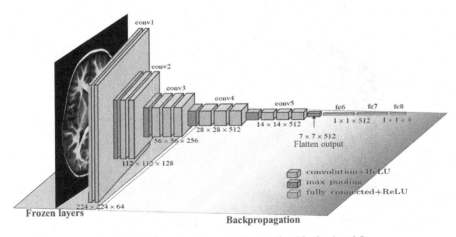

Fig. 4. Feature extraction with fine-tuning blocks 4 and 5.

3.2   Approach 02: Progressive Transfer Learning

In this Approach, we first trained 4 models on a brain cancer dataset one by one (the first model is not refined), and then we used progressive transfer learning to keep both generic and specific learned features. For this, we freeze the first layers of the pre-trained model on natural images and the last layers of the pre-trained model on brain MRI-scans respectively. Then the model is trained on Alzheimer's dataset following the same steps applied in the previous approach (without fine-tuning, fine-tuning block 5, fine-tuning block 4 and 5) (Fig. 5).

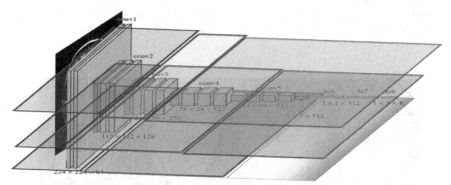

Freezing layers on image net weights

Freezing layers on brain cancer weights

Freezing layers on AD datasets

Fig. 5. Feature extraction with fine-tuning blocks 4 and 5 (Progressive transfer learning)

3.3   Approach 03: Configuration of MobileNetV2, Xception and Inception Model on VGG Pre-trained Model

To use progressive transfer learning from step 3 of approach 2 (fine-tuning blocks 4 and 5), we test three different configurations: MobileNetV2, Inception V3 and Xception. We use the same configuration as the pre-trained VGG16 for the three aforementioned architectures, transferring the pretrained models' training first on ImageNet while freezing 40% of the feature extractor's total number of layers, and then obtaining the rest from the cancer dataset. We then retrain the remaining 40% of the feature extractor layers of this final model using the Alzheimer dataset after freezing 60% of the layers.

To conduct our experiments, we used a computer with the following resources: Processor (Intel® Core™ i3-3120M CPU @ 2.50 GHz × 4), Memory (RAM = 8.00 Gb), Hard disk (500.1 GB), OS (Ubuntu 20.04.4 LTS).

4   Experiments and Results

Three approaches were conducted to examine the effectiveness of the proposed models, all using different datasets. At this stage of the study, we explain each of the Approaches and we give the produced results and the relevant findings.

4.1 Datasets

Public datasets contain many relevant benchmarks. It is therefore imperative to research how to use public dataset to increase the classification accuracy of target dataset. We used two types of MRI medical images from Kaggle to extract specific brain features: Alzheimer Dataset with FCIE includes four classes, and Multi Cancer dataset contains eight forms of cancer collected just from brain cancer images.

1. *Alzheimer Dataset:* contains 6400 images classified into four categories: mildly malignant (mildly demented), 64 moderately malignant (moderately demented), 3110 healthy (non-demented), and 3240 very malignant (very-mildly demented). It is divided as follows: 75% train, 15% validation, 10% test (Fig. 6).

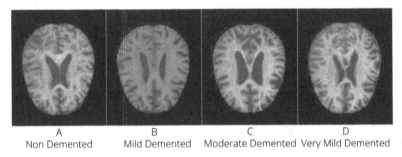

| A | B | C | D |
| Non Demented | Mild Demented | Moderate Demented | Very Mild Demented |

Fig. 6. Examples of dataset images relating to Alzheimer's disease.

2. *Multi Cancer Dataset:* contains 15000 images composed of 3 classes, including 5000 images of Glioma, 5000 images of Meningioma and 5000 images of Pituitary Tumor (Fig. 7).

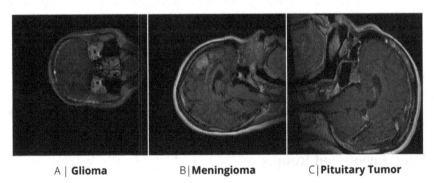

A | **Glioma**   B | **Meningioma**   C | **Pituitary Tumor**

Fig. 7. Examples of images of the multi cancer dataset.

4.2   Results of Approach 01

Retraining the model on the Alzheimer dataset yields an accuracy of 64.48% (f-measure = 62%) on the test images. By unfreezing the final block of the model, these values increased to an accuracy of 83.56% (f-measure = 83.49%). These values obtained show that this level of fine-tuning produced a better outcome than the initial scenario. By further freezing the final layers (blocks 4 and 5) with an accuracy of 69.79% (F1-measure = 68.63%), these values have marginally increased.

4.3   Results of Approach 02

In this approach, we trained four models using the brain cancer dataset, refined additional blocks (the first model is not refined), and summarized the results in Table 1. The results indicate that the fourth model performed the best during the test-set evaluation. It has an accuracy of 99.87% (recall = 99.99%, precision = 99.99%, and F-measure = 99.99%). The fourth model has been selected to be employed in the progressive learning of Alzheimer's categorization as a result.

Table 1. Results and evaluations metrics of Approach 02

Approach	Evaluation	Loss	Accuracy	Recall	Precision	AUC	F1-Score
1	Without fine-tuning	0,013	92	91,53	92,27	98,62	91,89
2	Fine-tuning B5	0,075	97,2	97,2	97,39	99,83	97,29
3	Fine-tuning B4-B5	0,02	99,13	99	99	99,99	99
4	Fine-tuning B3-B4-B5	0,01	99,87	99,99	99,99	99,99	99,99

Without fine-tuning, the application of progressive learning on Multi Cancer Dataset then on Alzheimer dataset gave an accuracy of 94.68% (f-measure = 95.0%). With fine-tuning block 5, the accuracy rose to 98.44% (f-measure = 98%). After one more block of refinement (block 5 and 4), the model's accuracy reached 99.22% (f-measure = 99%).

4.4   Results of Approach 03

Progressive transfer learning on the Alzheimer dataset, then the Multi Cancer dataset, without any fine-tuning, produces accuracy of 94.68% (f-measure = 95.0%) on test images. The accuracy increased to 98.44% (f-measure = 98.4%) with fine-tuning block 5. The model's accuracy is 99.22% after one additional refining (Block 5 and block 4).

As reported in Table 2, all models achieved an AUC of 99.99% when evaluated on predicting the final AD datasets, a precision of 99% for both MobileNetV2 and InceptionV3 and 98% for the Xception model. Furthermore, the accuracy measure allows to set the difference between the three models and recorded 99,37%, 98,9% and 98,28% for InceptionV3, MobileNetV2 and Xception respectively.

Table 2. Results and evaluation metrics of Approach 03.

Approach	Evaluation	Loss	Accuracy	Recall	Precision	AUC	F1-Score
1	mobileNetV2:unfreezing 09 blocks on BC then AD	0,03	98,9	99	99	99,99	99
2	Inception V3:freeze 40% andtrain model on BC then retrainit on AD en by freezing 60%	0,03	99,37	99	99	99,99	99
3	Xception: freeze 40% and trainmodel on BC then re-train it on AD by freezing 60%	0,05	98,28	98	98	99,99	98

4.5 Discussion

For many reasons, hidden, sluggish, non-lethal, and other features, AD images are exceedingly complicated. Furthermore, it might be challenging to obtain clinical samples. Therefore, there is a significant challenge in obtaining high-efficiency models with small size and dispersed samples. According to this literature and the principle of deep learning, to avoid inadequate training and overfitting caused by a small number of samples, we retrain the pre-trained model on ImageNet on a brain cancer MRI image dataset, which is larger than that of Alzheimer's disease. This dataset includes 5000 images from each of the three classes previously mentioned, totaling roughly 15,000. We primarily proposed a method based on deep transfer learning using MRI images of AD dataset. To this purpose, our tests showed that the performance results of this approach produced a high accuracy of 99.22% using VGG16, 98.90% using MobileNetV2, 98.28% using Xception and 99.37% using the Inception model. The AD classification results are shown in Table 3. Depending on the configuration used, the most accurate models for classifications were produced in the last approach.

Table 3. Results and evaluation metrics of three Approaches.

	Evaluation metrics	Loss	Accuracy (%)	Recall (%)	Precision (%)	AUC (%)	F1-score
1st Approach	Without fine tuning	0,73	64,48	57	68	90	62
	Fine-tuning B5	0,42	83,56	83	84	97	83,49
	Fine-tuning B5-B4	0,71	69,79	64	74	91	68,63

(continued)

Table 3. (*continued*)

	Evaluation metrics	Loss	Accuracy (%)	Recall (%)	Precision (%)	AUC (%)	F1-score
2nd Approach	Without fine tuning	0,0135	92	91,53	92,27	98,62	91,89
	Fine-tuning B5	0,0749	97,2	97,2	97,39	99,83	97,29
	Fine-tuning B5-B4	0,02	99,13	99	99	99,99	99
3rd Approach	mobileNetV2	0,03	98,9	99	99	99,99	99
	Inception V3	0,03	99,37	99	99	99,99	99
	Xception	0,05	98,28	98	98	99,99	98

The results summarized in Table 3 show that, even when beginning from a very dissimilar dataset, such as the ImageNet dataset, choosing the appropriate architecture and fine-tuning results in better adaptability and specialization of the pre-trained model (intended for the origin of the classification of 1000 object types).

According to the results obtained using Approach 01 and Approach 02 on the VGG16, MobileNetV2, InceptionV3, and Xception architectures (Approach 03), where the last approach has recorded high-performing results. This underlines the power and the performance of progressive deep transfer learning on a small dataset. We come to the conclusion that when we want to transfer learning and we train the model on a large dataset with images similar to those of the target domain dataset, then we move on to train it on the target domain, this gives better results than the classic method of transfer learning and fine-tuning. It appears to build a bridge between natural datasets and the target dataset, where the performance of the model depends on the nature of the bridge dataset used, it must be similar (Table 4).

Table 4. Comparison of accuracy values (%) of TL approaches used for Alzheimer classification

Ref	MRI 's training set	Dataset	Model used	Accuracy
[12]	5,120	OASIS	Inception V4	96.25%
[13]	2560	ADNI	VGG	*99.36%*
[14]	306	ADNI	Resnet	88.30%
[17]	48000	ADNI	VGG (2D)	97.11%
			VGG (3D)	97.36%
[18]	411 × 2 from ADNI 30 × 2 Local DS	ADNI + Clinical DS	ELA algorithm	88%

(*continued*)

Table 4. (*continued*)

Ref	MRI 's training set	Dataset	Model used	Accuracy
Ours	4800 AD 11250 BC	Kaggle[1]	InceptionV3	**99.37%**
			Mobile net V2	98.90%
			Xception	98.28%

When we compare our results to the most advanced state-of-the art approaches relative to AD disease in terms of accuracy, we can see that our approach with the three-configurations outperforms all the other approaches except for [13], our results are very close to theirs. [13] freezes groups of convolutional layers while using the image entropy of each slice to reduce the training set. So, it may decrease the model's capability for generalization across other datasets. Whereas in our approach, the use of multi-level feature extraction: InceptionV2, outperforms every model and increases the accuracy of the model.

The most significant effect of transfer learning in our approach can be seen when we freeze a group of layers progressively and take depth results of each block to be used for another smaller target dataset to progressively transfer the largest informative training images. This is what makes our approach more significant in obtaining high-efficiency models with small dataset and dispersed samples without using data augmentation.

While comparing with the other fine-tuned deep models, we can see that our approach has very good performance with a comparatively smaller training dataset.

Unlike the classic TL approach, our approach combines the advantages of two TL strategies: feature extraction for similar datasets and low-level conservation of generic features for dissimilar datasets. However, it lacks the strength and optimization to converge speedily, especially for other tasks like small object detection and segmentation tasks.

5   Conclusion

In this study, several methods were used for extracting discriminatory features from structural MRI in order to classify Alzheimer's disease. The experimental results on the Brain tumors and Alzheimer's disease dataset showed that the proposed progressive transfer learning offered high accuracy for AD classification. The best results were obtained by using the parameters of the first layers obtained on natural images (ImageNet dataset) and the last layers from the pre-trained model, obtained on Brain cancer dataset (which contains images similar to those of the AD dataset) to protect both generic and specific features from the first and last layers, respectively.

As the approaches were only tested on ILSVRC models, another future direction for this methodology is to configure these approaches on more recent CNN models. Explore additional models, such as transformer networks. Transfer learning's convergence may

---

[1] https://www.kaggle.com/datasets/gautamgc75/dataset-alzheimer-with-fcie.

be improved by the use of more advanced optimizers. Compare them and decide which can be completed quickly and efficiently for our model.

Finally, as previous works related to the integration of deep learning in medical imaging, including Alzheimer's disease, and in collaboration with clinicians to collect controls and surveys, we really hope to target real conditions and causes of this disease that clinicians and even analysts have not yet been able to identify. This assists the government in coming up with new strategies for fighting Alzheimer's disease and for finding new cures.

References

1. Lama, R.K., Gwak, J., Park, J.-S., Lee, S.-W.: Diagnosis of Alzheimer's disease based on structural MRI images using a regularized extreme learning machine and pca features. J. Healthc. Eng. **2017**, 1–11 (2017). https://doi.org/10.1155/2017/5485080
2. Prepared by: Office of Communications and Public Liaison: Neurological Diagnostic Tests and Procedures Fact Sheet. Neurological Diagnostic Tests and Procedures", NINDS. NIH Publication No. 19-NS-5380. 10 Apr 2019
3. Akkus, Z., Galimzianova, A., Hoogi, A., Rubin, D.L., Erickson, B.J.: Deep learning for brain MRI segmentation: state of the art and future directions. J. Digit. Imaging **30**(4), 449–459 (2017)
4. Arafa, D.A., Moustafa, H.E.D., Ali-Eldin, A.M., Ali, H.A.: Early detection of Alzheimer's disease based on the state-of-the-art deep learning approach: a comprehensive survey. Multimedia Tools Appl. **81**, 23735–23776 (2022)
5. Ding, Y., et al.: A deep learningmodel to predict a diagnosis of Alzheimer disease by using 18f-fdg pet of the brain. Radiology **290**(2), 456–464 (2019)
6. Hosseini-Asl, E., Gimel'farb, G., El-Baz, A.: Alzheimer's disease diagnostics by a deeply supervised adaptable 3d convolutional network. arXiv preprint arXiv:1607.00556 (2016)
7. Zamir, A.R., Sax, A., Shen, W., Guibas, L.J., Malik, J., Savarese, S.: Taskonomy: Disentangling task transfer learning. In: Proceedings of the IEEE Conference on Computer Vision and Pattern Recognition, pp. 3712–3722 (2018)
8. Zhang, L., Wang, M., Liu, M., Zhang, D.: A survey on deep learning for neuroimaging-based brain disorder analysis. Front. Neurosci. **14**, 779 (2020)
9. Acharya, H., Mehta, R. and Kumar Singh, D.: Alzheimer disease classification using transfer learning. In: 2021 5th International Conference on Computing Methodologies and Communication (ICCMC), pp. 1503–1508 (2021). https://doi.org/10.1109/ICCMC51019.2021.9418294
10. Taeho, J., Kwangsik, N., Andrew, J.S.: Deep learning in Alzheimer's disease: diagnostic classification and prognostic prediction using neuroimaging data. Front. Aging Neurosci. **11**, 220 (2019). https://doi.org/10.3389/fnagi.2019.00220
11. Agarwal, D., Marques, G., de la Torre-Díez, I., Franco Martin, M.A., García Zapiraín, B., Martín Rodríguez, F.: Transfer learning for Alzheimer's disease through neuro-imaging biomarkers: a systematic review. Sensors **21**(21), 7259 (2021)
12. Hon, M., Khan, N.M.: Towards Alzheimer's disease classification through transfer learning. In: 2017 IEEE International Conference on Bioinformatics and Biomedicine (BIBM), pp. 1166–1169. IEEE (2017)
13. Khan, N.M., Abraham, N., Hon, M.: Transfer learning with intelligent training data selection for prediction of Alzheimer's disease. IEEE Access **7**, 72726–72735 (2019)

14. Oktavian, M.W., Yudistira, N., Ridok, A.: Classification of alzheimer's disease using the convolutional neural network (CNN) with transfer learning and weighted loss. arXiv preprint arXiv:2207.01584 (2022)
15. Zhu, X., et al.: Transfer learning for cognitive reserve quantification. NeuroImage **258**, 119353 (2022). https://doi.org/10.1016/j.neuroimage.2022.119353
16. Zhao, X., Zhao, X.M.: Deep learning of brain magnetic resonance images: a brief review. Methods **192**, 131–140 (2021)
17. Helaly, H.A., Badawy, M., Haikal, A.Y.: Deep learning approach for early detection of Alzheimer's disease. Cognitive Comput. 1–17 (2021)
18. Tan, X., et al.: Localized instance fusion of MRI data of Alzheimer's disease for classification based on instance transfer ensemble learning. Biomed. Eng. Online **17**(1), 1–17 (2018)
19. Kandel, I., Castelli, M.: Transfer learning with convolutional neural networks for diabetic retinopathy image classification. A review. Appl. Sci. **10**(6), 2021 (2020). https://www.mdpi.com/2076-3417/10/6/2021
20. Wang, X., Lu, Y., Wang, Y., Chen, W.B.: Diabetic retinopathy stage classification using convolutional neural networks. In: 2018 IEEE International Conference on Information Reuse and Integration (IRI), pp. 465–471. IEEE (2018)
21. Lam, C., Yi, D., Guo, M., Lindsey, T.: Automated detection of diabetic retinopathy using deep learning. AMIA Summits Transl. Sci. Proc. **2018**, 147–155 (2018)
22. Li, X., Pang, T., Xiong, B., Liu, W., Liang, P., Wang, T.: Convolutional neural networks based transfer learning for diabetic retinopathy fundus image classification. In: 2017 10th International Congress on Image and Signal Processing, BioMedical Engineering and Informatics (CISP-BMEI), pp. 1–11 (2017)
23. Mohammadian, S., Karsaz, A., Roshan, Y.M.: Comparative study of fine-tuning of pre-trained convolutional neural networks for diabetic retinopathy screening. In: 2017 24th National and 2nd International Iranian Conference on Biomedical Engineering (ICBME), pp. 1–6 (2017)

An Optimized MSER Using Bat Algorithm for Skin Lesion Detection

Khadidja Belattar[1(✉)] ⓘ, Mohamed Ait Mehdi[2] ⓘ, Maroua Ridane[1],
and Loubna Ahmed Chaouch[1]

[1] Faculty of Sciences, Department of Computer Science,
University of Algiers, Algiers, Algeria
`k.belattar@univ-alger.dz`, `khadidja.belattar@gmail.com`
[2] LRIA, USTHB, Algiers, Algeria
`maitmehdi@usthb.dz`

Abstract. Detecting regions of interest in skin lesion images is of great significance in dermatological image analysis. In this article, we present a novel approach for the skin lesion detection in melanoma images based on bat algorithm and maximally stable extremal regions. The purpose is to better localize the regions of interest. To evaluate the proposed approach, the detection process is tested on the skin lesion images (melanoma, nevus) from MED-NODE and Atlas (dermIS, dermQUST) databases. The obtained qualitative and statistical results demonstrate the superiority of the proposed detector.

Keywords: Maximally stable extremal regions · Bat algorithm · Regions of interest · Skin lesion detection

1 Introduction

Medical imaging systems [28] provide more advantages over the clinical methods. Due to its efficiency and an improved accuracy, image recognition systems [5] are broadly established in the medical field. They also improve the health care quality and assist the clinicians. In this respect, the image recognition refers to a set of algorithms enabling the identification and detection of the objects or the features in the input image. The most common medical recognition applications include: malaria parasite detection [16], ophthalmic disease recognition [29], pneumonia recognition [23], colon cancer detection [2], alzheimer's disease recognition [31], tuberculosis disease detection [22], breast cancer recognition [24], COVID-19 detection [1], and the skin lesion classification [17].

Like any other medical recognition system, feature detection is a crucial phase in the skin lesion change evolution, severity assessment or computer-aided diagnosis systems. The detected features mainly describe the relevant discriminative skin lesion patterns. However, computing these features is very challenging due to the high intra-class variations and the artifact occlusion.

To this end, a performant detection process is required to achieve recognition quality improvement. Many algorithms have been developed in the literature to locally identify and describe the image features [12]. The basic ones

M. Salem et al. (Eds.): ICAITA 2022, CCIS 1769, pp. 79–93, 2023.
https://doi.org/10.1007/978-3-031-28540-0_7

are: Scale Invariant Feature Transform (SIFT) [19], Speeded Up Robust Feature (SURF) [3], Features from Accelerated Segment Test (FAST) [25], HARRIS [9], Binary Robust Independent Elementary Features (BRIEF) [4], Oriented FAST and Rotated BRIEF (ORB) [26], Hessian-Laplace and Hessian-Affine [21], SUSAN [32] and Maximally Stable Extremal Regions (MSER) [20]. They are currently used in the object detection applications for their robustness against different geometric transformations, an improved accuracy and fast recognition.

In this context, we propose to adapt bat algorithm with MSER algorithm for the skin lesion detection in melanoma images. The optimal set of MSER parameters is reached through the bat algorithm and used to identify the skin lesion region of the input image. We also utilize the intersection over union objective function for better detection results.

The remainder of the article is structured as follows. Section 2 presents a general introduction to the local feature detection and the bat algorithm. Section 3 describes the developed skin lesion detection system. Section 4 reports the achieved experimental results with the discussion. Section 5 outlines the conclusion and research perspectives.

2 Background

2.1 MSER-Based Local Feature Detection Feature

Feature detection refers to the process of selecting and localizing (by positions) the relevant features from the input image based on a given criterion. Local features could be edges, lines, intersections, corners, points or regions. They include the feature detector and descriptor. Feature detector is an algorithm that finds the salient features (known also interest points) from the given image, whereas descriptor is a vector computed to describe the image patches. In the literature, various algorithms for corner, blob and region detection were proposed. These detectors use gradient, intensity, segmentation or template techniques to locate the points of interest.

When it comes to medical images, the region detectors could be the suitable choice for the medical applications as the detected regions correspond well to the semantic image structures. Furthermore, the images are not so substantially distinct and lacking of sharp details. Accordingly, we can refer mainly to the MSER-based region detectors. In brief, the algorithm incrementally identifies small stable regions within the input image for different intensity thresholds. The detected region is a connected area characterized by uniform intensity and relatively constant area over a given range of thresholds. Nine parameters including: stability score (delta), max variation, min area, max area, min diversity, max evolution, area threshold, min margin and edge blur size, could be adjusted to optimize the detection quality. The main advantages of the MSER over region detectors are: (1) The accuracy of the feature localization, (2) Invariance to the rotation, scale and affine transformations, (3) High repeatability, and (4) The efficiency of the recognition. However, it fails to detect the relevant regions if the MSER parameters are not well selected.

Since our approach relies on the adaptation of the MSER detector, we provide here a literature review of the existing MSER-based methods in the medical field.

Previous studies include the work of Korotkov [15]. The author developed an automatic technique to detect the change in multiple pigmented skin lesions using MSER algorithm. The idea is to establish the correspondence between the skin lesions maps in three-dimensional space. Once the image registration is done, the change detection using the image patches around the moles is performed. However, the MSER parameters were tuned empirically during the experiments.

Following this work, Zhu et al. [37] improved the MSER algorithm for the ultrasound liver segmentation process. The principle here is to extract the lesions from the liver images using the generalized MSER detector. After the region detection, the edge detection and merging strategies are employed to determine edges and refine the edge map of the liver lesions. The last stage is the segmentation of the liver lesion according to the refined contour. The improved MSER algorithm is assessed on 88 ultrasound liver lesion images and it achieves the Dice similarity index of 0.6.

Jyotiyana and Maheshwari [13] applied successive Otsu's algorithm into hierarchical levels to define segmentation threshold for MRI tumor images. The segmented MRI images are introduced to MSER detector to extract the tumor area.

Another work [10] has been carried out using MSER detector. Hassan et al. (2019) suggested a breast cancer masses detection approach in mammograms images. The detection process includes the following steps: preprocessing the input mammogram image to improve the image quality. Then, the MSER regions are identified from the original and enhanced images using MSER detector. Finally, the mass area is identified by applying the feature matching process between the extracted MSER regions. The approach is evaluated on 300 mammogram images and produces an average detection accuracy of 95%.

Karthika et al. [14] performed the cyst boundary extraction in dental x-ray images based on MSER detector.

Based on the above-reported works, the application of the MSER in the medical analysis tasks is limited. This is may be due to the number of the parameters to be fine-tuned. If they are not well selected, the performed task will be affected. So, the MSER algorithm requires more adaptation in the medical tasks. Consequently, this work proposes the skin lesion detection based on the MSER algorithm with bat algorithm.

2.2   Bat Algorithm

Bat Algorithm (BA) is a bio-inspired metaheuristic that was first applied in [35] to solve hard continuous optimization problems. The algorithm is inspired by the echolocation behavior observed in microbats foraging for prey. During the prey search, the microbats randomly fly as they observe their environment. They emit sound with varying pulse properties and receiving the echoes reflected back from the surrounding objects. When the prey is detected, the microbats increase the rate of pulse emission, and reducing the loudness to overtake their prey. Inspired by this behaviour, the artificial bat algorithm is outlined as follows.

An artificial bat is determined by a position x_i, velocity v_i, frequency f_i, loudness A_i and pulse emission rate r_i at a time step t in a d-dimensional search space. The position of a bat i represents a potential solution to an optimization problem, and a set of n solutions constitute the population. Initially, the objective function is to be selected and the initial population is generated randomly.

During the search process, every bat updates its position x_i and velocity v_i according to the Eqs. (1), (2) and (3).

$$f_i = f_{min} + (f_{max} - f_{min})\beta \tag{1}$$

$$v_i^t = v_i^{t-1} + (x_i^t - x_{best})f_i \tag{2}$$

$$x_i^t = x_i^{t-1} + v_i^t \tag{3}$$

where β is a random vector drawn uniformly from the interval $[0, 1]$, x_{best} is the current global best solution, f_{min} and f_{max} are the lower and upper bounds of the frequency.

To enhance the solution quality, a local random walk is employed. A new solution around the best solution is obtained and it is computed based on the equation (4).

$$x_{new} = x_{best} + \varepsilon A^t \tag{4}$$

where ε is a random number generated within the range $[-1, 1]$, A^t stands for the average loudness of all bats at the time step t.

A random solution is also created and compared to the best non visited neighboring solution according to the objective function. If the recent randomly obtained solution has a better objective function value, then the loudness A^t and pulse emission rate r_i^t are updated. The formal definition of these parameters is given by the Eqs. (5) and (6).

$$r_i^t = r_i^0[1 - e^{-\gamma t}] \tag{5}$$

$$A_i^t = \alpha A_i^{t-1} \tag{6}$$

where γ is a constant, r_i^0 are the initial pulse rates and α is a user-defined constant depending on the problem.

Algorithm 1 presents the main steps involved in the standard bat algorithm. Since its development, the bat algorithm was applied to a number of problems, including numerical optimization problems [27], feature selection [18], scheduling tasks [30], data mining [11], clustering [38] classification [8]. Except for a very limited number of studies, identifying relevant regions of interest in the skin lesion imagery based on the MSER has not been addressed in the existing literature. In this work, combining the MSER detector with bat algorithm can substantially contribute in resolving the problem of the skin lesion detection.

Algorithm 1. Bat algorithm

Input: test skin lesion image

Output: a set of optimized MSER parameters

1: Select an objective function $f(x)$ with $x = (x_1, ..., x_d)^T$

2: Initialize randomly the bat population x_i and v_i, $i = 1...n$

3: Initialize empirical parameters pulse rates r_i^0, loudness A_i^0, pulse frequency f_i^0 and maximum iterations T_{max}

4: Compute the fitness value for each bat and find the best solution x_{best} in the population

5: **while** $t < T_{max}$ **do**

6:     Update the bat positions x_i and velocities v_i following Eq.(1) to (3)

7:     **if** $rand(0,1) > r_i$ **then**

8:         Select a random solution among the best solutions

9:         Generate a local solution around the best solution, according to the Eq.(4)

10:     **end if**

11:     Generate a new random solution

12:     **if** $rand(0,1) > r_i$ **then**

13:         Accept the new solution

14:         Increase r_i and reduce A_i

15:     **end if**

16:     Rank the bats and find the current best solution

17: **end while**

18: Return the best set of MSER parameters.

3 The Proposed Bat Algorithm for Skin Lesion Detection

As said above, the present study utilizes a bat algorithm for the skin lesion detection purposes. Using the bat algorithm is motivated by its efficiency compared to different state-of-the art metaheuristics algorithms when applied to several optimization contexts [36]. In this respect, the skin lesion detection is formulated as an optimization problem, where the output should be a best combination of parameters of the MSER of the input preprocessed skin lesion image. The best set of parameters is achieved through the bat algorithm. In fact, the exhaustive search of the better combination of the MSER parameters would be too costly in time, which justifies the utilization of metaheuristics.

In this regard, considered as an optimization problem, the skin lesion detection problem involves the specification of the solution representation, the objective function and the search process which will be next presented in this section.

Representation of Solutions. In this work, we adopted a vectorial representation of solutions as it has been employed in [6]. It defines a solution to the problem as a vector of five dimensions. Figure 1 illustrates the used representation. Such vectors contained real numbers of the MSER parameters within different ranges, as indicated in Table 1. The specified parameter ranges are empirically set based on several tests.

Delta	Min area	Max area	Max variation	Min diversity

Fig. 1. Representation of solution.

Table 1. Setting the ranges of the MSER parameters.

Parameter	Parameter range
Delta	$[1, 10]$
Min area	$[500, 1500]$
Max area	$[20000, 30000]$
Max variation	$[0.15, 0.3]$
Min diversity	$[0.1, 0.6]$

Objective Function. The solution quality in bat algorithm is assessed by the intersection over union (IoU) metric. It measures the overlap between the predicted (by MSER) and ground truth bounding boxes B1 and B2, respectively. The larger overlap, the higher detection will be obtained.

$$F = max(\frac{B1 \bigcap B2}{B1 \bigcup B2}) \tag{7}$$

The Search Process. In this work, the skin lesion detection based an optimized MSER is accomplished by bat algorithm metaheuristic. The latter aims to search for the best combination of optimal parameters by maximizing the IoU objective function. Having an input skin lesion image to be processed, the developed system operates as shown in the chart of Fig. 2. The required steps are summarized in the following:

1. **Initialization**: A set of solutions is created as the initial bat population. It consists of n real vectors of five dimensions. All the bat parameters are also initialized and the fitness values for all solutions are calculated. Once the initialization step is performed, the algorithm reiterates the following steps till a stop condition (i.e. pre-determined number of iterations) is met.
2. **Generation of new solutions**: In this step, each bat generates a new solution by updating the corresponding frequency, velocity and position.
3. **Solution quality evaluation**: The recently created solutions are assessed according to the IoU objective function.
4. **Generation of a local solution**: A new solution in the neighborhood of the best solution is generated and evaluated.
5. **Selection of a random solution**: A random solution is selected from the population and then evaluated.

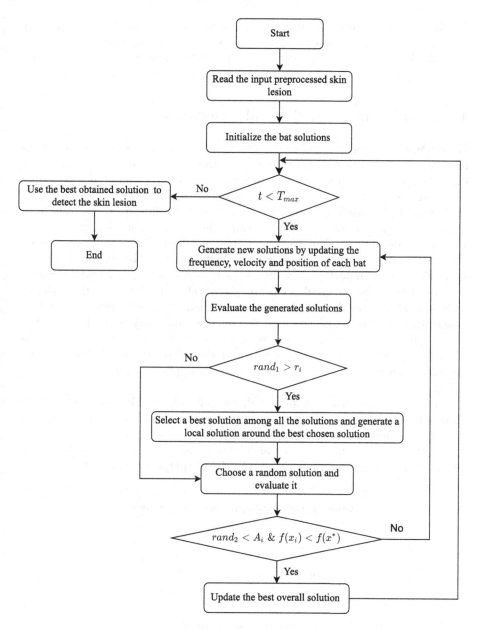

Fig. 2. The proposed bat algorithm for skin lesion detection.

6. **Solutions comparison and update**: In this stage, a fitness comparison is established between the local and random generated solutions : if the fitness value of the former solution is better than that of the latter, the loudness is to be increased and the pulse emission is reduced. The best obtained solution is kept and used for the next iteration.

7. **Check the stopping criterion**: if the stop criterion is reached, the best global solution found so far is introduced to the MSER to detect the skin lesion of the input preprocessed image, which results an image with the detected skin lesion.

4   Experimental Results and Discussion

To demonstrate the efficacy of the proposed approach, we used Python programming language and Google Colaboratory platform. We have also utilized LabelImg [34] tool to annotate the used images. It generates ground truth bounding box around each skin lesion appearing in each image.

Regarding the used dataset, we collected a set of clinical images from both MED-NODE [7] dataset and Atlas (dermIS, dermQUST[1]) site images. The collection contains 224 melanoma and nevus images of different resolutions, between 720×960 and 2045×1165 pixels. Having high image resolution may limit the skin lesion detection performance due to the high cost of the computation. So, we first resize the melanoma and nevus images to 300×300 pixels and the resulting images will focus on the skin lesions. In addition, we preprocessed the resized images trough Otsu-based adaptive thresholding and inpainting algorithms to get an enhanced (artifact-free) skin lesion images. Since the skin lesions are characterized by different features (colors and sizes), dynamic thresholds based on the histogram of the input skin lesion images is applied in the adaptive thresholding.

Once the image dataset is prepared, we split it into train and test sets containing 179 images and 45 images for training and testing sets, respectively.

For the empirical parameters setting concerns, we conducted extensive tests to tune the bat algorithm hyperparameters to values that yield the best detection results. Table 2 presents the adapted empirical hyperparameters setting.

Table 2. Empirical hyperparameter values.

Hyperparameter	Value
Number of iterations	50
Population size	50
Velocity alpha	0.5
Gamma (r)	0.5
The minimum frequency	0.0
The maximum frequency	2.0

Qualitatively speaking, Fig. 3 provides the obtained visual results from the application of the MSER detector with default parameters, the genetic algorithm and

---

[1] http://www.dermweb.com/photo_atlas/.

the proposed bat algorithm on some test images. So, the figure shows each image with colored bounding boxes and the blue detected skin lesion region. The red one corresponds to that of the ground truth annotation while the green represents the predicted bounding box.

For what is specific to the genetic algorithm, the population size is set to 100, the number of iterations is chosen to be 50, the mutation probability is 0.5, the probability of crossover is 0.1 and finally, the replacement rate is set to 0.1.

From these results, we can observe that the proposed bat algorithm and genetic algorithm detector have successfully identified the skin lesions compared to those of the MSER detector applied with default parameters. In addition, the figure shows that the skin lesion detection fails when applying the default MSER in 7 of ten cases.

	Original images	Default MSER	Genetic algorithm	Bat algorithm
1				
2				
3				

When comparing, visually, our results to those of the genetic algorithm, we remark that both algorithms yield almost identical results. So, a quantitative comparison seems to be needed. Table 3 represents the statistical comparison of the IoU metric, for the above-showed images between the default MSER detector, genetic algorithm and the proposed bat algorithm. It shows clearly that the proposed bat algorithm outperforms the genetic algorithm and the default MSER detector in terms of IoU. We have also employed the Wilcoxon rank-sum test [33]. It is a non parametric test that determines whether two populations are statistically different from each other based on ranks of the values in the used dataset. The statistical significance is expressed as a p-value.

Original Images	Default MSER	Genetic algorithm	Bat algorithm

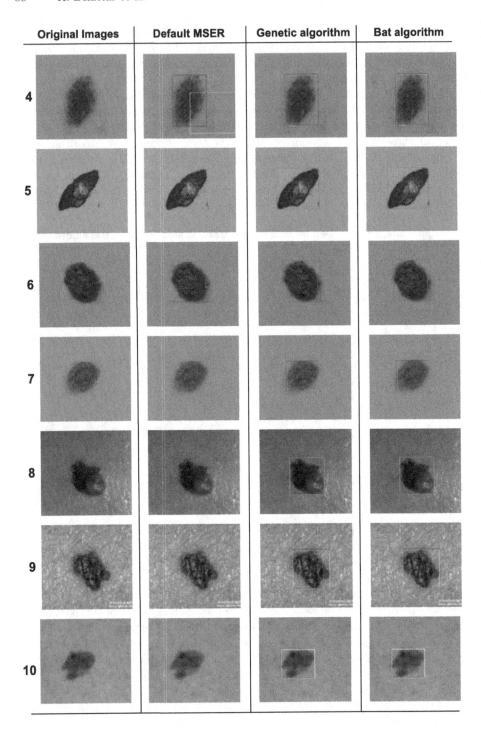

Fig. 3. Visual-based evaluation of the skin lesion detection approaches.

Table 3. Comparing the IoU values for the skin lesion detection approaches.

The skin lesion images	Default MSER detector	Genetic detector	The proposed bat detector
Image 1	0.01057	0.8386	**0.8584**
Image 2	0.0113	0.6681	**0.8622**
Image 3	0.0032	**0.9901**	**0.9901**
Image 4	0.0032	0.9432	**0.9433**
Image 5	0.0049	**0.3085**	**0.3085**
Image 6	0.0062	**0.8157**	**0.8157**
Image 7	0.0000	**0.9569**	**0.9569**
Image 8	0.0000	0.9605	**0.9643**
Image 9	0.0038	0.9694	**0.9740**
Image 10	0.0000	**0.9836**	**0.9836**

By measuring the IoU values of the test images for the detection approaches and assuming the null hypothesis is true, we determine the p-values associated with the used test. The obtained results over all test images are presented in the form of multiple box plots with respect to the IoU metric. These box plots allow us to respond to the following null hypothesis: does the proposed bat algorithm method tend to give higher IoU than of the genetic algorithm and default MSER?

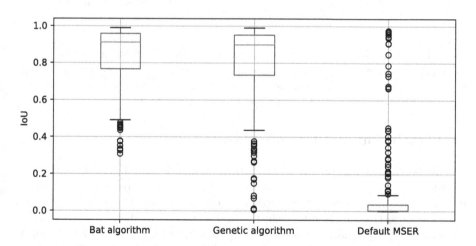

Fig. 4. IoU comparison with genetic algorithm and default MSER.

The proposed bat algorithm method gives the best IoU; having the median set higher than that of the genetic algorithm and default MSER. However, the genetic algorithm provides a large variation range of IoU values. Furthermore, MSER generates an important number of IoU outliers compared with the genetic and bat algorithms. Concerning the Wilcoxon rank-sum test, it yields 0.04391, 0.01386 as p-values for the proposed bat and genetic algorithms, respectively, which are smaller than our significance level (0.05). Therefore, we can conclude that our proposed bat and genetic algorithms better detect the skin lesions.

Table 4 summarises the obtained results of the detection methods for the images in terms of the IoU mean value, IoU standard deviation value (STD) and execution time obtained over the 30 runs.

Table 4. Comparing the IoU values for the bat and genetic algorithm.

Images	The proposed bat detector		Genetic detector	
	Mean ± STD	Execution time	Mean ± STD	Execution time
Image 1	85.1522 ± **42.4488**	13 min 25 s	**85.2624** ± 42.6395	**5 min 30 s**
Image 2	96.2608 ± **47.9595**	6 min 48 s	**96.2195** ± 48.1000	**5 min 11 s**
Image 3	98.1520 ± **48.9397**	10 min 30 s	**97.8215** ± 48.9833	**4 min 44 s**
Image 4	98.3480 ± **48.98522**	16 min 55 s	**98.3085** ± 49.1282	**9 min 49 s**
Image 5	**96.9745** ± **48.3165**	9 min 40 s	97.0251 ± 48.5024	**6 min 37 s**
Image 6	85.5944 ± 46.5897	11 min 20 s	**83.4412** ± **41.259**	**6 min 54 s**
Image 7	56.548 ± **30.5978**	16 min 12 s	**45.5698** ± 48.369	**12 min 26 s**
Image 8	93.2099 ± 48.3169	9 min 36 s	**91.4417** ± **47.392**	**5 min 12 s**
Image 9	**94.4471** ± 42.6488	6 min 36 s	94.8960 ± **42.056**	**3 min 50 s**
Image 10	**90.8310** ± 48.697	13 min 41 s	**90.8310** ± 48.269	**9 min 55 s**

As noticed from the table of IoU comparison, the bat algorithm has higher STD values in almost of cases which are presented in bold. So, the genetic algorithm is more numerically stable than bat algorithm. Furthermore, it outperformed the bat algorithm in terms of the IoU mean values and execution time.

5   Conclusion and Perspectives

In this work, we adopted a bat algorithm for the skin lesion detection of melanoma images. To this end, we used the MSER detector where the optimal set of parameters is metaheuristically gotten by utilizing the bat algorithm with IoU objective function to yield the best detection quality.

To test and evaluate the proposed approach, we used a set of melanoma images collected from the MED-NODE and Atlas image databases. The qualitative and numerical outcomes reveal the performance of the proposed approach compared to the basic MSER and genetic algorithm-based detectors.

As far as we know, this is the first time a skin lesion detection approaches based on an optimized MSER is proposed and one of the few works destined for the medical image analysis, which gives this work its originality.

Further work will focus on enhancing the proposed bat approach in terms of the processing time. To this end, we plan to establish the parallel implementation of bat algorithm using GPUs for the skin lesion detection problem. Other objective functions could be also developed to improve the skin lesion detection performance. Another interesting direction would be to expoit the proposed bat algorithm for the multiple skin lesion detection problem.

References

1. Altan, A., Karasu, S.: Recognition of covid-19 disease from x-ray images by hybrid model consisting of 2d curvelet transform, chaotic salp swarm algorithm and deep learning technique. Chaos Solitons Fractals **140**, 110071 (2020). https://doi.org/10.1016/j.chaos.2020.110071
2. Aurelia, J.E., Rustam, Z., Wibowo, V.V.P., Setiawan, Q.S.: Comparison between convolutional neural network and convolutional neural network-support vector machines as the classifier for colon cancer. In: 2020 International Conference on Decision Aid Sciences and Application (DASA), pp. 812–816 (2020). https://doi.org/10.1109/DASA51403.2020.9317103
3. Bay, H., Ess, A., Tuytelaars, T., Van Gool, L.: Speeded-up robust features (surf). Comput. Vis. Image Underst. **110**(3), 346–359 (2008). https://doi.org/10.1016/j.cviu.2007.09.014. Similarity Matching in Computer Vision and Multimedia
4. Calonder, M., Lepetit, V., Strecha, C., Fua, P.: BRIEF: binary robust independent elementary features. In: Daniilidis, K., Maragos, P., Paragios, N. (eds.) ECCV 2010. LNCS, vol. 6314, pp. 778–792. Springer, Heidelberg (2010). https://doi.org/10.1007/978-3-642-15561-1_56
5. Campilho, A., Karray, F., Wang, Z. (eds.): ICIAR 2020. LNCS, vol. 12132. Springer, Cham (2020). https://doi.org/10.1007/978-3-030-50516-5
6. Davis, J.E., Bednar, A.E., Goodin, C.T.: Optimizing maximally stable extremal regions (mser) parameters using the particle swarm optimization algorithm. Technical report ERDC (2019)
7. Giotis, I., Molders, N., Land, S., Biehl, M., Jonkman, M.F., Petkov, N.: Med-node: a computer-assisted melanoma diagnosis system using non-dermoscopic images. Expert Syst. Appl. **42**(19), 6578–6585 (2015)
8. Gupta, D., Arora, J., Agrawal, U., Khanna, A., de Albuquerque, V.H.C.: Optimized binary bat algorithm for classification of white blood cells. Measurement **143**, 180–190 (2019). https://doi.org/10.1016/j.measurement.2019.01.002
9. Harris, C., Stephens, M.: A combined corner and edge detector. In: Proceedings of the Alvey Vision Conference, pp. 23.1–23.6. Alvety Vision Club (1988). https://doi.org/10.5244/C.2.23
10. Hassan, S.A., Sayed, M.S., Abdalla, M.I., Rashwan, M.A.: Detection of breast cancer mass using MSER detector and features matching. Multimedia Tools Appl. **78**(14), 20239–20262 (2019). https://doi.org/10.1007/s11042-019-7358-1
11. Heraguemi, K.E., Kamel, N., Drias, H.: Multi-objective bat algorithm for mining numerical association rules. Int. J.Bio-Inspired Comput. **11**(4), 239–248 (2018). https://doi.org/10.1504/IJBIC.2018.092797
12. Joshi, K., Patel, M.I.: Recent advances in local feature detector and descriptor: a literature survey. Int. J. Multimedia Inf. Retrieval **9**(4), 231–247 (2020). https://doi.org/10.1007/s13735-020-00200-3

13. Jyotiyana, P., Maheshwari, S.: Maximal stable extremal region extraction of MRI tumor images using successive Otsu algorithm. In: Fong, S., Akashe, S., Mahalle, P.N. (eds.) Information and Communication Technology for Competitive Strategies. LNNS, vol. 40, pp. 687–700. Springer, Singapore (2019). https://doi.org/10.1007/978-981-13-0586-3_67

14. Karthika Devi, R., Banumathi, A., Sangavi, G., Sheik Dawood, M.: A novel region based thresholding for dental cyst extraction in digital dental X-ray images. In: Smys, S., Iliyasu, A.M., Bestak, R., Shi, F. (eds.) ICCVBIC 2018, pp. 1633–1640. Springer, Cham (2020). https://doi.org/10.1007/978-3-030-41862-5_167

15. Korotkov, K.: Automatic change detection in multiple pigmented skin lesions. TDX (Tesis Doctorals en Xarxa) (2014). https://dugi-doc.udg.edu/handle/10256/9276

16. Lee, Y.W., Choi, J.W., Shin, E.H.: Machine learning model for predicting malaria using clinical information. Comput. Biol. Med. **129**, 104151 (2021). https://doi.org/10.1016/j.compbiomed.2020.104151

17. Li, H., Pan, Y., Zhao, J., Zhang, L.: Skin disease diagnosis with deep learning: a review. Neurocomputing **464**, 364–393 (2021). https://doi.org/10.1016/j.neucom.2021.08.096

18. Li, Y., Cui, X., Fan, J., Wang, T.: Global chaotic bat algorithm for feature selection. J. Supercomput. (2022). https://doi.org/10.1007/s11227-022-04606-0

19. Lowe, D.: Object recognition from local scale-invariant features. In: Proceedings of the Seventh IEEE International Conference on Computer Vision, vol. 2, pp. 1150–1157 (1999). https://doi.org/10.1109/ICCV.1999.790410

20. Matas, J., Chum, O., Urban, M., Pajdla, T.: Robust wide-baseline stereo from maximally stable extremal regions. Image Vis. Comput. **22**(10), 761–767 (2004). https://doi.org/10.1016/j.imavis.2004.02.006. British Machine Vision Computing 2002

21. Mikolajczyk, K., Schmid, C.: An affine invariant interest point detector. In: Heyden, A., Sparr, G., Nielsen, M., Johansen, P. (eds.) ECCV 2002. LNCS, vol. 2350, pp. 128–142. Springer, Heidelberg (2002). https://doi.org/10.1007/3-540-47969-4_9

22. Oltu, B., Güney, S., Dengiz, B., Ağıldere, M.: Automated tuberculosis detection using pre-trained CNN and SVM. In: 2021 44th International Conference on Telecommunications and Signal Processing (TSP), pp. 92–95 (2021). https://doi.org/10.1109/TSP52935.2021.9522644

23. Ortiz-Toro, C., García-Pedrero, A., Lillo-Saavedra, M., Gonzalo-Martín, C.: Automatic detection of pneumonia in chest x-ray images using textural features. Comput. Biol. Med. **145**, 105466 (2022). https://doi.org/10.1016/j.compbiomed.2022.105466

24. Ravikumar, M., Rachana, P.G.: Study on different approaches for breast cancer detection: a review. SN Comput. Sci. **3**(1), 1–6 (2021). https://doi.org/10.1007/s42979-021-00898-w

25. Rosten, E., Drummond, T.: Machine learning for high-speed corner detection. In: Leonardis, A., Bischof, H., Pinz, A. (eds.) ECCV 2006. LNCS, vol. 3951, pp. 430–443. Springer, Heidelberg (2006). https://doi.org/10.1007/11744023_34

26. Rublee, E., Rabaud, V., Konolige, K., Bradski, G.: Orb: an efficient alternative to sift or surf. In: 2011 International Conference on Computer Vision, pp. 2564–2571 (2011). https://doi.org/10.1109/ICCV.2011.6126544

27. Saha, S.K., Kar, R., Mandal, D., Ghoshal, S.P., Mukherjee, V.: A new design method using opposition-based bat algorithm for IIR system identification problem. Int. J. Bio-Inspired Comput. **5**(2), 99–132 (2013). https://doi.org/10.1504/IJBIC.2013.053508

28. Santosh, K., Dey, N., Antani, S., Guru, D.: Medical imaging: artificial intelligence, image recognition, and machine learning techniques. CRC Press LLC (2019)
29. Sengupta, S., Singh, A., Leopold, H.A., Gulati, T., Lakshminarayanan, V.: Ophthalmic diagnosis using deep learning with fundus images - a critical review. Artif. Intell. Med. **102**, 101758 (2020). https://doi.org/10.1016/j.artmed.2019.101758
30. Shareh, M.B., Bargh, S.H., Hosseinabadi, A.A.R., Slowik, A.: An improved bat optimization algorithm to solve the tasks scheduling problem in open shop. Neural Comput. Appl. **33**(5), 1559–1573 (2020). https://doi.org/10.1007/s00521-020-05055-7
31. Singh, A., Kharkar, N., Priyanka, P., Parvartikar, S.: Alzheimer's disease detection using deep learning-CNN. In: Hu, Y.C., Tiwari, S., Trivedi, M.C., Mishra, K.K. (eds.) Ambient Communications and Computer Systems, pp. 529–537. Springer Nature Singapore, Singapore (2022). https://doi.org/10.1007/978-981-16-7952-0_50
32. Smith, S.M., Brady, J.M.: Susan-a new approach to low level image processing. Int. J. Comput. Vision **23**(1), 45–78 (1997). https://doi.org/10.1023/A:1007963824710
33. Smucker, M.D., Allan, J., Carterette, B.: A comparison of statistical significance tests for information retrieval evaluation. In: Proceedings of the 16th ACM Conference on Conference on Information and Knowledge Management, pp. 623–632 (2007)
34. Tzutalin: Labelimg. Free Software: MIT License (2015). https://github.com/tzutalin/labelImg
35. Yang, X.S.: A new metaheuristic bat-inspired algorithm. In: González, J.R., Pelta, D.A., Cruz, C., Terrazas, G., Krasnogor, N. (eds.) Nature Inspired Cooperative Strategies for Optimization (NICSO 2010), pp. 65–74. Springer, Berlin (2010). https://doi.org/10.1007/978-3-642-12538-6_6
36. Yang, X.S., He, X.: Bat algorithm: literature review and applications. Int. J. Bioinspired Comput. **5**(3), 141–149 (2013)
37. Zhu, H., Sheng, J., Zhang, F., Zhou, J., Wang, J.: Improved maximally stable extremal regions based method for the segmentation of ultrasonic liver images. Multimedia Tools Appl. **75**(18), 10979–10997 (2015). https://doi.org/10.1007/s11042-015-2822-z
38. Zhu, L.F., Wang, J.S., Wang, H.Y., Guo, S.S., Guo, M.W., Xie, W.: Data clustering method based on improved bat algorithm with six convergence factors and local search operators. IEEE Access **8**, 80536–80560 (2020). https://doi.org/10.1109/ACCESS.2020.2991091

Accurate Detection of Brain Tumor Using Compound Filter and Deep Neural Network

Praveen Kumar Ramtekkar$^{(\boxtimes)}$, Anjana Pandey ,
and Mahesh Kumar Pawar

Rajiv Gandhi Proudyogiki Vishwavidyalaya, Bhopal 462033, Madhya Pradesh, India
pramtekkar@rediffmail.com

Abstract. The unexpected growth of nerves inside the human brain that interferes with the normal function of the brain is referred to as a brain tumor. Magnetic Resonance Imaging (MRI) is used to provide images of better resolution of the brain. This paper proposes a system that applies a compound filter, along with a Convolution Neural Network (CNN) and Support Vector Machine (SVM) for the detection of brain tumors in MRI. For finding tumors this proposed system has been divided into the following sections: preprocessing, segmentation, feature extraction and tumor detection. This system employs a compound filter for preprocessing that is made up of Gaussian, mean and median filters. Threshold and histogram-based techniques have been applied for image segmentation and Grey Level Co-occurrence Matrix (GLCM) for feature extraction. For tumor detection, the SVM and CNN classifiers were employed. CNN is a Deep Neural Network (DNN) based classifier. The tumor detection accuracy of CNN and SVM classifiers have been estimated at 98.06% and 93.28%, respectively. The proposed system concludes that the accuracy of CNN is superior to SVM.

Keywords: Brain tumor · CNN · GLCM · MRI · SVM

1 Introduction

The brain is the main organ that controls all the activities of the human body. Brain Tumor is an abnormal development of brain cells, which affects the normal function of the brain and might be fatal for life. UltraSound, Single-Photon Emission Computerized Tomography (SPECT), Computed Tomography (CT) Scan, X-Rays, Positron Emission Tomography (PET) and MRI techniques have been used for scanning of the human brain [1]. MRI is one of the commonly used imaging techniques that generate a good contrast of medical images and malignant tissues [2]. The task included in this research paper concentrates on the detection of brain tumors through image-processing approaches by using a compound filter, SVM and CNN classifiers. The compound filter has been designed using Gaussian, mean and median filters. Well-timed detection of a

M. Salem et al. (Eds.): ICAITA 2022, CCIS 1769, pp. 94–109, 2023.
https://doi.org/10.1007/978-3-031-28540-0_8

brain tumor could assist physicians to offer better treatment. As soon as a brain tumor is detected the next task of radiologists is to repair the effect of the tumor in nearby parts of the brain and to ensure the size, location, type and grade of the tumor. This information may also help to decide the methods of treatment for patients such as surgery, radiation, therapy and chemotherapy. The possibility of saving the life of a brain tumor-infected person could be enhanced significantly if it is discovered correctly in its starting phase [3]. Therefore, the research work on tumor detection in MRI is significant.

Some of the important contributions of this article are:

- It develops a system that employs a compound filter, SVM and CNN classifier to detect brain tumors in MRIs automatically and accurately.
- One of the factors for the increase in accuracy is the use of a compound filter.
- In this system the CNN assists automatically in feature extraction and segmentation in addition to tumor detection.
- This article deeply explains the step-by-step procedure for tumor detection in MRI.

This article has been divided into seven stages. Stage 1, Introduction, summarizes this article. Stage 2, Literature Review, presents past work in brain tumor detection. Stage 3, Proposed Methodology, includes details of datasets, proposed algorithm, preprocessing techniques, compound filter, GLCM and brain tumor detection. Stage 4 and Stage 5, describe the working of SVM and CNN classifiers, respectively. Stage 6, Experimental Results, Analyzes the results obtained from GLCM, SVM and CNN classifiers. Stage 7, Performance Comparison, explains the accuracy comparison of the proposed and existing work. At last, a conclusion has been drawn.

2   Literature Review

Mahmoud Khaled Abd-Ellah [4] illustrated two-phase automatic computer-aided detection mechanisms to detect and classify tumors in MRIs. In this paper, segmentation has been performed using K-means clustering, feature extraction by Discrete Wavelet Transform (DWT), feature reduction using Principal Component Analysis (PCA) and classification by SVM. This system depicts the accuracy of 100% for tumor detection and classification using 120 MRIs.

Heba Mohsen [5] described DNN classifiers to categorize the dataset of 66 MRIs into four groups metastatic bronchogenic carcinoma, sarcoma, glioblastoma and normal tumor. In this system, DNN was combined with DWT and PCA which reported a classification accuracy of 96.97%.

Sanjeev Kumar [6] proposed a hybrid approach, in which DWT was used to extract image features. GA was applied for feature reduction and SVM has been used to classify the tumor as benign or malign. For the analysis of images, the parameters like entropy, kurtosis, smoothness, Root Mean Square (RMS) and correlation were applied.

Tonmoy Hossain [7] applied CNN for MRI classification into abnormal and normal pixels. This paper has used features based on statistics and texture to implement CNN using Keras and Tensor flow. The classification accuracy was achieved at 97.87% using CNN.

Manjunath [8] suggested a system that determines the grades of tumors of the appropriate class. In this paper, the practical was performed by using Back Propagation Neural Networks (BPNN) and CNN classifiers. This system has estimated a classification accuracy of 86.48%.

A. Isin [9] lighted on the use of DL techniques for the segmentation of the MRIs. This paper has reviewed various MRI segmentation techniques like KNN and SVM. This work also examined the performance of different DL methods in segmenting glioma tumors.

Selvakumar Raja [10] performed automatic tumor detection in the brain MRI by applying CNN classifiers. In this system, CNN uses a small-size filter with a very small neuron weight. The use of the CNN classifier in this work obtained an accuracy of 97.5%.

Janardhana Swamy [11] proposed a system that detects the brain tumor at an earlier stage. This model was initially developed using CNN and fits the CNN into images. Obtained results have been passed through various layers of the network for training. The trained system has made better predictions about the presence of tumors.

Chirodip Lodh Choudhury [12] described a work that involves a DNN-based approach with CNN to detect and classify the MRI images as tumor or non-tumor. This model has scored an average accuracy of 96.08% and an F-score up to 97.3.

Praveen Kumar Ramtekkar [13] proposed an automatic system that works on brain MRI images to detect and classify tumors using a CNN classifier and DL techniques. This article has reviewed different techniques of brain MRI classification and concluded that DL techniques have been preferred over other state-of-the-art brain tumor detection and classification methods since they offer higher accuracy.

3   Proposed Methodology

This section describes the materials and methods used in this proposed system.

3.1   Dataset

For performing the experimental task of this proposed work, brain MRI images have been collected online from Kaggle dataset (https://www.kaggle.com) for the identification of brain tumors. The dataset has been divided into two classes - Yes and No. Class Yes contains 155 tumor images and Class B has 99 non-tumor images. All these images are MRIs of different modalities like T1, T2 and Fluid Attenuated Inversion Recovery (FLAIR) [14]. The size of each image is

(128, 128) exactly in axial view. There is a total of 253 images, out of 253, 177 (i.e. 70% of 253) were used for the training of CNN and SVM network, 38 (i.e. 15% of 253) for testing and again 38 have been utilized for validation. All these images are available at: https://www.kaggle.com/navoneel/brain-mri-images-for-brain-tumor-detection.

3.2 Proposed Algorithm

This article suggested an algorithm that follows the set of instructions as given below.

Algorithm: Implementation of proposed system

- Input← Inputting MRI.
- Grey← Transform Input into Grayscale image form.
- ReSize← Reduction of the dimension of Gray image to 128 × 128.
- Filter← Eliminate noise of Gray image using a compound filter, which includes Gaussian, mean and median filters.
- Segment← Apply threshold and histogram segmentation operations to segment Filter image.
- Feature← Determine features of Segment image, like contrast, correlation, homogeneity, entropy, energy, shape, color, texture and intensity using GLCM.
- Detect← Detection of brain tumor using Feature by the SVM and CNN classifier.
- Output← MRI image with tumor

3.3 Conversion of MRIs into gray-scale

The RGB images contain a lot of information which are not necessary for their processing. This information can be eliminated by transforming RGB images to gray-scale [15]. Any image can be represented by RGB scale in three channels B, G, R. Each channel has 8 bits and an image holds different intensity levels for each B, G, R component. For any color image, we have intensities for each scale. Therefore, RGB requires a huge information to store and manipulate. Figure 1 depicts the conversion of RGB image into gray-scale.

3.4 Preprocessing

MRI Processing is a difficult job. It is compulsory to eliminate unnecessary artifacts from MRI. After the removal of extra noise, MRI becomes useful for further processing [16]. Preprocessing includes gray-scale conversion, resizing, noise removal and image reconstruction. Gray-scale conversion of the image is the general preprocessing practice [17]. After gray-scale conversion additional noise is removed by applying filtering techniques. This proposed system designs a compound filter by combining the Gaussian, mean and median filters. The

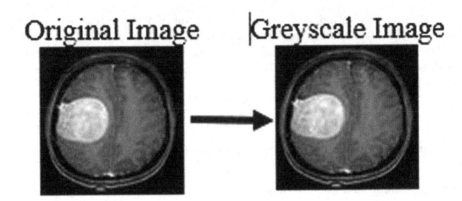

Fig. 1. Conversion of MRI into Grayscale Image

compound filter eliminates the Gaussian, salt and pepper and speckle noises from the gray-scale image [18]. It also preserves the edges and boundaries in MRI [19]. The algorithm for the compound filter is as follows:

Algorithm: Filtering of the gray-scale image using a compound filter

- Grey← Inputting Grayscale Image.
- ReSize← Reduction of the dimension of the Gray image to 128×128.
- Gaussian← Apply Gaussian Filter.
- Mean← Apply Mean Filter.
- Median ← Apply Median Filter.
- Compound ← Median Filter.
- Output← Filtered MRI image

A sample result of preprocessing using the compound filter is shown in Fig. 2.

Image segmentation is the division of the image into a number of distinct, non-overlapping sections. The image is segmented into groups of pixels that are more significant and simpler to analyze. The process of segmentation is used to find the edges or objects in an image and the generated sections collectively cover the entire image [20]. The similarity and discontinuity of image intensity are the two features that segmentation techniques focus on [21]. There are different segmentation methods that are utilized for image segmentation, including threshold-based, histogram-based, region-based, edge-based and clustering methods [22,23]. For the segmentation of tumor regions in MRI, this work uses threshold- and histogram-based segmentation algorithms [24,25]. The result obtained by applying the threshold- and histogram-based segmentation techniques have displayed in Fig. 3.

Feature extraction refers to the reduction of the number of resources required to define a large set of data correctly. At the time of analysis of a large data set, problems arise when a large number of variables are used. Analysis of a large number of variables requires more memory and computation time. Feature extraction creates groups of variables to solve these problems while describing

Grayscale Image   Filtered Image

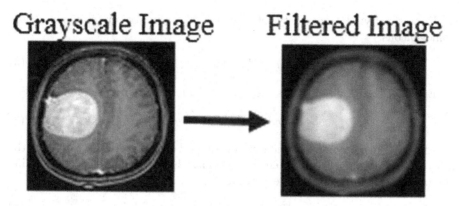

Fig. 2. Preprocessing using compound filters

Filtered Image Threshold Segmentation   Histogram Segmentation

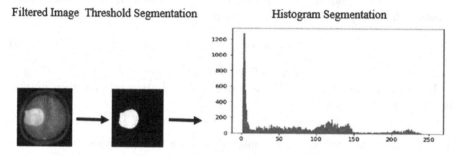

Fig. 3. Results of threshold and histogram segmentation

the data with enough accuracy. In this paper, the GLCM [26] has been used to obtain both statistical- and texture-based features of an image.

The number of rows and a number of columns in GLCM [27] remains the same as the number of gray levels in the image. Let $Fi,j(\delta x, \delta y)$ be an element of GLCM. Where, $Fi,j(\delta x, \delta y)$ is the relative frequency with which two pixels, which are separated by a pixel distance $(\delta x, \delta y)$ occur within a given neighborhood, where i and j are the intensity of the first and second pixel, respectively. The matrix element $(Fi,j : d, \theta)$ has the second-order statistical probability values for the changes between gray level i and j at a displacement distance d and angle θ. The features extracted using GLCM [28,29], are tabulated in Table 3.

Contrast: The darkest and lightest parts of an image are separated by contrast.

$$Contrast = \Sigma_{i=0}^{n-1} Fi,j(i-j)^2 \tag{1}$$

Correlation: Correlation is determined as the correlation coefficient between -1 and $+1$.

$$Correlation = \Sigma_{i=0}^{n-1} Fi,j((i-\mu)(j-\mu))^2/\sigma^2 \tag{2}$$

Homogeneity: Homogeneity stands for the quality or state of being homogeneous.

$$Homogeneity = \Sigma_{i=0}^{n-1} Fi, j(j/(1 + (i - j))^2 \tag{3}$$

Entropy: Entropy represents the degree of uncertainty in a random variable.

$$Entropy = \Sigma_{i=0}^{n-1} Fi, j(logFi, j) \tag{4}$$

Energy: Energy computes the total of squared elements. It also estimates the homogeneity. A high degree of energy specifies that the image has outstanding homogeneity or image pixels are very similar.

$$Energy = \Sigma_{i=0}^{n-1} Fi, j^2 \tag{5}$$

Smoothness: Smoothness is an amount of grey-level contrast which is used to establish descriptors of relative smoothness.

$$Smoothness = 1 - (1/(1 + \alpha^2)) \tag{6}$$

Here, α is the standard deviation of an image.

Kurtosis: It calculates the flatness of the distribution relative to a normal distribution.

$$Kurtosis = \sigma^{-4} \Sigma_{j=0}^{n-1} Fi, j(1 - \mu)^4 - 3 \tag{7}$$

Here, Fi, j, is the pixel value at point(i, j) and μ and σ are mean and standard deviation, respectively.

Root Mean Square (RMS): RMS computes the value of each row or column of the input matrix of the given dimension of the input or whole input.

$$RMS = \sqrt{Fi, j^2} \tag{8}$$

4   SVM

SVM classifier is applied for the classification of the image or pattern. It divides the group of images into two different classes. In this proposed work linear SVM has used [30]. In linear SVM, input data is mapped on high-dimension space using a linear kernel. Suppose n dimensional inputs x_i for $i = 1$ to n, belongs to Class-I or Class-II and the associated labels $y_i = 1$, for Class-I and $y_i = -1$, for Class-II. The decision function for linear SVM is $D(x) = w^t x_i + b$, Where w is an n-dimensional vector and b is a scalar. The separating hyperplane satisfies the condition: $(w^t x_i + b) \geq 0$, for $i = 1$ to n. The margin is the distance between the separating hyperplane $D(x) = 0$ and the training datum closest to the hyperplane. The hyperplane $D(x) = 0$, with the highest margin is known as the optimal separating hyperplane. The linear SVM function is defined by: $D(x) = (w^t x_i + b)$, where x_i is the training samples. This function creates two classes by inserting a hyperplane between the classes. The classification is performed by finding the hyperplane which differentiates the two classes very

Table 1. SVM parameters

Parameters used	Value
Kernel function	Linear
Degree of the kernel	3
Tolerance of the termination criteria	0.001
C parameter of C-SVC	1
Missing value replacement	off

well [31,32]. The SVM forms a hyperplane based on a kernel function D. Feature vectors on the left of the hyperplane have a value of -1 and they represent Class-I. Similarly, the values on the right of hyperplane is $+1$ and belong to Class-II as shown in Fig. 4. The values used for linear SVM parameters are tabulated in Table 1.

5 CNN

CNN [33] is frequently used in image processing to identify and classify brain tumors. In this work, CNN was applied and practically implemented using Python programming language to detect brain tumors. Figure 5 explains how CNN works.

In Fig. 5 An input image I of size (128, 128, 3) has entered CNN, which follows the sequence of CNN layers. One extra dimension represented as "None" was added to process multiple batches in one epoch of CNN. After the addition of a new dimension, the size of the input image becomes (None, 128, 128, 3).

The zero Padding layer has been used to pack the image border with 0s. This layer with a pool size of 2 * 2 has been applied to perfectly fit the input image by padding image I with 0s. After padding the new image size (132, 132, 3) becomes the input of the convolution layer.

Convolution of image size (132, 132, 3) uses filter size (7, 7), stride 1, dilation rate 1 and valid padding, to produce the size of the image: $(132 - 7 + 1, 132 - 7 + 1) = (126, 126)$. The convolution layer also applies 32 filters, therefore the output image becomes (126, 126, 32). A filter has been applied to detect the presence of features in the input image I. Output (126, 126, 32) of the convolution layer has fed to the batch normalization layer.

After performing a batch operation the batch normalization layer produces an image of size (126, 126, 32), which becomes the input of the activation layer. Further, the output of the activation layer (126, 126, 32) is fed to max pooling layer 0. This layer chooses the biggest element from the rectified feature map. It applies 4 filters, 4 strides and a pool size of (4, 4) for its processing. The formula for the output image is $(N - F)/S + 1$, where N is the dimension of the pooling layer, F is the dimension of the filter and S is the stride. Then the size of an output image is: $((126 - 4)/4 + 1, (126 - 4)/4 + 1) = (31, 31)$.

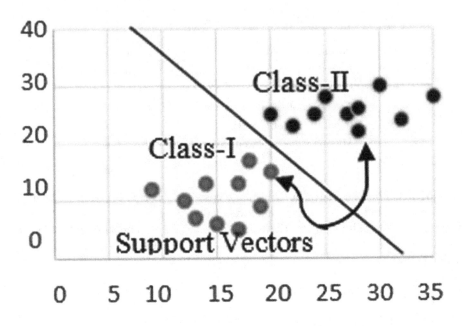

Fig. 4. Classification using SVM classifier

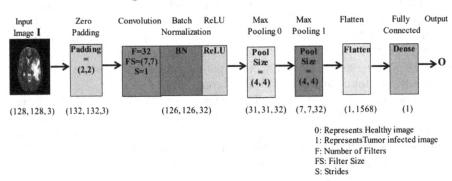

0: Represents Healthy image
1: RepresentsTumor infected image
F: Number of Filters
FS: Filter Size
S: Strides

Fig. 5. Working flow of CNN

There is another max pooling layer 1 that has been used by CNN to reduce the computation cost. The stride indicates the shifting of a number of pixels in the input matrix. The value of stride is 1, which allows moving filters 1 pixel at a time, if it is 2, means moving the filters 2 pixels at a time, and so on.

The image size of (31, 31, 32) has input to max pooling layer 1. This layer uses 4 filters, 4 strides and a pool size of (4, 4). It produces an output image of size $((31 - 4)/4 + 1, (31 - 4)/4 + 1) = (7, 7)$, The image size of (7, 7, 32) is fed to the next layer, flatten.

An activation function converts the input signal into output. The ReLU layer works as an activation function. The Nonlinear ReLU layer is used to apply nonlinear operation in CNN since non-negative linear values are needed to learn CNN. The output of ReLU is: $f(x) = max(0, x)$, if $x < 0$ then $f(x) = 0$, and if $x > 0$ then $f(x) = x$.

After pooling, The flattened layer has applied to transform the three dimensional matrix, i.e. input images into a one dimensional vector. Thereafter, it has been fed to the neural network to process further. Flatten layer takes all the pixels and channels of an image which creates a one dimensional vector without batch size. The input size of the image (7, 7, 32) is flattened to 7732 = 1562 values.

At last, a fully connected layer with one neuron and sigmoid activation function is referred to as the dense layer. It has been used for binary classification. The final output of our model is 1, it outputs 1 value per sample in the batch.

Each layer contains trainable and non-trainable parameters. Trainable parameters can be updated with CNN and non-trainable remain static. The number of parameters of the convolution layer is estimated by the formula $H * W * NIC * NOC + NOC$ (if bias is issued). Where, H: Height of kernel, W: Width of the kernel, NIC: Number of input channels and NOC: Number of output channels.

For this convolution layer, 6433 parameters have been calculated. Out of 6433 6369 are trainable and 64 are non-trainable. The detailed output of each CNN layer has shown in Table 2.

Table 2. Layer-wise output of CNN

Layer (type)	Output shape	No. of parameters
Input Layer	(None, 128, 128, 3)	0
Zero Padding	(None, 132, 132, 3)	0
Convolution	(None, 126, 126, 32)	4736
Batch Normalization	(None, 126, 126,32)	128
Activation	(None, 126, 126, 32)	0
Max Pooling 0	(None, 31, 31, 32)	0
Max Pooling 1	(None, 7, 7, 32)	0
Flatten	(None, 1568)	0
Dense	(None, 1)	1569
Total parameters: 6433		
Trainable parameters: 6369		
Non-trainable parameters: 64		

6   Experimental Process

Out of 253 images, eight GLCM textural features such as contrast, correlation, homogeneity, entropy, energy, smoothness, kurtosis and RMS are calculated for each image. These features are tabulated in Table 3.

The parameter Tumor presents two values, "Yes" and "No". Value "Yes" indicates that the image is tumor infected and "No" is healthy. Out of these 253 images, 98 are found healthy and 155 are identified as tumor infected.

Table 3. Extracted features wise result of MRI images

Image	Tumor	Contrast	Correlation	Homogeneity	Entropy	Energy	Smoothness	Kurtosis	RMS
I001	No	1.7326	0.1075	1.0987	2.4589	0.8927	0.9450	15.255	0.0845
I002	No	0.2414	0.1237	0.8876	3.8976	0.3686	0.9335	12.544	0.0933
I003	No	0.2725	0.1465	0.8745	3.1765	0.3644	0.9194	15.280	0.0919
I004	No	1.1327	0.1253	0.9178	2.8345	0.4168	0.9607	15.438	0.0860
I005	Yes	1.2434	0.1344	1.1984	2.6623	0.7424	0.9316	13.231	0.0931
I006	Yes	1.2723	0.1147	1.5873	3.4078	0.3683	0.9445	17.285	0.0944
I007	No	1.0326	0.1195	1.4742	3.4412	0.3246	0.9493	18.164	0.0849
I008	No	0.1417	0.0964	1.3177	2.4534	0.4268	0.9522	16.173	0.0852
I009	No	0.2736	0.1126	0.0985	3.2223	0.7923	0.9509	17.148	0.0950
I010	Yes	1.7325	0.1183	1.8875	2.5565	0.7680	0.9348	18.145	0.0934
I011	Yes	0.4414	0.0942	1.8744	3.7643	0.4649	0.6915	19.154	0.0891
I012	No	1.5720	0.1166	1.9172	3.1178	0.6165	0.9524	14.175	0.0952
I013	Yes	1.4411	0.1347	1.3329	2.9065	0.3423	0.3233	16.254	0.0823
I014	Yes	1.2526	0.1565	1.4378	2.6634	0.5685	0.9387	15.264	0.0938
I015	No	1.0625	0.1344	1.3246	2.8834	0.4243	0.9464	12.858	0.0946
I016	Yes	0.1414	0.1183	1.1274	2.4376	0.8262	0.9362	19.243	0.0936
I017	No	0.4723	0.1222	1.1180	3.6712	0.5527	0.9229	17.233	0.0922
I018	No	1.3326	0.1337	1.1871	2.4534	0.4485	0.9261	12.208	0.0926
I019	No	1.5415	0.1446	0.8743	2.8967	0.3684	0.9396	15.676	0.0939
I020	No	0.5924	0.1115	1.2174	3.2754	0.4163	0.9145	15.917	0.0914
I021	Yes	1.8413	0.1623	1.3985	2.5834	0.7422	0.9554	12.133	0.0955
I022	Yes	1.6727	0.1067	1.4576	2.9623	0.3680	0.9183	13.615	0.0918
I023	Yes	1.5328	0.1332	1.5747	3.1076	0.3241	0.9205	12.859	0.0920
I024	No	0.4459	0.1571	1.6173	3.1454	0.4263	0.9530	17.447	0.0953
I025	Yes	0.2445	0.1730	0.7984	2.7834	0.7924	0.9368	19.355	0.0936
I026	Yes	1.3424	0.1358	1.7875	3.1287	0.7685	0.9555	18.412	0.0855
I027	No	1.4667	0.1036	1.5749	2.5167	0.4646	0.9634	13.287	0.0863
I028	Yes	0.7726	0.1145	1.4170	3.7156	0.6167	0.9493	17.433	0.0949
I029	Yes	1.4313	0.1974	0.8328	3.1756	0.3424	0.9506	21.363	0.0950
I030	Yes	0.4722	0.1583	1.4377	2.5023	0.5683	0.9315	17.298	0.0931

The local variations in the GLCM are measured by contrast. Contrast 0 indicates that the image is constant. In Table 3 values of contrast vary from 0.1124 to 2.3219 which have been calculated using the formula in Eq. (1).

Correlation defines how a pixel correlates with its adjacent pixel. Values of correlation vary from −1 to 1. 1 represents a positively correlated image and −1 represents a negatively correlated image. For constant image value of correlation is NaN. It has been seen from Table 3 that the value of correlation varies from 0.0264 to 0.7633 and is calculated by the formula in Eq. (2).

Homogeneity is used to measure the closeness of the distribution of elements in GLCM to diagonal GLCM. Homogeneity has a value between 0 and 1. The value of diagonal GLCM is measured as 1. Table 3 represents homogeneity from 0.0985 to 2.5749 which was calculated using the formula in Eq. (3).

Entropy calculates the loss of information in an image. It also measures the image information. Entropy represents randomness, which defines the texture part of the input image which means distribution variation within an image. In Table 3 entropy varies from 2.284 to 3.9137 which is calculated using the formula in Eq. (4). The images I043, I002 and I011 have energy 3.9137, 3.8976 and 3.7643, respectively, which are the highest. This higher energy indicates that the loss of information in these images is high as compared to other images in the data set.

Energy measures the homogeneity. Higher energy indicates very good homogeneity and close similarity in pixels. Its range is between 0 and 1. For a constant image, its value is 1. In Table 3 energy varies from 0.1232 to 0.9566 which has been calculated using the formula in Eq. (5). Images I044, I001 and I046 have higher energy 0.8949, 0.8927 and 0.8651, respectively. Therefore, the pixels are very similar in these images.

Smoothness represents the measure of dissimilar gray levels, which are used to create descriptors of relative smoothness. Table 3 shows the range of smoothness from 0.3233 to 0.9650 which has been calculated by Eq. (6). For images 1013, I038 and I036 values of smoothness are 0.3233, 0.3835 and 0.6311 which are very low.

Kurtosis is a measure of how flat a distribution region is in comparison to a normal region. The value of kurtosis varies from 11.459 to 29.255 in Table 3, was computed using Eq. (7). Low kurtosis value indicates a flat top near the mean. Images I032, I021 and I018 have low values of kurtosis, which are 12.124, 12.133 and 12.208, respectively.

RMS represents the root mean square error which computes the RMS value of every row and column. The values of RMS are in the range between 0.0383 to 0.0988 as shown in Table 3 and have been computed using the Eq. (8). In Table 3 RMS values of images I038, I036 and I013 are 0.0383, 0.0631 and 0.0823 are very low.

253 samples of MRI images, which create the input vector of size 253 * 128 * 128 * 1 and the target output of size 253 * 128 * 128 * 1 were formed. The training algorithm performed in this practical work classifies the input data into 3 epochs with an average training time of 5.28 s. The result of this work and performance measure of SVM and CNN has been displayed in Table 4.

Table 4. Performance measure of SVM and CNN

Total images	Infected images	Healthy images	Accuracy of SVM (%)	Accuracy of CNN (%)
253	155	98	93.28	98.06

The confusion matrix provides an accuracy of detection of tumors of any classifier. A confusion matrix [34] is created by using actual and predicted values.

$$Accuracy = (TN + TP)/(TP + FN + TN + FP) \times 100 \qquad (9)$$

where: TP: indicates both the actual and predicted values are true. FP: implies predicted value is true but actual value is false, TN: represents both the actual and predicted values are false FN: denotes predicted value is false but the actual value is true.

253 MRI different images have been trained and tested for detecting the tumor in the human brain. The accuracy computed by CNN and SVM is 98.06% and 93.28%, respectively.

7   Performance Comparison

This section compares the performance of the proposed work with T. Hossain, et al. [7], who have worked on SVM and CNN classifiers to detect brain tumors, as shown in Table 5. T. Hossain, et al. [7], report 97.87% and 92.42% of brain tumor detection accuracy using CNN and SVM classifiers. The dataset used by T. Hossain, et al. [7], contains 217 images. The tumor detection accuracy obtained by this system is quite promising, however, the performance of the system degrades gradually while the number of images increases and the system is not workable for 3D images. Our proposed methodology provides an improvement in the accuracy of CNN to 98.06% and SVM to 93.28%. One of the reasons for this improvement in accuracy is the use of a compound filter, which has been designed using the Gaussian, mean and median filters. This compound filter removes the unwanted noises from MRI images, preserves the boundaries and edges of images and produces filtered images of high resolution for feature extraction. The second reason behind the improved accuracy is the usage of more GLCM features and a large number of images. A third reason for the increase in accuracy is the application of more layers in our CNN. We have used eight layers, including max pooling layer 0 and max pooling layer 1. These two max pooling layers also assisted in increasing accuracy.

Table 5. Performance comparison of proposed work and T. Hossain, et al. [7]

Proposed work		T. Hossain, et al. [7]	
Methodology	Accuracy (%)	Methodology	Accuracy (%)
SVM	93.28	SVM	92.42
CNN	98.06	CNN	97.87

Conclusion

This study presented a system that uses a compound filter, SVM and DNN-based classifier CNN to identify brain tumors in MRIs. The preprocessing, segmentation, feature extraction and detection stages were followed to detect brain tumors in MRIs by this proposed system. A compound filter, which is made up of using a combination of Gaussian, mean and median filters was used for pre-processing of MRI images. The use of a compound filter in preprocessing stage has eliminated the Gaussian, salt and pepper and speckle noises from the grayscale images. It also preserves the edges and boundaries in MRIs during the filtering process. For MRI image segmentation, threshold and histogram segmentation algorithms have been applied. Different image features were extracted by GLCM. Brain tumors have been detected by CNN and SVM classifiers. The brain tumor detection accuracy of CNN and SVM classifier was reported at 98.06% and 93.28%, respectively. These accuracies justify that the performance of CNN classifiers is more promising than SVM in tumor detection in MRI images. This accuracy can be improved if more numbers of MRIs and high-resolution images were used.

It is hoped that this work will be helpful for future researchers to collect the literature on brain tumor detection and also assist physicians in making their decision for further treatment.

References

1. Borole, V.Y.: Image processing techniques for brain tumor detection: a review. Int. J. Emerg. Trends Technol. Comput. Sci. **4**, 28–32 (2015)
2. Bahadure, N.B.: Image analysis for MRI-based brain tumor detection and feature extraction using biologically inspired BWT and SVM. Int. J. Biomed. Imaging **2017**, 1–12 (2017)
3. Coatrieux, G.: A watermarking-based medical image integrity control system and an image moment signature for tampering characterization. IEEE J. Biomed. Health Inform. **17**, 1057–1067 (2017)
4. Abd-Ellah, M.K., Awad, A.I., Khalaf, A.A.M., Hamed, H.F.A.: Design and implementation of a computer-aided diagnosis system for brain tumor classification. In: 28th International Conference on Microelectronics (2016)
5. Mohsen, H., El-Dahshan, E.A., El-Horbaty, E.M., Salem, A.M.: Classification using deep learning neural networks for brain tumors. Future Comput. Inform. J. **3**, 68–71 (2018)
6. Kumar, S., Dabas, C., Godara, S.: Classification of brain MRI tumor images: a hybrid approach. Procedia Comput. Sci. **122**, 510–517 (2017)
7. Hossain, T., Shishir, F.S., Ashraf, M., Nasim, M.A., Shah, F.M.: Brain tumor detection using convolution neural network. In: 1st International Conference on Advances in Science, Engineering and Robotics Technology (2019)
8. Manjunath, S., Pande, S.M.B., Raveesh, B.N., Madhusudhan, G.K.: Brain tumor detection and classification using convolution neural network. Int. J. Recent Technol. Eng. **8**, 34–40 (2019)

9. Isin, A., Direkoglu, C., Sah, M.: Review of MRI-based brain tumor image segmentation using deep learning methods. In: 12th International Conference on Application of Fuzzy Systems and Soft Computing, vol. 102, pp. 317–324 (2016)

10. Seetha, J., Raja, S.S.: Brain tumor classification using convolution neural networks. Biomed. Pharmacol. J. **11**, 1457–1461 (2018)

11. Swamy, G.J., Anusha, L., Meghashree, S.: Brain tumor detection using convolution neural network. Int. J. Res. Appl. Sci. Eng. Technol. **8** (2020)

12. Choudhury, C.L., Mahanty, C., Kumar, R., Mishra, B.K.: Brain tumor detection and classification using convolution neural network and deep neural network. In: International Conference on Computer Science, Engineering and Applications (2020)

13. Ramtekkar, P.K., Pandey, A., Pawar, M.K.: A proposed model for automation of detection and classification of brain tumor by deep learning. In: 2nd International Conference on Data, Engineering and Applications (2020)

14. https://case.edu/med/neurology/NR/MRI%20Basics.htm

15. Saravanan, C.: Color image to grayscale image conversion. In: 2nd International Conference on Computer Engineering and Applications, pp. 196–199 (2010)

16. Hebli, A.P., Gupta S.: Brain tumor detection using image processing: a survey. In: Proceedings of 65th IRF International Conference (2016)

17. Kapoor, L., Thakur, S.: A survey on brain tumor detection using image processing techniques. In: 7th International Conference on Cloud Computing, Data Science and Engineering Confluence, pp. 582–585 (2017)

18. Patil, R.C., Bhalchandra, A.S.: Brain tumor extraction from MRI images using MatLab. Int. J. Electron. Commun. Soft Comput. Sci. Eng. **2**, 1–4 (2012)

19. Gopal, N.N., Karnan, N.: Diagnose brain tumor through MRI using image processing clustering algorithms such as Fuzzy C Means along with intelligent optimization techniques. In: Computational Intelligence and Computing Research IEEE International Conference, pp. 1–4 (2010)

20. Janani, V., Meena, P.: Image segmentation for tumor detection using fuzzy inference system. Int. J. Comput. Sci. Mob. Comput. **2**, 244–248 (2013)

21. Acharya, J.: Segmentation techniques for image analysis: a review. Int. J. Comput. Sci. Manag. Res. **2**, 1218–1221 (2013)

22. Naik, D., Shah, P.: A review on image segmentation clustering algorithms. Int. J. Comput. Sci. Inf. Technol. **5**, 3289–3293 (2014)

23. Christ, S.A.: Improved hybrid segmentation of brain MRI tissue and tumor using the statistical feature. ICTACT J. Image Video Process. **1**, 34–49 (2010)

24. Bhima, K., Jagan, A.: Analysis of MRI-based brain tumor identification using segmentation technique. In: International Conference on Communication and Signal Processing, pp. 2109–2113 (2016)

25. Seerh, G.K., Kaur, R.: Review on recent image segmentation techniques. Int. J. Comput. Sci. Eng. **5**, 109–112 (2013)

26. Zulpe, N., Pawar, V.: GLCM textural features for brain tumor classification. Int. J. Comput. Sci. Issues **9**, 354–359 (2012)

27. Mohanaiah, P.: Image texture feature extraction using GLCM approach. Int. J. Sci. Res. Publ. **3**, 1–5 (2013)

28. Kumar, P., Kumar, B.V.: Brain tumour MR image segmentation and classification using by PCA and RBF kernel based support vector machine. Middle-East J. Sci. Res. **23**, 2106–2116 (2015)

29. Kumar, P., Vijayakumar, B.: An efficient brain tumor MRI segmentation and classification using GLCM texture features and feed forward neural networks. World J. Med. Sci. **13**, 85–92 (2016)

30. Vani, N.: Brain tumor classification using support vector machine. Int. Res. J. Eng. Technol. **4**, 1724–1729 (2017)
31. Kumar, T.S.: Brain tumor detection using SVM classifier. In: 3rd International Conference on Sensing, Signal Processing and Security (2017)
32. Srinivasaraghavan, A., Josef, V.: Machine Learning, vol. 181, 1st edn. Wiley India Pvt. Ltd. (2019)
33. Parihar, A.S.: A study on brain tumor segmentation using convolution neural network. In: International Conference on Inventive Computing and Informatics (2017)
34. Srinivasaraghavan, A., Josef, V.: Machine Learning. First ed. Wiley India Pvt. Ltd. 221 (2019)

Bi-ESRGAN: A New Approach of Document Image Super-Resolution Based on Dual Deep Transfer Learning

Zakia Kezzoula[1]([✉]), Djamel Gaceb[1], Zineddine Akli[1], Abdelouaheb Kahouli[1], Ayoub Titoun[2], and Fayçal Touazi[1]

[1] LIMOSE Laboratory, University M'Hamed Bougara of Boumerdes, Boumerdes, Algeria
{z.kezzoula,d.gaceb,f.touazi}@univ-boumerdes.dz
[2] National School of Computer Science (ESI), Algiers, Algeria
ia_titoun@esi.dz

Abstract. This paper proposes a new super-resolution approach for low-resolution document images based on dual deep transfer learning and GAN Architecture. It is an improvement of an already existing method but constrained by its poor caliber by its low quality on document images. In these images of complex types, it is necessary to preserve the most details and outlines of text and graphic areas. These constraints were the target of our contribution, which aims to improve the ESRGAN method. The proposed approach is called "Bi-ESRGAN". It is based on the combination of two ESRGAN networks. The networks act in double focal on two different image maps (full image and details on the contour map) with a collaborative decision. Our approach has been tested and compared on our document image dataset that we built from document images presenting different challenges, categories, complexity levels and degradation kinds. The experimental results carried out are encouraging and confirmed the superiority of our approach compared to more than sixteen existing approaches with and without learning.

Keywords: Image processing · Super-resolution · Documents images · Deep learning · Transfer learning · Edges detection

1 Introduction

In recent years, we are witnessing significant advancements in image processing and computer vision, which relates to the emergence and progress of deep learning models that can be adapted to the complexity and enhance the quality of the images. Image digitization has various applications in a growing number of fields as diverse as automatic mail and document sorting, medical diagnosis, telecommunications (compression) and military target recognition. This is due to the exponential growth of processor capacities. Accordingly, the development of digitization devices, cameras and scanners in particular, which are becoming diverse and more efficient. However, in spite of the known technical evolution of the digital devices, we can still have a document image that is not clear and the numerical version is not of acceptable or not close enough to the real image.

M. Salem et al. (Eds.): ICAITA 2022, CCIS 1769, pp. 110–122, 2023.
https://doi.org/10.1007/978-3-031-28540-0_9

Digitization of documents does not only mean to archive them, but also to process them (restore, binarize, compress, etc.) and to analyze them (Automatic Document Reading ADR, Optical Character Recognition OCR, Automatic Document Recognition ADR...).

The Single Image Super Resolution "SISR" aims to estimate a high-resolution image (HR) from a single low-resolution image (LR), the LR image is obtained by down sampling HR image with a slight modification [1]. This category of algorithms is divided into two subcategories of methods: the first one is based on direct transformation of the LR image (without learning) and the second one is based on machine learning.

The techniques based on direct transformation are classified into two main categories: interpolation and reconstruction.

The interpolation does not directly estimate the HR image, but a blurred version of this image. In the literature, there are two types of interpolation: non-adaptive (nearest neighbor [2], bilinear [3], bicubic [3], bspline [4], Mitchell-Netravali approach [5], Lanczos approach [5]), adaptive interpolation methods (NEDI [6], DDT [7], EDI [8], FCBI [9], ICBI [10], and DSDF [11]).

Techniques based on reconstruction use generally mathematical operators (masks, convolution, etc.) to find the HR image. Many of the techniques in this category are based on wavelet theory. This method includes several techniques such as expert domains [12], wavelet-based [13], SISR based on surveying adjustment [14], methods based on a Gaussian mixture model [15]. However, these methods introduce blurring and blending the edges and transitional parts, which leads to degradation of fine areas and text in the document images.

The second important category of SR techniques is learning-based method. These approaches are better than those of the first category; in particular those that use deep learning and more efficient on images of documents. In this paper, we propose a new SR method using a dual deep transfer learning, more suitable for document images. Our approach is based on a dual ESRGAN networks using transfer learning to improve document images in combination with edge detection. The proposed approach's performance was compared to that of sixteen approaches (without and with learning) which are most used in literature.

The rest of this article is organized as follows: the second section presents a bibliographic study of existing image SR approaches based on deep learning. Additionally, the third section is devoted to present the proposed approach. Accordingly, the last section presents the experiments carried out and the obtained results.

2 Existing Super-Resolution Approaches Using Deep Learning

Techniques based on deep learning are done in two phases:

a) Learning phase where the model is trained on dictionary. The dictionary that is composed of the pairs (LR, HR).
b) Prediction phase that makes it possible to reconstruct an HR image from a single LR image, based on the model learned in the learning phase.

For several years, artificial neural networks have been very effective in the field of super-resolution. Dong et al. proposed in [16] the SRCNN model (SR with conventional neural network), based on an architecture of three layers: one for patch extraction, non-linear mapping and reconstruction, with a mean square error (MSE) loss function. The SRResNet, SR with residual network is proposed in [17]. A sort of specialized neural network called the deep residual network (ResNet) aids in handling more complex deep learning tasks and models. ESPCN (Efficient Sub-Pixel Convolutional Neural Network) is a novel model for single image SR that uses DenseNet (Densely connected convolutional Networks) [18]. VDSR (Accurate image super-resolution using deep convolutional networks) is another architecture proposed in [19], it uses a very deep convolutional network inspired by the VGG-net. In the same context, the authors of [20] propose a deep CNN model (up to 52 layers), called DRRN (Image super-resolution via deep recursive residual network), which strives to create networks that are both deep and concise. Recursive learning is used to control model parameters while increasing depth. In [21], the authors proposes a new method of SR, WDSR (Wide Activation for Efficient and Accurate Image Super-Resolution), the intuition of the proposed work is that the expansion of the features before ReLU allows more information to pass all retaining the high nonlinearity of deep neural networks. Thus, low-level SR features from shallow layers can be easier to propagate to the final layer for better dense pixel value predictions. The article [22] presents a GLEAN (Generative Latent Bank for Large-Factor Image Super-Resolution) is a new technique for using pre-trained GANs for the problem of large-scale super-resolution, up to 64 upscaling factor. It was demonstrated that an encoder-decoder architecture can leverage a pre-trained GAN as a generative latent bank. In [23], self-attention generative adversarial networks (SAGANs) are proposed. Convolutions are accompanied by the self-attention module, which helps in modeling long-range. The authors of [24] present an improvement of a nonlinear contour stencil method published in [25] that is limited by its high complexity and low quality on images degraded by the introduction of neural networks. Authors of [26] present a new technique for removing noise that is based on learning for the image of surveillance camera. SRGAN (SR with generative adversarial network) is a deep residual network (ResNet) with skip connection [26]. Unlike previous work, which uses MSE as the only optimization target a new perceptual loss is defined, it uses high-level feature maps of the VGG network combined with a discriminator that encourages perpetually difficult solutions to be distinguished from reference images. The generative model's objective is to learn as much as possible about the distribution of the real data. Whereas, the discriminant model's objective is to attempt to accurately determine whether the input data is genuine data or from the generative model HR. The author of [27] extensively examined the three essential components of SRGAN: network structure, adversarial loss, and perceptual domain loss. In order to further enhance the visual quality, and to make each item better to obtain ESRGAN. Removed all BN layers to improve efficiency and minimize computing complexity in a variety of tasks, including super resolution and blur removal [28]. Exchange the original base block with the Residual-in-Residual Dense (RRDB) block, which combines a tiered residual network and density connections. Residual scaling [28, 29], i.e. Reduce residuals by multiplying a constant between 0 and 1 before adding them to the main path to avoid the instability of a smaller initialization. The Residual-in-Residual Dense Block network unit that ESRGAN suggests eliminates the BN layer. The Relativistic GAN concept is also incorporated into ESRGAN, allowing the discriminator to forecast if an image is real rather than whether it is a fake. Utilizing features prior to activation, ESRGAN also enhances perceptual domain loss and can provide more stringent supervision for the restoration of texture and brightness constancy. The Fig. 1 shows the ESRGAN architecture.

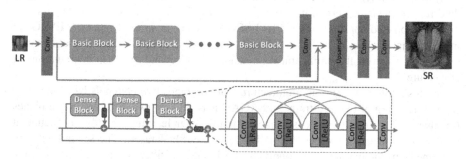

Fig. 1. ESRGAN Architecture [27].

These upgrades gave ESRGAN higher visual quality as well as more realistic and natural texturing. The network is pre-trained on the DIV2K and the Flickr2K datasets. The model is evaluated using benchmark datasets Set5, Set14, BSD100, Urban100 and the PIRM self-validation dataset provided in the PIRM-SR challenge [30]. The main advantage of the ESRGAN model is that it is deeper and shows superior performance with easier training than SRGAN [31]. Therefore, our approach improves this powerful network by introducing an edge detection mechanism, for better adaptation and robustness on document images of all types.

Our proposed approach addresses two important aspects that will be presented below: the transfer of deep learning from natural images to document images. Taking into account the contours that correspond to the details on a document image (often textual or graphical areas).

3 Proposed Method

For our super-resolution approach to document images, we adopt the ESRGAN model in combination with a contrast enhancement step. It focuses on two aspects; the transfer of deep learning from natural images to document images and taking into account the contours that correspond to the details (often textual or graphic areas) carried by a document image. Transfer Learning has been very successful with the rise of Deep Learning. Indeed, the models used in this field often require high computation times and important resources. However, by using pre-trained models as a starting point, Transfer Learning makes it possible to quickly develop high-performance models and efficiently solve complex problems in Computer Vision.

Edge detection via zero crossing of the Laplacian were proposed in 1976. They use the fact that the zero crossing of the Laplacian makes it possible to clearly highlight the extrema of the derivative. These methods also profit from the fact that the zeros of the second derivative constitute a network of closed lines (thus avoiding, in principle, the tracking and closing steps). It is the same for the network of the crest lines of the gradient, but the first is more easily detected from a simple labeling of the positive and negative zones. However, the estimation of the second derivative being very sensitive to noise, it is necessary to filter the image very strongly before measuring the Laplacian. The most used filters for these low-pass filtering are Gaussian filters. The Laplacian edge detector uses a single kernel. It computes second-order derivatives in a single pass.

Our proposed approach (called Bi-ESRGAN) consists in improving the ESRGAN model so that it is more adapted to the super resolution of images of digitized documents of all types (historical, administrative, manuscripts, etc.) and robust to degradations of any nature.

In this context, we use an architecture that combines two channels based on deep transfer learning of two pre-trained ESRGAN networks on two different image planes (or layers). The first network is trained using transfer learning directly on document images of a very rich dataset (SR_IVISION_LIMOSE dataset). The second network is trained on contour maps by applying the Laplacian filter on images of the same database SR_IVISION_LIMOSE. Learning the second network on the edge layer gives the network a better ability to process its detail areas (especially the text zones) with more precision, while being more robust to the presence of blur and background noise. The goal of edge detection is to represent points in a digital image that correspond to a sudden change in light intensity. These changes in document image properties usually reflect important events or changes in the properties of the image (text, drawing, etc.). Contour detection is therefore mainly used for the extraction of useful features i.e. characters and texts. The proposed method is therefore a combination between two approaches with the same model, which will give more readability to the resulting image and thus minimize errors due to noise.

The figure below illustrates the architecture of the proposed approach Bi-ESRGAN (Fig. 2):

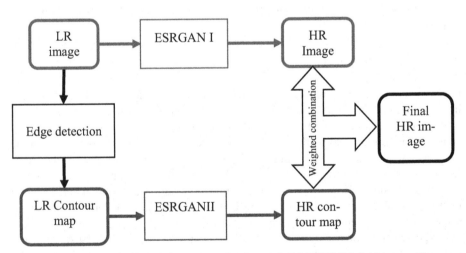

Fig. 2. Architecture and steps of the proposed SR approach (Bi-ESRGAN) for document images.

The ESRGANI model is trained using images pairs (128×128 LR image, 512×512 h image) from the SR_IVISION_LIMOSE document image dataset. Then, it builds patches on which it performs deep transfer learning.

For ESRGANII model, the learning step is carried out in the same way as the ESR-GANI model except that it learns on the basis filtered document image of dataset using Laplacian operator, taking pairs of LR 128×128 and HR 512×512 images which are images filtered by the Laplacian edge detector filter. Subsequently, the two networks ESRGAN II and I trained on the two image planes are combined to build the Bi-ESRGAN architecture. This combination gives our architecture the ability to produce sharper, more legible high-resolution document images that preserve better the quality of thin text or line areas. The combination is based on the following formula:

$$f = \alpha \times f_{ESRGAB1} + (1 - \alpha) \times f_{ESRGAN2} \tag{1}$$

where α is a weighting constant fESRGAN1 and fESRGAN2 are the output HR images of the two models.

4   Evaluation and Experimental Results

To evaluate our model we used two metrics: PSNR and SSIM. PSNR (abbreviation for Peak Signal to Noise Ratio): is a measure of distortion used in digital images, especially in image compression. It makes it possible to quantify the performance of coders by measuring the quality of reconstruction of the compressed image compared to the original image.

$$PSNR = 10 \log_{10}(\frac{Maxf^2}{\sqrt{MSE}}) \tag{2}$$

where,

$$MSE = \frac{1}{mn} \sum_{i=0}^{m-1} \sum_{j=0}^{n-1} \|I(x, y) - I'(x, y)\|^2 \tag{3}$$

(x) represents the matrix data of input original image (Reference HR image). $I'(x,y)$ represents the matrix data of the degraded image in question (HR image resulting from our approach). M and n represents the number of rows and columns of pixels in the image, "i" the index of these rows, and "j" the index of these columns. $Maxf$ is the maximum value of the signal that exists in the original image (knowing that the original image is in good quality).

SSIM (Structural Similarity Index): It was created to compare the visual quality of an original image to a compressed image. In contrast to PSNR, which measures pixel-to-pixel differences, SSIM measures the structural similarity between the two images. The fundamental presumption is that changes in image structure are more perceptible to the human eye.

$$SSIM(x, y) = \frac{(2\mu_x\mu_y + c_1)(2\sigma_x\sigma_y + c_2)(2cov_{xy} + c_3)}{(\mu_x^2 + \mu_y^2 + c_1)(\sigma_x^2 + \sigma_y^2 + c_2)(\sigma_x\sigma_y + c_3)} \tag{4}$$

where μ_x the mean of x; μ_y the mean of y; $\sigma_x{}^2$ variance of x; $\sigma_y{}^2$ variance of y; cov_{xy} the covariance of x and y; $c_1 = (k_1L)^2$, $c_2 = (k_2L)^2$ and $c_3 = c_2/2$.

To gauge the effectiveness of our strategy, We apply it to a test dataset and comparing it with traditional and modern techniques. (deep learning based method).

The choice of the training dataset is very important in the design of a powerful network because the bad choice of this dataset considerably reduces the model's effectiveness. Faced with the absence of SR datasets for deep learning on document images, we have created our new SR_IVISION_LIMOSE dataset of document images. This dataset makes it possible to test super-resolution deep learning approaches on images of scanned documents of all types: historical, administrative, manuscripts, medical, printed matter, degradation of all types (noise, blur, poor lighting, light ink, physical degradation, etc.). It presents challenges to assess the robustness of different approaches including the preservation of the quality of the layout. All folders are divided into two subfolders, TRAIN: 757 images and TEST: 757 images. The existing folders are:

- HR_GT_512 × 512: Contains the folder of HR images of size 512 × 512, these images will be compared with those resulting from super-resolution approaches to assess the quality of the result (ex. PSNR).
- LR_256 × 256: Contains LR images of size 256 × 256.

Fig. 3. Example of images of our SR_VISION_LIMOSE dataset.

– LR_128 × 128: Contains LR images of size 128 × 128 (Fig. 3).

The proposed approach is developed in computer with the following technical characteristics: AppleM1 with RAM = 8 GB, hard drive size 256 Gb and operating system: Mac OS 12.5 Monterey and PC2 with processor: Intel "M5, RAM = 8 GB, hard drive size 256 Gb and HP ELITE BOOK 1030 G1.

The architecture network is the same as ESRGAN with learning rate equals to $1e^{-8}$ and 10 epochs, with the VGG perceptual loss as loss function.

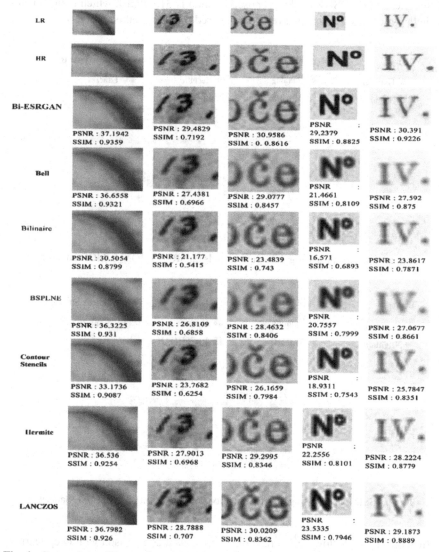

Fig. 4. Comparison of the quality of our Bi-ESRGAN method with conventional approaches.

The following figure (Fig. 4) shows an example of the performance of our method compared to conventional approaches (Bilinear, Bspline, Nearest, Mitchell, Median), with a focus on the errors (blurry, noise, false color) that the SR methods can generate.

We can see in Fig. 4 that the results of the proposed Bi-ESRGAN approach are closer to the HR reference image. The proposed method exhibits less blur and noise than the Mitchell and Median methods and show better quality than the remaining others which leave an image of poor quality which involves a risk of losing information.

The graphs of Fig. 5 show the values of PSNR from the application of the classic methods on the SR_IVISION_LIMOSE document image dataset. In this graph, we can see that the average PSNR of the proposed method is 31.045 dB, which is higher than Mitchell, which has the highest average compared to other methods (Fig. 6).

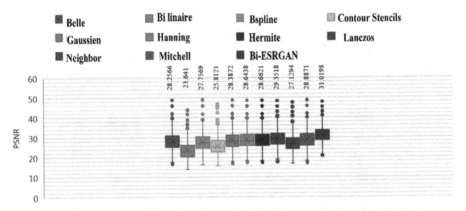

Fig. 5. PSNR values using different methods on SR_IVISION_LIMOSE dataset

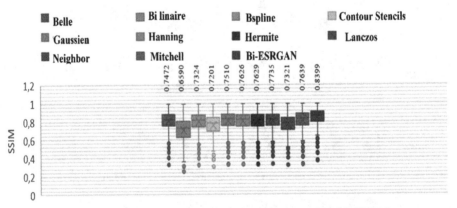

Fig. 6. SSIM values using different methods on SR_IVISION_LIMOSE dataset

In addition, the following figure (Fig. 7) presents a comparison with deep learning methods (SRGAN, WDSR, EDSR, SRCNN).

Image restoration through the deep learning models presented in the table above shows deterioration of the resulting images represented by blurring (SRGAN, WDSR) and emergence of unnecessary details (SRCNN, EDSR).

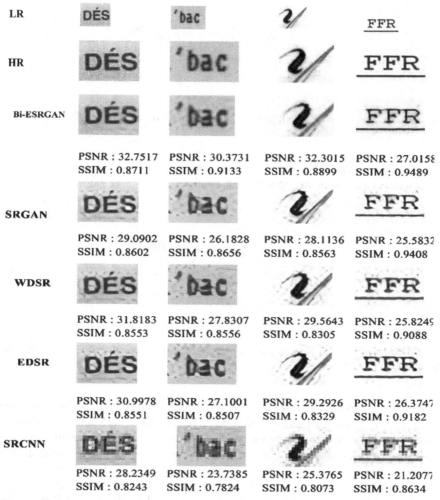

Fig. 7. Comparison of the quality of our Bi-ESRGAN method with existed learning methods

The following graph (Fig. 8) gives the average performance of the proposed model in comparison with other existing models based on deep learning (Fig. 9).

The graphs show an interesting improvement given by our Bi-ERSGAN method, indeed the average of our method with both PSNR and SSIM metrics is higher than the rest of the methods. This confirms that contouring improves the ability of the network to provide sharper and less degraded images of HR documents.

The comparison with classical methods has revealed the limits of classical interpolation techniques to increase the resolution of document images. They do not remove

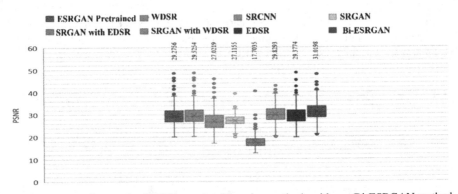

Fig. 8. Comparative graph of PSNR metric of learning methods with our Bi-ESRGAN method, on SR_IVISION_LIMOSE dataset.

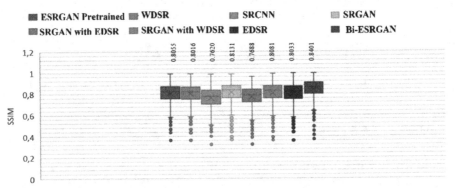

Fig. 9. Comparative graph of SSIM of learning methods with our Bi-ESRGAN method, on SR_IVISION_LIMOSE dataset.

all the degradations present at the level of the LR image, but sometimes these methods themselves cause degradations on the processed images. While the SSIM shows unsatisfactory similarity results for these methods. The comparison with deep learning methods shows the degree of performance of the used models. We notice through these images that the letters is well readable by the learning approaches compared to the classic interpolation methods, but with certain degradations in the contours and our approach has better traced these letters and well defined its contours. The PSNR and SSIM metrics demonstrate the superiority of our method over other models. The SRCNN model was the least preformed which shows these limitations in document and fine image processing. Same thing with the SRGAN model that indicates false colors and a bad tint of the image. All these results allow us to conclude that the developed method gives improved the results compared to these methods. We tried to combine the remaining methods by using the median and the mean to raising the resolution obtained and improved the metrics, which gave the better improvement with the median.

5 Conclusion

In this paper, we have proposed a new SR method, called Bi-ESRGAN, more adapted for the SR of document image, with the aim of improving the existing ESRGAN method based on Transfer Learning and edge detection. The taking into account of the contours allows a better blur reduction on the HR image on the informative zones, with more details. The proposed combination also offers better contrast enhancement and noise reduction on the HR image. Indeed, the advantages of the developed method compared to other approaches have been confirmed by experimental results. In perspective of this work, we intend to explore an approach based on the combination of other image layers with this ESRGAN network and to test our approach on more advanced bases. We also intend to elaborate the complementarity between the different approaches and to build, if necessary, a collaborative approach with weighting.

References

1. Huang, S., Tsai, R.Y.: Multiframe image restoration and registration. Adv. Comput. Vis. Image Process. 1(7), 317–339 (1984)
2. Prajapati, A., Naik, S., Mehta, S.: Evaluation of different image interpolation algorithms. Int. J. Comput. Appl 58(12), 6–12 (2012). https://doi.org/10.5120/9332-3638
3. Kawa, S., Kawano, M.: An overview. In: Umehara, H., Okazaki, K., Stone, J.H., Kawa, S., Kawano, M. (eds.) IgG4-Related Disease, pp. 3–7. Springer, Tokyo (2014). https://doi.org/10.1007/978-4-431-54228-5_1
4. Acharya, T., Tsai, P.-S.: Computational foundations of image interpolation algorithms. Ubiquity 2007, 1–17 (2007)
5. Burger, W., Burge, M.J.: Digital Image Processing: An Algorithmic Introduction Using Java. Springer London, London (2008)
6. Li, X., Orchard, M.T.: New edge-directed interpolation. IEEE Trans. Image Process. 10(10), 521–1527 (2001)
7. Su, D., Willis, P.: Image interpolation by pixel-level data-dependent triangulation. In: Computer graphics forum. 9600 Garsington Road, Oxford, OX4 2DQ. Blackwell Publishing Ltd., UK, pp. 189–201 (2004)
8. Jiji, C.V.: Chaudhuri S: Single-frame image super-resolution through contourlet learning. EURASIP J. Adv. Signal Process 11, 737–767 (2006)
9. Reddy, K.S., Reddy, K.R.L.: Enlargement of Image Based Upon Interpolation Techniques, Department of Electronics and Communication Engineering VITS, Karimna- gar India (2013)
10. Sun, J., Zheng, N.N., Tao, H., Shum, H.Y.: Image hallucination with primal sketch priors. In: Proceedings of IEEE Conference on Computer Vision and Pattern Recognition, vol. 2, pp. 729–736 (2003)
11. Ajit, K., Khobragade, S., Nalbalwar, S.: Review of image reconstruction by interpolation techniques. Int. J. Eng. Res. Technol 03, 198–202 (2014)
12. Roth, S., Black M.J.: Fields of experts: a framework for learning image priors. In: Proceedings of IEEE Conference on Computer Vision and Pattern Recognition, USA (2005)
13. Anbarjafari, G., Demirel, H.: Image super resolution based on interpolation of wavelet domain high frequency subbands and the spatial domain input image. ETRI J. 32(3), 390–394 (2010)
14. Jianjun, Z., Cui, Z., Donghao, F., Jinghong, Z.: A new method for super resolution image reconstruction based on surveying adjustment. J. Nanomaterials 2014, 931616 (2014)

15. Ogawa, Y., Ariki Y., Takiguchi T.: Super-resolution by GMM based conversion using self-reduction image. In: Proceedings of IEEE International Conference on Acoustics, Speech and Signal Processing, Japan, pp. 1285–1288 (2012)
16. Dong, C., Loy, C.C., He, K., Tang, X.: Learning a deep convolutional network for image super-resolution. In: Fleet, D., Pajdla, T., Schiele, B., Tuytelaars, T. (eds.) Computer Vision – ECCV 2014: 13th European Conference, Zurich, Switzerland, September 6-12, 2014, Proceedings, Part IV, pp. 184–199. Springer International Publishing, Cham (2014). https://doi.org/10.1007/978-3-319-10593-2_13
17. Fazli, S., Tahmasebi, M.: PSO and GA based neighbor embedding super resolution. Int. J. Tech. Phys. Problems Eng. 6, 17–21 (2014)
18. Shi, W., et al.: Real-time single image and video super-resolution using an efficient sub-pixel convolutional neural network. In: 2016 IEEE Conference on Computer Vision and Pattern Recognition (CVPR), pp. 1874–1883 (2016)
19. Kim, J., Lee, K.J., Lee, K.M.: Accurate image super-resolution using very deep convolutional networks. In: IEEE Conference on Computer Vision and Pattern Recognition (CVPR), pp. 1646–1654 (2016)
20. Qin, J., Sun, X., Yan, Y., Jin, L., Peng, X.: Multi-resolution space-attended residual dense network for single image super-resolution. IEEE Access 8, 40499–40511 (2020)
21. Yu, J., et al.: Wide activation for efficient and accurate image super-resolution (2018)
22. Chan, K.C.K., Wang, X., Xu, X., Gu, J., Loy, C.C.: Glean: Generative latent bank for large-factor image super-resolution. In: Proceedings of the IEEE/CVF conference on computer vision and pattern recognition, pp. 14245–14254 (2021)
23. Zhang, H., Goodfellow, I., Metaxas, D., Odena, A.: Self-attention generative adversarial networks. In: 36th International Conference on Machine Learning, ICML, pp. 12744–12753 (2019)
24. Kezzoula, Z., Faouci, S., Gaceb, Dj.: Neural approach for the magnification of low-resolution document images. In: IEEE, International Conference, AICCSA, pp. 1–8 (2018)
25. Getreuer, P.: Contour stencils for edge- adaptive image interpolation, vol. 7257. University of California Los Angeles, Mathematics Department, U.S.A (2009)
26. Akiyama, D., Goto, T.: Improving image quality using noise removal based on learning method for surveillance camera images. In: 2022 IEEE 4th Global Conference on Life Sciences and Technologies (LifeTech), pp. 325–326 (2022)
27. Wang, X., et al.: Esrgan: Enhanced super-resolution generative adversarial networks. In: Leal-Taixé, L., Roth, S. (eds.) Computer Vision – ECCV 2018 Workshops: Munich, Germany, September 8–14, 2018, Proceedings, Part V, pp. 63–79. Springer International Publishing, Cham (2019). https://doi.org/10.1007/978-3-030-11021-5_5
28. Nah, S., Kim, T.H., Lee, K.M.: Deep multi-scale convolutional neural network for dynamic scene deblurring. In: IEEE Conference on Computer Vision and Pattern Recognition, pp. 3883–3891 (2017)
29. Szegedy, C., Ioffe, S., Vanhoucke, V.: Inception-v4, inception-resnet and the impact of residual connections on learning. In: Thirty-First AAAI Conference on Artificial Intelligence (2016)
30. Zhu, F.: A review of deep learning based image super-resolution techniques. Comput. Vis. Pattern Recogn. 14, 5423 (2022)
31. Blau, Y., Mechrez, R., Timofte, R., Michaeli, T., Zelnik-Manor, L.: The 2018 PIRM challenge on perceptual image super-resolution. In: Proceedings of the European Conference on Computer Vision (ECCV) Workshops https://doi.org/10.48550/arXiv.1809.07517 (2018)

Evolutionary Algorithms Applications

A Novel Classification Method for Group Decision-Making Dimensions

Badria Sulaiman Alfurhood[1]([✉])[iD] and Marius Silaghi[2][iD]

[1] Department of Computer Sciences, College of Computer and Information Sciences,
Princess Nourah Bint Abdulrahman University,
P.O. Box 84428, Riyadh 11671, Saudi Arabia
bsalfurhood@pnu.edu.sa
[2] Florida Institute of Technology, 150 West University Blv., Melbourne, USA
msilaghi@fit.edu
https://faculty.pnu.edu.sa/bsalfurhood/Pages/default.aspx ,
https://www.fit.edu/faculty-profiles/s/silaghi-marius/

Abstract. This paper presents a novel and comprehensive classification mechanism that groups numerous dimensions associated with group
decision-making approaches. We identify three broad categorizations of
group decision-making literature. The classified dimensions are clustered
by whether they are intrinsic to the group of participants, the nature
of the addressed problems (topics), or a choice in the decision process
design. We highlight the unique challenges facing manual group decision-
making and those linked with intelligent support for group decision-
making. Manual rules governing traditional face-to-face meetings must
be adapted to suit virtual meetings where co-decisions must be made
electronically. Virtual group decision-making for large groups facilitated
by the extensive use of social media tools is an emergent area that
has proven to be a necessity that has had a tremendous influence on
society for the whole world during the recent Covid-19 pandemic. We
first present the associated challenges, followed by the potential solution
of computer-based group decision-making systems. We further discuss
binary and multi-features of group decision approaches with justifications
and examples. Communication has been identified as the main bottleneck, and various attacks have been revealed. This research opens up
several aspects of group decision-making that could be further studied.
For example, a cluster of dimensions concerns the truthfulness of the
information exchanged and how untruthfulness is handled, starting with
detection and how to react to detected lies. The scale of technology-
supported group decision-making has grown to the point where its influence has been accused or lauded in the last few US elections. There is a
persistent call for properties such as transparency and fairness in group
decision-making systems.

Keywords: Group decision-making classification · Group
decision-making communication · Group decision-making challenges ·
Unstructured problems · Group decision support systems · Group
dynamics

© The Author(s), under exclusive license to Springer Nature Switzerland AG 2023
M. Salem et al. (Eds.): ICAITA 2022, CCIS 1769, pp. 125–138, 2023.
https://doi.org/10.1007/978-3-031-28540-0_10

1   Introduction

Decision-making technology for large groups is undergoing significant advancement with the widespread use of social networks and online decision support tools. While group decision-making (GDM) is not new, virtual group decision-making (VGDM) is relatively new, but it showed huge advancement during the Covid-19 pandemic era [5]. Virtual group decision-making is the version of the process when decisions are conducted online. The Internet has helped to facilitate its vast growth with new concepts and techniques that makes it adequate for large groups of widely spread participants. However, challenges do exist. The recent US elections make it clear that this is an issue in group decision-making. The scale of technology-supported group decision-making has grown to the point where its influence has been accused or lauded in the last few US elections. There is a persistent call for properties such as transparency and fairness.

Efforts were made to design and analyze processes for group decision-making. These may refer to rules for face-to-face (FtF) meetings and their transcripts and processing or to architectures, mechanisms, and interaction constraints on computer-supported decision-making systems. Studies of the existing features of group decision-making are sparse in the literature [30]. Therefore, this work aims to develop a comprehensive classification that covers various aspects of group decision-making approaches. In reality spectra, there are very few binary dimensions, and some binary dimensions may have in-between possibilities. Even without considering that nowadays you can be face-to-face online, many feasibilities of hybrid meetings involve FtF and online sessions. For example, between the face-to-face and online meeting approaches, some people might be in one room and others in another, which opens up many possibilities for structuring communication and processes for group decision-making. For instance, having discussions in each room and then exchanging information between rooms with additional talks afterward.

This paper is organized as follows: we first review existing group decision-making difficulties, justifications, and the potential solution of the group decision support system, followed by a broad classification of dimensions grouped by whether they are intrinsic to the group of participants, the nature of the addressed problems (topics), or a choice in the decision process design. We finally provide a summary of the paper's contribution.

2   Group Decision-Making Challenges

As information technology has added more dimensions to problems tackled by humans, it is a challenge to make decisions in such a dynamic era [25]. Constraints affecting groups' decision-making processes include *cognitive limitations, collaboration requirements, information technology costs, and trust issues* [16]. Additionally, many alternatives entail *uncertainty*; therefore, researchers try to model the decision-making mental process taking into account inherent deficiencies and strengths [18]. There exist biases in how humans process information and norms that can provide some cues towards reducing them [14]. Moreover, cognitive biases may impact decision-making negatively, especially when assessing small probabilities [26].

2.1 Face-to-Face (FtF) Meetings Challenges and Robert's Rules of Order

Based on existing research, it is estimated that meetings for making group decisions consume 30–70% of employees' time [29]. Traditional face-to-face (FtF) group decision-making debate suffers from the superiority of some group members, minority intolerance, and the inaccessibility to required information [29]. Issues concerning group members' domination, responsibility dissemination, lack of self-distinction, group agreement stress, and organizational matters may lead to conventional, high-risk, and unreasonable decisions. Given coordination and structural issues, difficulties faced by FtF group decision-making include recording meeting progress, communication, ideas' presentation, and reassurance [29]. Some matters facing FtF meetings have been dealt with using Robert's Rules of Order, which sets proper rules for conducting meetings [28]. These rules are implemented in software tools [6] for easy retrieving. Nevertheless, Robert's Rules of Order are suboptimal for virtual meetings as they do not address the current opportunities for parallel and asynchronous interactions. The authors in [4] suggest shifting the focus on face-to-face versus virtual interactions to the alternative virtual team interactions as a post-pandemic consideration of changes.

2.2 Group Dynamics Challenges

Group dynamics is used to understand behavioral and psychological changes that happen, or take place, when a social group interacts or when new proposals and technologies emerge [3,20]. Some identified group dynamics challenges are: building trust, handling team efficiency, avoiding member isolation and detachment, balancing technical and interpersonal skills, and assessing performance [19].

2.3 Group Decision-Making Communication

Effective communication is generally assumed to lead to superior decisions. Still, communication is believed to be the current bottleneck of common group decision support systems [29]. Even without considering that nowadays, you can be face-to-face online, some people might be in one room and others in another, which opens up many possibilities for structuring communication and processes for group decision-making. For instance, having discussions first in each room and then exchanging information between rooms with additional talks afterward.

3 Solutions: Potential of Computer-Based GDSSs

An accurately designed computer-based group decision support system (GDSS) can decrease some of the FtF meetings' problems mentioned before as the GDSS structured methods can help group members to boost contribution, be on task, and focus on solutions [29]. Group decision-making systems open up different ways for structuring communications for group decision-making. Some of the decision-making participants can initiate the discussion by posting initial ideas anonymously, which is believed by [29] to solve the dominance and minority issues. The group's collaboration for generating the necessary data that assists wrong decisions is the answer to the information inaccessibility problem. There is an extra advantage for GDSS as it lessens or eliminates unneeded social interaction gestures or unrelated information when people tend to type pertinent details only. Group members use several social interactions to collaborate in decision-making, such as conversation, critique, negotiation, and persuasion [25].

4 Group Decision-Making Classifications

4.1 GDM Classifications in Literature

Communication in GDSSs has been classified in [29] along the following dimensions:

- face to face (FtF) vs. virtual meetings,
- synchronous vs. asynchronous interactions,
- closeby vs. geographically dispersed teams, and
- collaborative vs. competing participants.

Other decision-making dimensions raised in the literature include:

- individuals' vs. groups' decisions,
- unstructured vs. semi-structured problems, and
- anonymous vs. authenticated participants [9].

4.2 A Novel GDM Classifications

The above GDM features, together with more than a dozen of other dimensions that we identify, are listed in Tables 1, 2, and 3, respectively grouped by whether they are intrinsic to the group of participants, the nature of the addressed problems (topics), or a choice in the decision process design. Next, we mention significant efforts clarifying some distinctions along these dimensions, noting that the nature of some classifications may overlap.

Table 1. GDSS communication classification based on participant's properties

Individuals' decisions	vs.	Groups' decisions
Small groups	vs.	Large groups
Static groups	vs.	Dynamic groups
Collaborative participants	vs.	Competing participants
Advisory participants' roles	vs.	Constituent participants' roles
Anonymous participants	vs.	Authenticated participants
Closeby teams	vs.	Geographically dispersed teams
Single language	vs.	Multiple languages
Homogeneous participants	vs.	Heterogeneous participants

Table 2. GDSS communication classification based on intrinsic problems' (Topics') properties

Unstructured problems	vs.	Semi-structured problems
Controversial subjects	vs.	Non-controversial subjects
Sensitive matters	vs.	Ordinary matters
Highly secret matters	vs.	Public matters
Urgent matters	vs.	Non-urgent matters
Low certainty facts	vs.	High certainty facts
Global issues	vs.	Local issues

Table 3. GDSS communication classification based on process design

Synchronous interactions	vs.	Asynchronous interactions
Face-to-face meetings	vs.	Hybrid and virtual meetings
Access restricted meetings	vs.	Open meetings
Automated decision-making	vs.	Semi-automated and deliberative decision-making
Consistency/truthfulness enforcing	vs.	Inconsistency/lie tolerance
Egalitarian decision-making	vs.	Weighted decision-making
More information	vs.	Less information
Unlimited time	vs.	Nearly instantaneous time
Single participation	vs.	Multiple participations
Retractable participation	vs.	Permanent participation
Issue-based positions	vs.	Alternative-based positions

5   GDM Classifications: Nature of the Group's Participants

5.1   *Individuals' vs. Groups' Decisions*

Group decision-making is a social activity that seeks the competencies of decision makers and is inherently more complex than individuals' decision-making process [15]. However, group decisions do not consistently outperform individuals' decisions [17]. Although individuals and groups ultimately have to make decisions, the methods used to reach a decision may differ [24]. For example, a debate or a discussion can be added as a social component to the process of group decision-making [23]. As a consequence, the mechanism design is in charge of ensuring reasonable outcomes and responsibility.

5.2   *Small vs. Large Groups*

The number of possible or allowed participants can vary from a tiny group like in the group decision for some medical treatment to a vast group of members that vote for a US president. System design varies accordingly to consider the issues related to handling large groups.

5.3   *Static vs. Dynamic Groups*

Groups may have a fixed composition or be dynamic, with members joining or leaving during the deliberation process. Groups can be considered static with deliberation processes that are relatively short compared to the churning rate of participants. The dynamism of groups has to be accounted for with deliberation processes that are continuous or that take a long time [1].

5.4   *Collaborating vs. Competing Participants*

An example of a *collaborative* application of GDSSs is the Forest Management Decision Support System, where participants *brainstorm ideas* and try to discover solutions by composing their knowledge [27,31]. Nevertheless, *competing* decision-making is expected in voting, politics, and religion. Competing participants mainly try to persuade others about their preconceived dogmas, but they may have some degree of openness that gives a leeway to reach a compromise. A GDSS can aid competing participants in discovering their leeway to find possible common areas of agreement and make compromises accordingly.

5.5   *Advisory vs. Constituent Roles of Participants*

With *advisory* citizen review boards and deliberative polls, for example, groups of citizens obtained by random sampling from the set of the whole eligible population are asked to meet and deliberate to propose solutions to community problems [13]. Henceforth, the obtained solutions or arguments can be presented to the decision-makers or the public ahead of a referendum. The logical soundness of the proposed solutions and the corresponding justifications are more important than the vote count since the real decision-makers have to be persuaded to follow up eventually. Nonetheless, with meetings of *constituents*, the votes are more important than the logic of the provided explanations. For this reason, and in some cases, rationality cannot be guaranteed, and it is up to constituents to support irrational conclusions.

5.6 *Anonymous vs. Authenticated Participants*

In various Internet forums, users can use pseudonyms that might have been verified for eligibility but where other participants do not know the actual identity. The identity of users may still be accessible to a system administrator, except if hidden using cryptographic anonymization mechanisms such as Tor. At the other extreme, there are cases where the identity of participants is validated using a tool that publishes the names known to an authenticated organization. Intermediary situations appear in Facebook groups where most people use their real names and are known by their peers, but some users may still join with pseudonyms or false identities.

5.7 *Closeby vs. Geographically Dispersed Teams*

An example of a closeby team is a neighborhood association where the participants, homeowners, are grouped based on their addresses and may identify each other using their house numbers [32,34]. Face-to-face meetings are most appropriate for a nearby team. Other organizations such as Facebook diasporas, for instance, enable people dispersed worldwide to make decisions concerning their communities virtually.

5.8 *Single vs. Multiple Languages*

In some platforms, deliberations on issues can be restricted to one language. Other platforms, like Loomio, support multi-language deliberation, which allows virtual participants to cooperate to reach a decision using their languages. However, rules and tools must be set to facilitate communication among various users, such as instant translation.

5.9 *Homogeneous vs. Heterogeneous Participants*

Some researchers have tried classifying group decision problems based on levels of similarity between participants. Various traits can define the similarity between humans. The following list is just a glimpse into the tremendous number of possible dimensions: *age, gender, race, studies, qualifications, experience, knowledge, civil status, military status, health status, ethnic identification, nationality, tribe, number of citizenships, religious affiliation, political affiliation, specific beliefs, dress code, time zone, latitude, altitude, climate, and geographical area.* Some efforts were also made to classify along a more general concept of culture [11]. Every person is unique. Hence, associating one discrete metric with cultural differences needs a careful scientific and ethical basis to avoid racism or other types of unethical stereotyping.

6   GDM Classifications: Nature of the Addressed Problems (Topics)

6.1   *Unstructured vs. Semi-structured Problems*

Unstructured problems are open-ended problems. For instance, class or political debates. GDSSs can support search management and visualization of information for such problems. The semi-structured problems exist where rules exist, but a human component is still maintained to guide the process in essential situations. Semi-structured and unstructured problems are the cases where it is difficult to uniquely and automatically identify and assess all the feasible alternatives [12].

6.2   *Controversial vs. Non-Controversial Subjects*

Some collaborative platforms, like Wikipedia, are known to be appropriate for non-controversial subjects such as science and remote historical events. However, such media are being less trusted in matters of political and religious issues, which might raise operation difficulty in some countries. Loomio is an excellent example of a platform that supports deliberation on controversial topics. Some interaction mechanisms, such as the common poll platforms, can be designed for controversial subjects at the expense of other properties.

6.3   *Sensitive vs. Ordinary Matters*

Decision platforms stress accuracy and responsibility with sensitive matters more, while comfort and simplicity are essential in handling common issues. For example, GDM platforms should be designed to handle restrictions regarding age, material, and sensitive national issues.

6.4   *Highly Secret vs. Public Matters*

Access control to data, authentication and anonymization for privacy are the foci of systems involved with highly certain matters, such as those involved in medical, political, and military areas. However, indexing and efficiency are the primary metrics for public issues.

6.5   *Urgent vs. Non-urgent Matters*

Access to multimedia and multimodal interfaces is an essential part of the platforms involved in handling disasters, while data structuring is more straightforward in non-urgent matters as there may not be a need for extra documents and related data.

6.6 Low Certainty vs. High Certainty Facts

Various expert systems support mechanisms to assert doubts and propagate probabilities involved in engineering and science materials, while certainties are a hallmark for political and religious systems.

6.7 Global vs. Local Issues

For global issues affecting a considerable number of people, like dealing with climate change, finding and illustrating the opinions of outsiders (stakeholders) are essential in making decisions. Nevertheless, local issues such as choosing a logo or dress code for a company or the type of food to eat at a particular gathering are topics that have lower impacts on public participants.

7 GDM Classifications: Choices of Decision Process Design

7.1 Synchronous vs. Asynchronous Interactions

Video conferences and electronic implementations of Robert's Rules of order induce mechanisms for synchronous group decision-making processes. Robert's Rules of Order sets proper rules for conducting meetings [28]. Other group decision-making mechanisms support asynchronous interactions where users can simultaneously submit ideas, comments, and votes.

7.2 Face to Face vs. Hybrid and Virtual Meetings

While face-to-face is the classical way of conducting meetings, virtual meetings can be facilitated by the exclusive use of technology. Using web technologies and mobile communication, Giorgio [7] suggests a distributed group decision support system to enable widely dispersed individuals to form "virtual teams" who cooperate to make real-time co-decisions. There exist hybrid systems where meetings could be conducted in which some participants meet in-person or face-to-face online, with multiple possibilities of participants being in the same building or on different continents.

7.3 Access Restricted vs. Open Meetings

Some examples of open group decision-making systems that enable people to join without significant verification of eligibility include YourView[1] and We the People (the White House petition system[2]). The White House petition system allows initiating a petition, and the matter is eligible to be viewed by the White

---

[1] yourview.org.au.

[2] petitions.whitehouse.gov.

House officials when the petition acquires more than a threshold number of signatures. In contrast, Confluence[3] and Choicla have access restricted to previously vetted participants only. An intermediary approach in DebateDecide allows new users to join based on a distributed verification of eligibility [32] with subjective results.

7.4   Automated vs. Semi-automated and Deliberative Decision-Making

This dimension defines a whole spectrum, from very little support (like just being an electronic blackboard) to fully automated systems having agents acting for all the humans (naturally provided with the goals of the humans) and deciding without further involvement of humans. There are a lot of possibilities in-between. In automated systems such as *Choicla*, humans input their preferences, and the computer aggregates them without further consultation. At the other extreme, systems such as *YourView*[4] or *ReframeIt* provide support tools that help humans in ranking and structuring information. However, they left the decision itself in the hands of the group. The use of multi-choice voting systems like *Borda count* leads to situations where the decision is partly automated in the sense of not being directly linked to a vote by the group members on isolated pairs of choices.

Since user preferences are not stable, mechanical adoption of the decisions based on mathematical aggregation strategies does not lead to perfect predictions of the happiness of groups [8]. Therefore, it is suggested to incorporate discussion groups where the computer has to be a facilitator rather than a rigid mediator [22]. For example, the STSGroup system is based on the *Orientation-Discussion-Decision-Implementation* (ODDI) model.

7.5   Consistency/Truthfulness Enforcing vs. Inconsistency/Untruthfulness Tolerance

A cluster of dimensions represented here is the truthfulness of the information exchanged and how untruthfulness is handled (starting with detection and how to react to detected lies). The recent US election makes it clear that this is an issue in group decision-making! In essence, humans are not always representing their beliefs as a consistent set of logical statements; instead, logic could be just a local justification of their emotional preferences. Moreover, some comments are incorrect or exaggerated not due to conscious intentions of lying but because of the unequal importance given to components of "truth". In some platforms, logic reasoning or deliberation is used to identify inconsistent positions of contributors, such as claiming priority for nonviolence principles and supporting specific violent actions. In these situations, detecting inconsistencies

---

[3] atlassian.com/software/confluence.
[4] yourview.org.au.

can help people improve their understanding of their emotions, but they cannot always accuse of bad intentions. In other words, deliberation or automatic logic consistency verification systems can help flag subjectivity and improve the objectivity of users. In contrast, an automated consistency enforcement technique would flag the user as lying or being inconsistent. However, given humans' lack of complete consistency in their reasoning, it is impossible to enforce consistency for provided statements without risking representing the participants' minds incorrectly. Deliberation systems, like *ReframeIt*, implicitly detect partial inconsistency through peer-review; however, the discrepancies and lies may remain in the conclusions. Argumentation frameworks [10] is focused on rooting out any inconsistency but may fail to represent members' positions fully.

7.6 *Egalitarian vs. Weighted Decision-Making*

In some systems, all participants have an equal vote (like in polls and elections). In contrast, in other tools like (DirectDemocracyP2P [33,34]), users can have weights highlighting their relevance to the decision being made: advisors, stakeholders, or constituents.

7.7 *More vs. Less Information*

While it has been shown that providing more information to participants is helpful, it remains to be decided when and how one should give this information. Information provided early and before the participants have made up their minds leads to the phenomena of *anchoring* where participants have incentives to accept a decision made by others without putting the effort to think for themselves. In addition, research shows that informing participants about others' preferences may negatively impact [21,35].

7.8 *Unlimited vs. Nearly Instantaneous Time*

The time dimension is around the time available for making a decision: from unlimited to having to be nearly instantaneous, being done in one round or several, and so on. Time constraints may affect group decision-making in different ways. Some applications give the ability to make decisions almost instantaneously, such as in *Choicla*, while others allow users to improve their choices in rounds, like in *ReframeIt*.

7.9 *Single vs. Multiple Participation*

Participants can submit only one vote and comment in certain group decision support procedures, such as polls. For other types of collaborative modes, participants can submit any number of comments [2].

7.10   *Retractable vs. Permanent Participation*

With some group decision-making systems, such as forums or wikis, participants can retract their opinions statements or vote when they change their minds. Loomio is an example of such a systems. With other methods, the submitted comments remain permanent, such as in the case of email-based communications.

7.11   *Issue-Based vs. Alternative-Based Positions*

In a common group decision-making architecture known as an Issue-based Information System (IBIS), each statement only has to be associated with the discussed issue. In the Alternative-based Information System model, statements have to be in support of some decisions [33].

8   Conclusions

This paper contributes to scientific research by developing a new classification of the group decision-making literature. We identify three broad categorizations of group decision-making clustered around various dimensions. In addition, a set of new classification dimensions is proposed, e.g., advisory vs. constituent participants, retractable vs. permanent participation, secret vs. public matters, global vs. local issues, egalitarian vs. weighted decision-making, unlimited vs. instantaneous time, issue-based vs. alternative-based positions, and low vs. high certainty facts.

Virtual group decision-making is essential for several applications where face-to-face meetings are not pertinent. Online group decision-making could impact people's meaningful choices in different ways. Communication threats in group decision-making may affect critical decisions, and specific attributes are highly required. Ongoing Internet technology advancement could mediate existing communications' limitation and resolves threat issues. Several aspects of group decision-making could be further studied. Examples include solutions to challenges facing significant group decision-making aspects, the impact of decision-makers roles in decision-making, and the tools that best serve their different needs. This research opens up several parts of group decision-making that could be further studied. For example, a cluster of dimensions concerns the truthfulness of the information exchanged and how untruthfulness is handled, starting with detection and how to react to detected lies. The recent US elections make it clear that this is an issue in group decision-making!

Acknowledgment. I would like to thank Princess Nourah bint Abdulrahman University Researchers Supporting Project number (PNURSP2023R359), Princess Nourah bint Abdulrahman, University, Riyadh, Saudi Arabia.

References

1. Alfurhood, B.S., Silaghi, M.C.: A survey of group decision making methods and evaluation techniques. In: FLAIRS Conference (2018)
2. Alqahtani, A., Silaghi, M.: Evaluations techniques for argumentation architectures from the perspective of human cognition. In: Florida Conference on Artificial Intelligence (FLAIRS) (2016)
3. Backstrom, L., Huttenlocher, D., Kleinberg, J., Lan, X.: Group formation in large social networks: membership, growth, and evolution. In: Knowledge Discovery and Data Mining, pp. 44–54 (2006)
4. Bauer, T.D., Humphreys, K.A., Trotman, K.T.: Group judgment and decision making in auditing: research in the time of Covid-19 and beyond. AUDITING: J. Pract. Theory **41**(1), 3–23 (2022)
5. Cui, C., Li, B., Wang, L.: The selection of Covid-19 epidemic prevention and control programs based on group decision-making. Complex Intell. Syst. **8**(2), 1653–1662 (2022)
6. Dahlstrom, D.B., Shanks, B.: Software support for face-to-face parliamentary procedure. Online Deliberation Des. Res. Pract. 213–220 (2009)
7. Michelis, G.: Co-decision within cooperative processes: analysis, design and implementation issues. In: Humphreys, P., Bannon, L., McCosh, A., Migliarese, P., Pomerol, J.-C. (eds.) Implementing Systems for Supporting Management Decisions. ITIFIP, pp. 124–138. Springer, Boston (1996). https://doi.org/10.1007/978-0-387-34967-1_9
8. Delic, A., et al.: Observing group decision making processes. In: Proceedings of the 10th ACM Conference on Recommender Systems, pp. 147–150. ACM (2016)
9. Dennis, A.R., Quek, F., Pootheri, S.K.: Using the Internet to implement support for distributed decision making. In: Humphreys, P., Bannon, L., McCosh, A., Migliarese, P., Pomerol, J.-C. (eds.) Implementing Systems for Supporting Management Decisions. ITIFIP, pp. 139–159. Springer, Boston, MA (1996). https://doi.org/10.1007/978-0-387-34967-1_10
10. Dung, P.M.: On the acceptability of arguments and its fundamental role in nonmonotonic reasoning, logic programming and n-person games. Artif. Intell. **77**(2), 321–357 (1995)
11. Eisenberg, C.: Integrating a distributed and heterogeneous organisation using constraint programming. Challenge 3, 4 (2000)
12. Filip, F.: Computer-aided decision-making: systems, applications, and modern solutions. In: IMACS Multiconference on Computational Engineering in Systems Applications, vol. 1, p. PL-11. IEEE (2006)
13. Fishkin, J., Farrar, C.: Deliberative polling: from experiment to community resource. In: The Deliberative Democracy Handbook: Strategies for Effective Civic Engagement in the Twenty-First Century, pp. 68–79 (2005)
14. Gordon-Becker, S.E., Lee, J.D., Liu, Y., Wickens, C.D.: An introduction to human factors engineering (2004)
15. Hitzler, S., Messmer, H.: Group decision-making in child welfare and the pursuit of participation. Qual. Soc. Work. **9**(2), 205–226 (2010). https://doi.org/10.1177/1473325010372156
16. Holsapple, C.W., Whinston, A.B.: Decision support systems: a knowledge-based approach. Stud. Inf. Control **10**(1), 73–76 (2001)
17. Hsieh, C.-J., Fifić, M., Yang, C.-T.: A new measure of group decision-making efficiency. Cogn. Res.: Principles Implications **5**(1), 1–23 (2020). https://doi.org/10.1186/s41235-020-00244-3

18. Kendler, J.: Visual artifacts as decision analysis and support tools. Wiklund Research and Design (2004)
19. Kirkman, B.L., Rosen, B., Gibson, C.B., Tesluk, P.E., McPherson, S.O.: Five challenges to virtual team success: Lessons from Sabre, Inc. Acad. Manag. Exec. **16**(3), 67–79 (2002)
20. Levi, D.: Group Dynamics for Teams. Sage Publications (2013)
21. Mojzisch, A., Schulz-Hardt, S.: Knowing others' preferences degrades the quality of group decisions. J. Pers. Soc. Psychol. **98**(5), 794 (2010)
22. Nguyen, T.N., Ricci, F.: Supporting group decision making with recommendations and explanations. In: 24th Conference on User Modeling, Adaptation and Personalization (UMAP) (2016)
23. O'Connell, T.S., Cuthbertson, B.: Group dynamics in recreation and leisure: creating conscious groups through an experiential approach. Human Kinetics (2009)
24. Ozuem, W., Bowen, G.: Competitive Social Media Marketing Strategies, p. 317. IGI Global (2016)
25. Pohl, J.: Cognitive elements of human decision making. In: Phillips-Wren, G., Ichalkaranje, N., Jain, L.C. (eds.) Intelligent Decision Making: An AI-Based Approach, pp. 41–76. Springer, Cham (2008). https://doi.org/10.1007/978-3-540-76829-6_2
26. Pomerol, J.C., Adam, F.: Understanding human decision making – a fundamental step towards effective intelligent decision support. In: Phillips-Wren, G., Ichalkaranje, N., Jain, L.C. (eds.) Intelligent Decision Making: An AI-Based Approach. SCI, vol. 97, pp. 3–40. Springer, Heidelberg (2008). https://doi.org/10.1007/978-3-540-76829-6_1
27. Reynolds, K.M., Schmoldt, D.L.: Computer-aided decision making. In: Shao, G., Reynolds, K.M. (eds.) Computer Applications in Sustainable Forest Management, vol. 11, pp. 143–169. Springer, Dordrecht (2006). https://doi.org/10.1007/978-1-4020-4387-1_8
28. Robert, H.M., Robert, S.C., Honemann, D.H.: Robert's Rules of Order Newly Revised. Da Capo Press (2011)
29. Rothi, J., Yen, D.: Group Decision Support Systems: Data Communications Considerations and Requirements. Auerbach Publications (2016)
30. Shakre, E., et al.: Decentralized decision making in dynamic groups for distributed free and open-source software updating. In: Florida Conference on Artificial Intelligence (FLAIRS) (2017)
31. Shao, G., Reynolds, K.M.: Computer applications in sustainable forest management (2006)
32. Silaghi, M., Qin, S., Matsui, T., Yokoo, M., Hirayama, K.: Bayesian network-based extension for PGP, estimating petition support. In: Florida Conference on Artificial Intelligence (FLAIRS) (2016)
33. Silaghi, M., Roussev, R., Alfurhood, B.: Why do they vote that? In: Florida Conference on Artificial Intelligence (FLAIRS) (2017)
34. Silaghi, M.C., et al.: Directdemocracyp2p - decentralized deliberative petition drives. In: 2013 IEEE Thirteenth International Conference on Peer-to-Peer Computing (P2P), pp. 1–2. IEEE (2013)
35. Stettinger, M., Felfernig, A.: Choicla: intelligent decision support for groups of users in the context of personnel decisions. In: RecSys, pp. 28–32 (2014)

Intelligent Cache Placement
in Multi-cache and One-Tenant Networks

Sara Ameghchouche[1]([✉]) [iD], Mohammed Amine Mami[2] [iD],
and Mohamed Fayçal Khelfi[3] [iD]

[1] University of Oran1 Ahmed Ben Bella, Oran, Algeria
amgsara63@gmail.com
[2] Department of Computer Science, University of Oran1 Ahmed Ben Bella,
Oran, Algeria
[3] RIIR Lab University of Oran1 Ahmed Ben Bella, Oran, Algeria

Abstract. The increasing of the network devices number could generate
a massive traffic, which causes overloads of the edge network or core network due to the large number of requests per second. The Caching is one
of network and telecommunications' services that has several advantages
for: ISPs, enterprise networks and Final users. This optimal caching system in hierarchical and multi-cache networks is an emerging challenge
since it's not easy due to its high computational complexity. So, our
problem is related to the multi-objective optimization issues which we
seek to optimize several parameters such as minimizing average distance
ratio, content provider load and increasing content diversity. However, we
suggest solving with the use of a pure-ordered approach of the GRASP
meta-heuristic (Greedy Randomized Adaptive Search Procedure) and
going to compare the results with the Pareto front's solution under the
LCE (leave copy everywhere) placement strategy to find the set of optimal cache distributions.

Keywords: Caching · Intelligent placement · GRASP ·
Multi-objective optimization

1 Introduction

The Implementing of an optimal caching system within a network infrastructure
is too challenging and remains to open question that is rarely addressed in the
literature. While single-objective placement has been addressed in the existing
field, there is still a lack of studies considering multi-objective placement. In
addition, optimal cache placement is not easy to achieve due to its high computational complexity. Indeed, the different Internet Service Providers (ISPs)
involved in the deployment have no interest in revealing their infrastructure,
which makes this task even more difficult.

In previous work with the single-tenant case, they studied caching at the network edge (which brings content closer to the final users and retrieves it much

M. Salem et al. (Eds.): ICAITA 2022, CCIS 1769, pp. 139–146, 2023.
https://doi.org/10.1007/978-3-031-28540-0_11

faster) and caching at the core (which it reduces content redundancy and thus dependence on peering links, ensuring lower costs for ISPs). But in other studies such as [1], the results show that the solution is not to put all resources at the edge or core of the network, as some studies claim, but to distribute them judiciously. In our study, we will extend the theory as a multi-objective optimization problem in which we aim to optimize various metrics such as minimizing content provider load, average distance ratio and increasing content diversity. We're going to using the GRASP meta-heuristic, this improved our understanding of the effect of cache distribution on network performance and how our solution can adapt to the different goals set in a multi-cache system.

2   Related Work

In [2], the authors proposed a proactive cache management approach for ISP networks in a scenario with multiple content providers by setting the problem as an integer linear program (ILP) to derive the optimization theoretical placement and routing strategy. The ILP approach can store content in the cache associated with the requesting edge node, resulting in a bandwidth utilization of 0. However, as proactive content placement on a specific node involves transferring the entire movie to that node, adjusting the placement every time the demand pattern changes can result in significant overhead in a real scenario.

Subsequently, the same authors proposed in [3]; a hybrid cache management approach that combines proactive content placement with traditional reactive caching strategies. The ISP leases the cache space in its network to one or more content providers. Each content provider specifies the amount of caching capacity it wants to reserve to store a portion of its content catalog, while the ISP decides which content items will be stored and where the proactive placement algorithm is run periodically by a core manager to allocate the proactive portion of the capacity across the network based on the predicted value of content popularity and geographic distribution. In contrast, the reactive approach (LRU) is applied to each cache independently and serves as a buffer to react locally to unpredicted changes in popularity. The performance of this approach is highly dependent on the accuracy of the prediction of future content requests.

Some evaluation of the performance of the NDN architecture as a function of the number of content stores in the network has been proposed in [4]. The number of content stores in the NDN network has an impact on its performance. The performance of the cache network increases up to about 40% of content stores in the network and then is not improved. Their results show that the performance of NDN caches (cache hit) is optimal with only 50% of the nodes equipped with Content Stores and this is important for network providers as it shows that the NDN architecture can be implemented at a reduced infrastructure cost.

A summary packet defined in [5], by using a Bloom filter and a method to share the summary. When a data packet is received, a router makes a decision whether or not to save the data based on the results of querying the summaries of the neighboring routers' cache. If one of the neighboring routers has the data, it has not necessarily been saved. When an Interest packet is received, a router can forward the Interest to a neighboring router that has the requested content by checking the summaries. Each router is able to avoid storing duplicate content that neighboring routers might have. Therefore, a greater diversity of content can be achieved, resulting in a better cache hit rate and a reduction in the average content delivery time. However, duplicate copy prevention is not advantageous from a content retrieval time perspective.

In [6], caching on Named Data Network must be able to store content dynamically. It should be selectively select content that is eligible to be stored or deleted from the content storage based on certain considerations. The results obtained by this approach are: Reducing the latency, maximizing the diversity of content and increase the ease and speed of consumers in accessing data. But there are some limitation like: The algorithms are more complex if there are many parameters, The prediction may be incorrect and the internal calculation of router is more complex.

3   Methodology to Solve the Cache Placement Problem

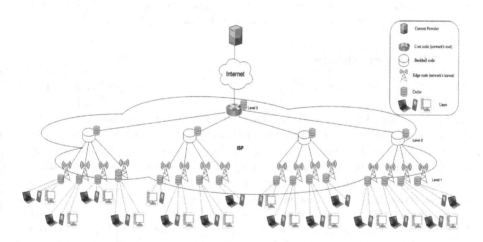

Fig. 1. Network architecture of one-tenant and multi-cache nodes

3.1   Network Architecture

Our Network Architecture as show in Fig. 1 is defining by three levels (one node for the 1st level, four nodes for the 2nd level and 16 nodes of the last one). All the network's nodes have the same cache size (X) and permanent copies of the available contents are hosted on one repository attached to the only node of level

3 (root node of the network) and the users are attached to the nodes of level 1 (network's leaves). User requests originate at a particular leaf node and proceed to the root node until the requested content is found. If the requested content cannot be found, even at the root node, the request is forwarded to the content provider (repository) that contains the content. When the content is found, it is forwarded to the client via the back path. Each cache on the back path decides whether to cache the content or not according to a chosen strategy like LCE (Leave copy everywhere) used in our study.

3.2   The Simulation Environment

Our experiments were performed on a hierarchical network that contains 21 nodes with the same distance and latency between each two adjacent nodes. And a content catalog (repository) with a size of 10000 contents or chunks. For the nodes cache we tested different size (Tc) which is a proportion of the catalog content size (10%, 25%, 30% from 10000). Following Zipf's law (content is ranked in a decreasing manner according to its popularity, from one to the total number of items in the catalog) and testing with different values of the Zipf's law skewness parameter α ranging from 0.8 to 1.2 (as many studies show [7], these two values correspond to the Zipf's popularity exponent in the case of user-generated content (UGC) and video-on-demand (VoD), respectively).

3.3   The Approach Used

We use the GRASP metaheuristic to evaluate two network performance measures: the content provider load $f1(X)$ and the average distance ratio $f2(X)$, where GRASP is a multi-iteration or multi-start process, in which each iteration was composed of two phases: construction and local search. The construction phase constructs a feasible solution, whose adjacency is examined until a local minimum is obtained during the local search phase. The method chosen to generate solutions based on the objective functions $f1(X)$ and $f2(X)$ in GRASP is called the pure ordered approach, using at each iteration one of the objective function ($f1(X)$ or $f2(X)$) in the construction phase. Then the other one is applied in the local search. Then, we alternate the functions chosen for each phase in the next iteration (e.g., if we chose $f2(X)$ for the construction phase and $f1(X)$ for the local search phase in the first iteration, then, in the second iteration, we will use $f2(X)$ in the local search and $f1(X)$ in the construction phase).

4   The Simulation Results and Analysis

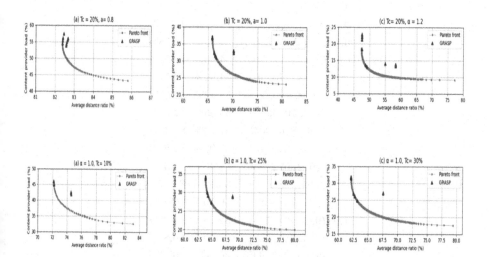

Fig. 2. Comparison of GRASP and Pareto front solutions under the LCE scheme strategy

4.1   Performance Evaluation

In Fig. 2, we compare the two network performance metrics (content provider load and average distance ratio) of different cache distribution results generated by GRASP with the Pareto front using the separate evaluation functions approach f1(X) and f2(X). Where the Content provider load is defined as the ratio of requests not served by intermediate network caches and so obtained from the content provider. And the Average distance ratio represents the average gain in terms of the distance an interest travels before finding a copy of the requested object. We can see and notice on the graphs the good quality of the GRASP metaheuristic where its results produced in different scenarios are very close to the set of efficient points.

4.2   Metrics Analysis

Effect of the Zipf Parameter (α): We notice a high values of the content provider load (57.40%) and the average distance ratio (82.66%) as shown in Table 1 when we fixed α at 0.8 as opposed to α when it equal to 1.2 there is an evident decrease on network performance metrics (13.00% and 47.49%, respectively). We explain this by: a small value of α like 0.8 means a smaller difference in popularity between different contents, which makes it more difficult to construct cache distribution solutions using GRASP that approximate the best performance.

Table 1. Cache distribution solutions using GRASP with two separate evaluation f1(X) and f2(X) function under the LCE strategy with various Zipf parameter (α)

Tc = 20%			
α	X	$f1(X)$	$f2(X)$
0.8	((1072, 912, 16))	54.23%	82.58%
	(1072, 672, 256)	54.46%	82.37%
	(1024, 240, 16)	54.90%	82.61%
	(1008, 720, 272)	55.42%	82.38%
	(976, 1008, 16)	55.56%	82.64%
	(880, 816, 304)	57.40%	82.44%
	(944, 1040, 16)	56.00%	82.66%
	(1072, 672, 256)	54.46%	82.37%
	(1104, 880, 16)	53.79%	82.57%
	(960, 752, 288)	56.16%	82.40%
1.0	(992, 464, 544)	32.34%	66.20%
	(704, 544, 752)	36.69%	65.80%
	(1104, 448, 448)	30.96%	66.57%
	(688, 576, 736)	36.97%	65.80%
	(976, 1008, 16)	32.37%	70.12%
	(704, 592, 704)	36.69%	65.80%
	(928, 1056, 16)	32.92%	70.11%
	(704, 576, 720)	36.69%	65.80%
	(1056, 432, 512)	31.54%	66.39%
	(688, 576, 736)	36.97%	65.80%
1.2	(1008, 976, 16)	13.73%	58.28%
	(320, 848, 832)	21.27%	47.53%
	(976, 960, 64)	13.96%	55.03%
	(144, 1040, 816)	22.77%	47.49%
	(1040, 416, 544)	13.53%	48.74%
	(544, 576, 880)	18.46%	47.50%
	(1120, 368, 512)	13.00%	49.13%
	(224, 960, 816)	22.13%	47.51%
	(1072, 912, 16)	13.30%	58.33%
	(224, 960, 816)	22.13%	47.51%

Effect of the Cache Size (Tc): With different cache size, we see an obvious impact on both network performance metrics. As shown in Table 2, the content provider load decreases from 46.02% to 26.00%, when the percentage of total size increases from 15% to 30%, which means that a large size of the intermediate cache nodes reduces the overload of the repository node. And the same

Table 2. Cache distribution solutions using GRASP with two separate evaluation
f1(X) and f2(X) function under the LCE strategy with various cache size

$\alpha = 1.0$			
Tc	X	$f1(X)$	$f2(X)$
15%	(464, 512, 16)	41.90%	74.55%
	(368, 288, 336)	44.95%	72.16%
	(432, 544, 16)	42.62%	74.55%
	(368, 288, 336)	44.95%	72.16%
	(448, 528, 16)	42.26%	74.55%
	(368, 336, 288)	44.97%	72.20%
	(464, 512, 16)	41.90%	74.55%
	(336, 320, 336)	46.02%	72.15%
	(464, 512, 16)	41.90%	74.55%
	(368, 304, 320)	44.96%	72.17%
25%	(1280, 1200, 16)	28.84%	68.70%
	(864, 720, 912)	34.01%	63.76%
	(1264, 1216, 16)	28.99%	68.70%
	(864, 720, 912)	34.01%	63.76%
	(1264, 576, 656)	29.15%	64.22%
	(864, 720, 912)	34.01%	63.76%
	(1184, 560, 752)	30.00%	64.06%
	(864, 720, 912)	34.01%	63.76%
	(1456, 560, 480)	27.29%	64.81%
	(880, 736, 880)	33.78%	63.77%
30%	(1440, 1536, 16)	27.14%	67.51%
	(1040, 864, 1088)	31.58%	62.08%
	(1504, 656, 832)	26.81%	62.51%
	(1056, 864, 1072)	31.39%	62.09%
	(1456, 1520, 16)	27.01%	67.51%
	(1040, 864, 1088)	31.58%	62.08%
	(1744, 688, 560)	24.85%	63.13%
	(1024, 880, 1088)	31.77%	62.08%
	(1600, 720, 672)	26.00%	62.73%
	(1056, 848, 1088)	31.39%	62.08%

for the average distance ratio, which decreases from 74.55% to 24.85%, when
the percentage of the total size increases from 15% to 30%, which means that
placing more content in the intermediate cache nodes converges it to the users
and minimizes the distance.

5 Conclusion

Our work allows us to study the optimal cache distribution in multi-cache network with one-tenant and evaluate its performance metrics by minimizing both the content provider load and the average distance ratio. As our problem is a multi-objective optimization problem we use the GRASP meta-heuristic to achieve this and the results were too close to pareto sets.

References

1. Ben-Ammar, H., Hadjadj-Aoul, Y.: A GRASP-based approach for dynamic cache resources placement in future networks. J. Netw. Syst. Manage. **28**(3), 457–477 (2020). https://doi.org/10.1007/s10922-020-09521-4
2. Claeys, M., et al.: Towards multi- tenant cache managem ent for ISP networks. In: EuCNC 2014 - European Conference on Networks and Communications (2014). https://doi.org/10.1109/EuCNC.2014.6882692
3. Claeys, M., Tuncer, D., Famaey, J., Charalam bides, M., Latré, S., Pavlou, G., De Turck, F.: Hybrid multi-tenant cache management for virtualized ISP networks. J. Netw. Comput. Appl. (2016). https://doi.org/10.1016/j.jnca.2016.04.004
4. Aubry, E., Silverston, T., Chrisment, I.: Green growth in NDN: deployment of content stores. In: IEEE Workshop on Local and Metropolitan Area Networks (2016). https://doi.org/10.1109/LANMAN. 2016.754 8850
5. Cache sharing using bloom filters in named data networking. J. Netw. Comput. Appl. https://doi.org/10.1016/j.jnca.2017.04.0 11
6. Yovita, L.V., Syambas, N.R.: Caching on named data network: a survey and future research. Int. J. Electr. Comput. Eng. 4456–4466 (IJECE) (2018). https://doi.org/10.11591/ijece.v8i6
7. Fricker, C., Robert, P., Roberts, J., Sbihi, N.: Impact of traffic mix on caching performance in a content-centric network. In: 2012 Proceedings IEEE INFOCOM Workshops, pp. 310–315. IEEE (2012)

Enhanced Grey Wolf Optimizer for Data Clustering

Ibrahim Zebiri[1]([✉]) [ID], Djamel Zeghida[1,2] [ID], and Mohammed Redjimi[1] [ID]

[1] Department of Computer Science, LICUS Laboratory,
Université 20 Août 1955, 21000 Skikda, Algeria
`i.zebiri@univ-skikda.dz`
[2] Department of Computer Science, LISCO Laboratory, University Badji Mokhtar,
23000 Annaba, Algeria

Abstract. Data clustering is an unsupervised learning method used to extract knowledge from data, it is an NP-Hard (nondeterministic polynomial time) problem; there is no known deterministic technique that can find the optimal solution with an appropriate time complexity. Metaheuristics are powerful tools used to find good solutions (near to the best one) in a feasible time. The objective of this work is to improve the quality of one of recent metaheuristic clustering-based algorithms, which is grey wolf optimizer metaheuristic (GWO) by proposing an enhanced version of GWO called Enhanced Grey Wolf Algorithm-based Clustering (EGWAC), GWO is applied to find the best cluster centers. The optimization is essentially done in the updation of wolves position. The assessment of the results is measured by three measures; precision, recall and G-measure. The enhanced version of GWO algorithm for data clustering showed the impressive effect of the optimizations.

Keywords: Data clustering · Hard clustering · Unsupervised classification · Metaheuristics · Grey wolf optimizer (GWO) · Optimization

1 Introduction

Data clustering is a technique of unsupervised learning. It aims to partition a dataset into finite similar groups known as clusters. Data (objects) of the same cluster are supposed to be similar, alike, or closer. Objects from different clusters are more dispersed or less alike. Clustering has been used in different domains including data mining, machine learning, bioinformatics, Image processing, statistics, and many other fields [7,27,43,49]. Using exact methods to find the optimal solution of clustering is useless. With a dataset of just 50 objects, applying Stirling partition number, there are 119 649 664 052 358 809 649 152 different ways to cluster it on just three clusters. For a computer that verifies a million solutions per second, it will take more than 119 649 664 052 358 809 s to verify all possible solutions, i.e., it will take more than 3.794 billion years. Thus, the use of approached methods is mandatory to detect good solutions for this problem. Metaheuristics are one of the approached methods.

M. Salem et al. (Eds.): ICAITA 2022, CCIS 1769, pp. 147–159, 2023.
https://doi.org/10.1007/978-3-031-28540-0_12

Metaheuristics are methods that mimic systems in nature. They can discover good solutions near to the optimal one in a feasible time. There are many different metaheuristics used to solve several problems, including clustering, such as Genetic algorithms (GA) [24,28], Tabu Search (TS) [21,22], Particle Swarm Optimization (PSO) [33], Ant Colony Optimization (ACO) [13–15], Gravitational Search Algorithm (GSA) [48], Black Hole Mechanics Optimization (BHMO) [32], and many others [8,9,16]. Grey Wolf Optimizer (GWO) [42] is one of the recently proposed metaheuristics, it has been applied to solve several problems, for instance, it was applied for gene expression data [55], clustering [34], path planning [61], biomedical research [36,51], knapsack problem [57], etc. The fundamental contribution of this paper is:

- Propose an enhanced approach of GWO for clustering as explained in Sect. 4.2.
- Applied it on several real datasets.
- Compare it to the original GWO and other algorithms.

The reminder of this paper is structured like this, Sect. 2 gives a short introduction to the data clustering problem and mentions some previous related works to metaheuristic-based clustering. Section 3 presents briefly the GWO metaheuristic. In Sect. 4 we present the enhanced approach of GWO for data clustering. Data sets, parameters, experimental results, and discussion are Sect. 5.

2   Background

2.1   Data Clustering

Data clustering is a process of dividing a collection of unlabeled data D into finitely similar groups known as clusters $C = \{C_1, C_2, ..., C_k\}$ with the result that objects of any given cluster are similar to each other, yet dissimilar to objects from another cluster. In hard clustering, each object $x \in D$ belongs to one and only one cluster $C_i \cap C_j = \varnothing \mid i \neq j$ and $\cup_{i=1}^k C_i = D$. Each cluster contains at least an object $C_i \neq \varnothing \mid \forall i \in \{1, ..., k\}$. Data are clustered in a d-dimensional space, which represents the set of variables or features $x = \{x_1, x_2, ..., x_d\}$ [19, 20,27,56]. The main feature expected from any accurate clustering technique is the ability to provide an adequate number of clusters, where members of any cluster are more similar or nearer, while members from distinct clusters are less similar or farther away. That is minimizing the intra-cluster distance and maximizing the inter-cluster's [20,43]. There is a large variety of proximity or distance measures. The most popular one is Euclidean distance. For two data points x and y the Euclidean distance is defined as [6,56]:

$$d_{Euc}(x,y) = \sqrt{(x_1 - y_1)^2 + (x_2 - y_2)^2 + ... + (x_d - y_d)^2}.$$

$$d_{Euc}(x,y) = \left(\sum_{i=1}^d (x_i - y_i)^2 \right)^{\frac{1}{2}}. \tag{1}$$

In addition to d_{Euc}, the squared distance is also used [6]:

$$d^2_{Euc}(x,y) = \sum_{i=1}^{d}(x_i - y_i)^2. \tag{2}$$

Or else, other measures are presented in [4,20,56].

To evaluate the results of clustering, there are two main frequently used categories of clustering validation. The first one uses external knowledge, such as the classification of the dataset by an expert, this one is called external indexes. The second one is the internal indexes, in this category, there is no external knowledge, which means the quality of results is calculated from the data itself [5,25]. An example of internal indices can be the sum of intra-cluster distances. Precision, Recall, G-measure are examples of external indices. Further information can be found in [5,20,25,26,56].

2.2 Related Works

Metaheuristics have been extensively used in solving the problem of data clustering. The fundamental approach is [54], which uses simulated annealing (SA). In [1] Al-Sultan developed the first Tabu Search (TS) based clustering algorithm, where it was compared with the TS, and K-means. The approach outperforms k-means in almost all cases. With reference to SA the results were equal in two out of eight benchmarks, yet TS outperformed SA on the others. Genetic algorithms (GA) have been widely applied in data clustering problem [11,39,40,46]. In [46] the authors intended to attain more accurate results excluding the requirement of user inputs. The approach has been tested on over 20 databases. Other metaheuristics used for data clustering are Particle Swarm Optimization (PSO) [12,41], and Ant colony Optimization (ACO) [29,38,50]. In [30,31,60] the Artificial Bee Colony (ABC) was applied to clustering. In [30] the authors proposed an ABC clustering based on K-Modes algorithm, they tested it on 6 databases with three measures, accuracy, precision, and recall and compared it with some other techniques. The results of the proposed algorithm were the best in almost all cases. Santosa et al. [52] tackled data clustering using Cat Swarm Optimization (CSO) in order to classify benchmarks in [17]. An improved CSO to overcome diversity and local optima problems was proposed in [35]. Cheng et al. [10] applied Fish Swarm Algorithm (FSA), which mimics the schooling behavior of fish, for clustering. The Cuckoo Search Algorithm, which mimics Cuckoos behavior of breeding, i.e., laying their eggs in other birds' nests, has been used for clustering in [23]. In [3] Ibrahim Aljarah et al. used Multi-Verse Optimizer (MVO) for clustering, which compared to other algorithms and shows good results. Authors in [59], applied the recent metaheuristic Rat Swarm Optimizer (RSO) to data clustering. The algorithm was compared to MVO and other algorithms where RSO shows the best results. In [34] the authors proposed a Grey wolf algorithm (GWA) for clustering. Ibrahim Aljarah et al. [2] hybridized GWO and TS techniques, where the new version showed best results than other algorithms compared with, including GWO, and TS.

3   Preliminaries

3.1   The GWO

The GWO is a population-based metaheuristic, proposed by Mirjalili et al. [42]. This method has shown a good adaptation in several problems [2,34,37,47,53, 62]. It is unique as it imitates the grey wolves social hierarchy and their natural hunting behavior [42,45].

Grey wolves live in packs. At the head of the social hierarchy comes the alpha (α), he is the leader of the group. He has the authority to choose the place and time for hunting, and make other decisions. Next to alpha comes beta wolves (β). Beta wolves aid the alpha in decisions making and dominate other wolves. They are probably the best ones that can take the place of the alpha if he became older or passed away. At the next level comes the delta wolves (δ), these wolves are the guardians and the protectors of the group. At the lowest rank comes omega wolves (ω) they are the scapegoat, and they should submit to other dominant wolves [2].

In GWO algorithm, each wolf constitutes a solution. But there is no pre-designated hierarchy. The three best solutions are considered as alpha, beta, and delta. The other solutions are considered as omega wolves [42]. Tracking, encircling, and attacking the prey are Grey wolves hunting mechanisms [42].

3.2   Encircling a Prey

Wolves in nature encircle their prey, this behavior was modeled as follows:

$$\vec{X}(t+1) = \vec{X}_p(t) - \vec{A}.\vec{D}. \tag{3}$$

where $\vec{X}(t+1)$ is the next position of the wolf. t indicates current iteration. $\vec{X}_p(t)$ is the position of the prey. $\vec{D}$ represent the distance depends on the position of current wolf and the prey. It is calculated as follows:

$$\vec{D} = \left| \vec{C}.\vec{X}_p(t) - \vec{X}(t) \right|. \tag{4}$$

$\vec{X}(t)$ is the current wolf's position. $\vec{A}$ and $\vec{C}$ are random vectors calculated as follows:

$$\vec{A} = 2.\vec{a}.\vec{r}_1 - \vec{a}. \tag{5}$$

$$\vec{C} = 2.\vec{r}_2. \tag{6}$$

where $\vec{a}$ is a vector that its components decrease linearly from 2 to 0 during the process. $\vec{r}_1$ and $\vec{r}_2$ are two random vectors in the interval [0, 1]. $\vec{a}$ is calculated as follows:

$$\vec{a} = 2 - t\left(\frac{2}{T}\right). \tag{7}$$

where the maximum number of iterations is represented by T, and the present iteration by t. By the two Eqs. 3 and 4, wolves can move to different positions around the prey.

3.3   Hunting the Prey

Hunting is generally led by alpha. In an abstract search space there is no idea about prey position. To model this behavior, it is supposed that the three best wolves have a good idea about prey position. Wolves' position update depends on the positions of the three best ones. It is calculated as follows:

$$\vec{X}(t+1) = \frac{\vec{X}_1 + \vec{X}_2 + \vec{X}_3}{3}. \tag{8}$$

where,

$$\vec{X}_1 = \left| \vec{X}_\alpha - \vec{A}_1.\vec{D}_\alpha \right|$$

$$\vec{X}_2 = \left| \vec{X}_\beta - \vec{A}_2.\vec{D}_\beta \right|$$

$$\vec{X}_3 = \left| \vec{X}_\delta - \vec{A}_3.\vec{D}_\delta \right|$$

Here $\vec{X}_\alpha$, $\vec{X}_\beta$ and $\vec{X}_\delta$ are respectively the positions of the three best solutions and $\vec{A}_i$ are random vectors calculated with the Eq. 5. $\vec{D}_\alpha$, $\vec{D}_\beta$, and $\vec{D}_\delta$ are calculated as follows:

$$\vec{D}_\alpha = \left| \vec{C}_1.\vec{X}_\alpha - \vec{X} \right|$$

$$\vec{D}_\beta = \left| \vec{C}_2.\vec{X}_\beta - \vec{X} \right|$$

$$\vec{D}_\delta = \left| \vec{C}_3.\vec{X}_\delta - \vec{X} \right|$$

where $\vec{C}_i$ are random vectors calculated with the Eq. 6. and $\vec{X}$ is the position of the current wolf.

Grey wolves research depends on the three best wolves. They diverge searching preys and converge attacking. Attacking prey is manifested whenever $|\vec{A}| < 1$, whereas $|\vec{A}| > 1$ oblige wolves to search far from prey position. These two behaviors can represent exploitation and exploration. Another parameter which favor the exploration is $\vec{C}$ which takes random values between 0 and 2. Hence GWO shows a random behavior in course of optimization process which favor exploration and avoid stagnating in a local optima.

4   Enhanced Version of GWA for Clustering

The idea here is to use GWA with K-means, that is, using GWA to explore the space of solutions, and two iterations of K-means to cluster the data. The approach starts by initializing parameters and generating random k cluster centers $\mu_j (j = 1, \ldots, k)$, each one can be represented in a $d - dimensional$ space (whither each attribute is represented by an axe), $\mu_j = (v_{j1}, v_{j2}, \ldots, v_{jd})$, then group each data point with the nearest center. Wolves' population $WP = (w_1, w_2, \ldots, w_{|WP|})$ represents the set of solutions. Each wolf w_i represents a solution. A solution possesses k data points, which represent the centers of clusters $w_i(\mu_{i1}, \mu_{i2}, \ldots, \mu_{ik} | i = 1, \ldots, |WP|)$. Thus, a wolf can be represented as follows:

$$w_i = ((v_{i11}, v_{i12}, \ldots, v_{i1d}), (v_{i21}, v_{i22}, \ldots, v_{i2d}), \ldots, (v_{ik1}, v_{ik2}, \ldots, v_{ikd}))$$

where v_{ijm} is the feature (variable) number m, of the jth center of the wolf number i. With $i = 1, \ldots, |WP|, j = 1, \ldots, k$, and $m = 1, \ldots, d$.

After the random generation of cluster centers for each wolf, each data point joins the cluster the nearest center. An assessment is done by an objective function. The three best solutions are considered as alpha, beta and delta. The update of parameters is done by Eqs. 5, 6, and 7. The position update for the next step is done by Eq. 8. This process; from the clustering task till the end, is repeated a max number of iterations T which is defined at the beginning of the process.

4.1 Objective Function

The sum of intra-clusters distances is the objective function used for evaluating the results quality. This measure is defined as follows:

$$\sum_{C_i \in C} \sum_{x \in C_i} d^2(x, \mu_i) \tag{9}$$

μ_i here is the cluster i center. And the function $d^2(\ldots)$ is the squared Euclidean distance (Eq. 2).

4.2 Positions Update

The simplest operation here, which is done in [34], is to add the first component of the first cluster center of $\overrightarrow{X}_1$ to the one of $\overrightarrow{X}_2$ and $\overrightarrow{X}_3$ and divide the result by three, the same with the other components. And the same with the other centers. However, this process may mislead the wolves far away from the prey, because, here, clusters centers position in the wolf are meaningless i.e. the clustering result of a wolf $w_i(\mu_1, \mu_2, \mu_3)$ will be the same as another wolf with the same centers $w_j(\mu_2, \mu_3, \mu_1)$ but in different positions. The ideal process here is to summate the closer center from $\overrightarrow{X}_2$ and $\overrightarrow{X}_3$ to the first center from $\overrightarrow{X}_1$ and divide the result by three. Same with the second center of $\overrightarrow{X}_1$, which will be added to the closer center from $\overrightarrow{X}_2$ and $\overrightarrow{X}_3$ without including the ones that are chosen before with the first center of $\overrightarrow{X}_1$ and so on. For example let's assume $\overrightarrow{X}_1 = ((1,3), (50,9), (7,5))$, $\overrightarrow{X}_2 = ((9,7), (2,3), (40,7))$ and $\overrightarrow{X}_3 = ((45,8), (8,6), (0,6))$. The position update here is done as follows:

$$\frac{((1,3) + (2,3) + (0,6)), ((50,9) + (40,7) + (45,8)), ((7,5) + (9,7) + (8,6))}{3}$$

Which gives: $\overrightarrow{X}_j(t+1) = ((1,4), (45,8), (8,6))$ The obvious method will give: $\overrightarrow{X}_j(t+1) = ((18.33,6), (20,6), (15.66,6))$. The two results are very distinct.

Besides the result of the technique used, which is closer to the three best solutions, the result of the obvious technique is farther than the supposed prey's position.

Algorithm 1 depicts the proposed EGWAC approach.

Algorithm 1: EGWAC

Set number of clusters k, population size, and max number of iterations T and the dataset
Initialize the grey wolf population $X_i (i = 1, ..., n)$
Initialize $\overrightarrow{a}$, $\overrightarrow{A}$ and $\overrightarrow{C}$ and set $t = 0$
Cluster data by each wolf
Calculate the fitness of each wolf
$\overrightarrow{X}_\alpha, \overrightarrow{X}_\beta, \overrightarrow{X}_\delta$ are the best three solutions
while $(t < T)$ **do**
    for *each wolf* **do**
        Update the position of the current wolf by the enhanced procedure explained previously (4.2)
        Cluster data by the current wolf
    Update $\overrightarrow{a}$, $\overrightarrow{A}$ and $\overrightarrow{C}$
    Calculate the fitness of all solutions
    if *a wolf has an empty cluster* **then**
        the current wolf position is updated to a random position from the dataset
    Update $\overrightarrow{X}_\alpha, \overrightarrow{X}_\beta, \overrightarrow{X}_\delta$
    $t = t + 1$
Cluster the dataset using $\overrightarrow{X}_\alpha$ and return the result

---

5   Experimental Results and Discussion

This section shows the experimentation and discusses enhanced approach results compared to several other algorithms. The benchmarks used here can be found in [17]. The Table 1 presents the characteristic of the used benchmarks.

The comparisons here are done through three indexes: precision, recall, and g-measure. They are defined as follows [18,44,58]:

$$Precision = \frac{TP}{TP + FP} \tag{10}$$

$$Recall = \frac{TP}{TP + FN} \tag{11}$$

$$G - Measure = \sqrt{Precision * Recall} \tag{12}$$

Table 1. Used datasets

Datasets	Number of instances	Number of features	Number of classes
Iris	150	4	3
Glass	214	9	6
Seeds	210	7	3
Haberman	306	3	2
Liver (Bupa)	345	6	2
CMC	1473	9	3

where,

TP (True Positives): The number of pairs from the same class and that are assigned to the same cluster.

FN (False Negatives): The number of pairs of the same class that are clustered in different clusters.

FP (False Positives): the number of pairs of different classes that are clustered in the same cluster. High values of the previous measures is required for best results.

Parameters of techniques compared with; K-means (KM), Genetic Algorithm-based clustering (GAC), Harmony search-based clustering (HSC), Modified Harmony search-based clustering (MHSC), Particle Swarm Optimization algorithm-based clustering (PSOC), Flower Pollination algorithm-based clustering (FPAC), Bat algorithm-based clustering (BAC), and Grey Wolf algorithm-based clustering (GWAC), are the same as mentioned in [34]. For the enhanced approach, the parameters are the same as GWAC, 30 for the size of population and 500 as a max number of iterations. The number of clusters is adjusted for each benchmark to be the same as the number of classes. Results are collected over more than 20 independent runs.

Figures (1, 2 and 3) shows the results quality of EGWAC.

For the datasets Iris, Wine, Glass, and Conceptive Method Choice (CMC), EGWAC outperforms the other algorithms in all measures with the least value of standard deviation. In Haberman, GWAC shows the best result in recall, however for precision and g-measure, EGWAC has the best values with the least standard deviation. For Bupa dataset, EGWA outperforms other algorithms in terms of recall and g-measure, yet FPAC has the best value in terms of precision. FPAC and BATC shows an impressive adaptation for Bupa dataset which calls the law of no free lunch. In general, EGWAC outperforms the other algorithms for almost all measures, and shows the least standard deviation. Comparing to GWAC, results shows the great effect of the improvements presented in the previous Sect. 4. Figures (1, 2 and 3) illustrate a visual comparison for more clarity. Globally, EGWAC dominated by outperforming other algorithms with a significant gap.

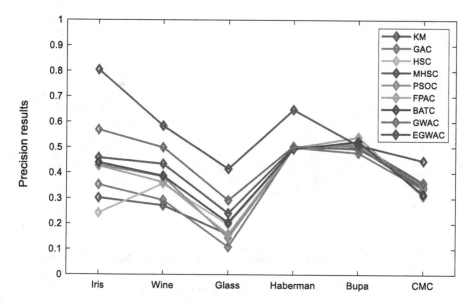

Fig. 1. Visual comparison of Precision

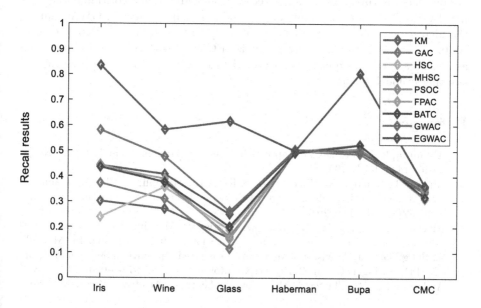

Fig. 2. Visual comparison of Recall

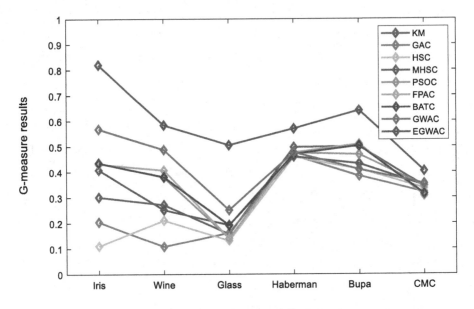

Fig. 3. Visual comparison of G-measure

6   Conclusion and Future Works

We presented in this work, an improved version of grey wolf algorithm for clustering (EGWAC). The results and analysis show the superiority of EGWAC among
the algorithms compared with in terms of quality of results and convergence.
As future studies, we will try to hybridize EGWAC with other techniques such
as cuckoo search and quantum genetic algorithm, to alleviate the drawback of
stacking in local optima.

References

1. Al-Sultan, K.S.: A tabu search approach to the clustering problem. Pattern Recogn.
   28(9), 1443–1451 (1995)
2. Aljarah, I., Mafarja, M., Heidari, A.A., Faris, H., Mirjalili, S.: Clustering analysis
   using a novel locality-informed grey wolf-inspired clustering approach. Knowl. Inf.
   Syst. **62**(2), 507–539 (2020)
3. Aljarah, I., Mafarja, M., Heidari, A.A., Faris, H., Mirjalili, S.: Multi-verse optimizer: theory, literature review, and application in data clustering. In: Mirjalili,
   S., Song Dong, J., Lewis, A. (eds.) Nature-Inspired Optimizers. SCI, vol. 811, pp.
   123–141. Springer, Cham (2020). https://doi.org/10.1007/978-3-030-12127-3_8
4. Anderberg, M.R.: Cluster Analysis for Applications. Academic Press, New York
   (1973)
5. Bagirov, A.M., Karmitsa, N., Taheri, S.: Partitional Clustering via Nonsmooth
   Optimization: Clustering via Optimization. Springer, Cham (2020). https://doi.
   org/10.1007/978-3-030-37826-4

6. Bailey, K.: Cluster analysis, pp. 59–128 (1974). In DR Heise (ed.) (1975)
7. Berkhin, P.: A survey of clustering data mining techniques. In: Kogan, J., Nicholas, C., Teboulle, M. (eds.) Grouping Multidimensional Data, pp. 25–71. Springer, Heidelberg (2006). https://doi.org/10.1007/3-540-28349-8_2
8. Bozorg-Haddad, O.: Advanced Optimization by Nature-Inspired Algorithms. Springer, Heidelberg (2018). https://doi.org/10.1007/978-981-10-5221-7
9. Bozorg-Haddad, O., Solgi, M., Loáiciga, H.A.: Meta-heuristic and Evolutionary Algorithms for Engineering Optimization. Wiley, Hoboken (2017)
10. Cheng, Y., Jiang, M., Yuan, D.: Novel clustering algorithms based on improved artificial fish swarm algorithm. In: 2009 Sixth International Conference on Fuzzy Systems and Knowledge Discovery, vol. 3, pp. 141–145. IEEE (2009)
11. Cowgill, M.C., Harvey, R.J., Watson, L.T.: A genetic algorithm approach to cluster analysis. Comput. Math. Appl. **37**(7), 99–108 (1999)
12. Cura, T.: A particle swarm optimization approach to clustering. Expert Syst. Appl. **39**(1), 1582–1588 (2012)
13. Dorigo, M.: Optimization, learning and natural algorithms [Ph. D. thesis]. Politecnico di Milano, Italy (1992)
14. Dorigo, M., Maniezzo, V., Colorni, A.: Ant system: optimization by a colony of cooperating agents. IEEE Trans. Syst. Man Cybern. Part B (Cybern.) **26**(1), 29–41 (1996)
15. Dorigo, M., Stützle, T.: The ant colony optimization metaheuristic: algorithms, applications, and advances. In: Glover, F., Kochenberger, G.A. (eds.) Handbook of Metaheuristics. International Series in Operations Research & Management Science, vol. 57, pp. 250–285. Springer, Boston (2003). https://doi.org/10.1007/0-306-48056-5_9
16. Du, K.L., Swamy, M., et al.: Search and optimization by metaheuristics. Tech. Algorithms Inspired Nat. (2016)
17. Dua, D., Graff, C.: UCI machine learning repository (2017). http://archive.ics.uci.edu/ml
18. Espíndola, R.P., Ebecken, N.F.: On extending F-measure and G-mean metrics to multi-class problems. WIT Trans. Inf. Commun. Technol. **35**, 25–34 (2005)
19. Everitt, B., Landau, S., Leese, M., Stahl, D.: Cluster Analysis. Wiley Series in Probability and Statistics. Wiley (2011). https://books.google.dz/books?id=WSayDAEACAAJ
20. Gan, G., Ma, C., Wu, J.: Data Clustering: Theory, Algorithms, and Applications. SIAM (2020)
21. Glover, F.: Future paths for integer programming and links to artificial intelligence. Comput. Oper. Res. **13**(5), 533–549 (1986)
22. Glover, F.: Tabu search—part I. ORSA J. Comput. **1**(3), 190–206 (1989)
23. Goel, S., Sharma, A., Bedi, P.: Cuckoo search clustering algorithm: a novel strategy of biomimicry. In: 2011 World Congress on Information and Communication Technologies, pp. 916–921. IEEE (2011)
24. Goldberg, D.E.: Genetic Algorithms in Search, Optimization and Machine Learning, 1st edn. Addison-Wesley Longman Publishing Co., Inc., USA (1989)
25. Halkidi, M., Batistakis, Y., Varzigiannis, M.: Cluster validity methods part I. ACM Sigmod Rec. **31**, 40–45 (2002)
26. Halkidi, M., Batistakis, Y., Vazirgiannis, M.: Clustering validity checking methods: part II. ACM SIGMOD Rec. **31**(3), 19–27 (2002)
27. Han, J., Pei, J., Kamber, M.: Data Mining: Concepts and Techniques. Elsevier, Amsterdam (2011)

28. Holland, J.H.: Adaptation in Natural and Artificial Systems. University of Michigan Press, Ann Arbor (1975). Second edition, 1992
29. İnkaya, T., Kayalıgil, S., Özdemirel, N.E.: Ant colony optimization based clustering methodology. Appl. Soft Comput. **28**, 301–311 (2015)
30. Ji, J., Pang, W., Zheng, Y., Wang, Z., Ma, Z.: A novel artificial bee colony based clustering algorithm for categorical data. PLoS One **10**(5), e0127125 (2015)
31. Karaboga, D., Ozturk, C.: A novel clustering approach: artificial bee colony (ABC) algorithm. Appl. Soft Comput. **11**(1), 652–657 (2011)
32. Kaveh, A., Seddighian, M., Ghanadpour, E.: Black hole mechanics optimization: a novel meta-heuristic algorithm. Asian J. Civ. Eng. **21**(7), 1129–1149 (2020)
33. Kennedy, J., Eberhart, R.: Particle swarm optimization. In: Proceedings of ICNN 1995 - International Conference on Neural Networks, vol. 4, pp. 1942–1948 (1995). https://doi.org/10.1109/ICNN.1995.488968
34. Kumar, V., Chhabra, J.K., Kumar, D.: Grey wolf algorithm-based clustering technique. J. Intell. Syst. **26**(1), 153–168 (2017)
35. Kumar, Y., Sahoo, G.: An improved cat swarm optimization algorithm based on opposition-based learning and Cauchy operator for clustering. J. Inf. Process. Syst. **13**(4), 1000–1013 (2017)
36. Li, Q., et al.: An enhanced grey wolf optimization based feature selection wrapped kernel extreme learning machine for medical diagnosis. Comput. Math. Methods Med. **2017** (2017)
37. Liu, H., Hua, G., Yin, H., Xu, Y.: An intelligent grey wolf optimizer algorithm for distributed compressed sensing. Comput. Intell. Neurosci. **2018** (2018)
38. Liu, X., Fu, H.: An effective clustering algorithm with ant colony. J. Comput. **5**(4), 598–605 (2010)
39. Liu, Y., Wu, X., Shen, Y.: Automatic clustering using genetic algorithms. Appl. Math. Comput. **218**(4), 1267–1279 (2011)
40. Maulik, U., Bandyopadhyay, S.: Genetic algorithm-based clustering technique. Pattern Recogn. **33**(9), 1455–1465 (2000)
41. Van der Merwe, D., Engelbrecht, A.P.: Data clustering using particle swarm optimization. In: The 2003 Congress on Evolutionary Computation, CEC 2003, vol. 1, pp. 215–220. IEEE (2003)
42. Mirjalili, S., Mirjalili, S.M., Lewis, A.: Grey wolf optimizer. Adv. Eng. Softw. **69**, 46–61 (2014)
43. Mirkin, B.: Clustering for Data Mining: A Data Recovery Approach. Chapman and Hall/CRC (2005)
44. Palacio-Niño, J.O., Berzal, F.: Evaluation metrics for unsupervised learning algorithms. arXiv preprint arXiv:1905.05667 (2019)
45. Panda, M., Das, B.: Grey wolf optimizer and its applications: a survey. In: Nath, V., Mandal, J.K. (eds.) Proceedings of the Third International Conference on Microelectronics, Computing and Communication Systems. LNEE, vol. 556, pp. 179–194. Springer, Singapore (2019). https://doi.org/10.1007/978-981-13-7091-5_17
46. Rahman, M.A., Islam, M.Z.: A hybrid clustering technique combining a novel genetic algorithm with K-means. Knowl.-Based Syst. **71**, 345–365 (2014)
47. Rashaideh, H., Sawaie, A., Al-Betar, M.A., Abualigah, L.M., Al-Laham, M.M., Ra'ed, M., Braik, M.: A grey wolf optimizer for text document clustering. J. Intell. Syst. **29**(1), 814–830 (2020)
48. Rashedi, E., Nezamabadi-Pour, H., Saryazdi, S.: GSA: a gravitational search algorithm. Inf. Sci. **179**(13), 2232–2248 (2009)
49. Romesburg, C.: Cluster Analysis for Researcher (2004). Lulu.com

50. Runkler, T.A.: Ant colony optimization of clustering models. Int. J. Intell. Syst. **20**(12), 1233–1251 (2005)
51. Sánchez, D., Melin, P., Castillo, O.: A grey wolf optimizer for modular granular neural networks for human recognition. Comput. Intell. Neurosci. **2017** (2017)
52. Santosa, B., Ningrum, M.K.: Cat swarm optimization for clustering. In: 2009 International Conference of Soft Computing and Pattern Recognition, pp. 54–59. IEEE (2009)
53. Sathiyabhama, B., et al.: A novel feature selection framework based on grey wolf optimizer for mammogram image analysis. Neural Comput. Appl. **33**, 14583–14602 (2021)
54. Selim, S.Z., Alsultan, K.: A simulated annealing algorithm for the clustering problem. Pattern Recogn. **24**(10), 1003–1008 (1991)
55. Vosooghifard, M., Ebrahimpour, H.: Applying grey wolf optimizer-based decision tree classifer for cancer classification on gene expression data. In: 2015 5th International Conference on Computer and Knowledge Engineering (ICCKE), pp. 147–151. IEEE (2015)
56. Xu, R., Wunsch, D.: Clustering, vol. 10. Wiley, Hoboken (2008)
57. Yassien, E., Masadeh, R., Alzaqebah, A., Shaheen, A.: Grey wolf optimization applied to the 0/1 knapsack problem. Int. J. Comput. Appl. **169**(5), 11–15 (2017)
58. Zafarani, R., Abbasi, M.A., Liu, H.: Social Media Mining: An Introduction. Cambridge University Press, Cambridge (2014)
59. Zebiri, I., Zeghida, D., Mohamed, R.: Rat swarm optimizer for data clustering. Jordan. J. Comput. Inf. Technol. (JJCIT) **08**(03), 297–307 (2022). https://doi.org/10.5455/jjcit.71-1652735477
60. Zhang, C., Ouyang, D., Ning, J.: An artificial bee colony approach for clustering. Expert Syst. Appl. **37**(7), 4761–4767 (2010)
61. Zhang, S., Zhou, Y., Li, Z., Pan, W.: Grey wolf optimizer for unmanned combat aerial vehicle path planning. Adv. Eng. Softw. **99**, 121–136 (2016)
62. Zhao, M., Wang, X., Yu, J., Bi, L., Xiao, Y., Zhang, J.: Optimization of construction duration and schedule robustness based on hybrid grey wolf optimizer with sine cosine algorithm. Energies **13**(1), 215 (2020)

Artificial Orca Algorithm for Solving University Course Timetabling Issue

Abdelhamid Rahali[1,2](✉), KamelEddine Heraguemi [1], Samir Akhrouf[1], Mouhamed Benouis[1], and Brahim Bouderah[1]

[1] M'Sila University, M'Sila, Algeria
{abdelhamid.rahali,KamelEddine.Heraguemi,Samir.Akhrouf,
Mouhemed.Benouis,Brahim.Bouderah}@univ-msila.dz
[2] Informatics and Mathematics Laboratory - Souk-Ahras University,
Souk Ahras, Algeria

Abstract. Timetabling problem for university courses (UCTP) is one of most traditional challenges that have been emphasized for a long time by many researches. This issue belong to NP-Hard problems, which are hard to solved with classical algorithms due to their complexity. Swarm intelligence become a trend to solve NP-hard problems, and also solve real life issues. This paper proposes a new based Artificial Orca Algorithm (AOA) solver for university courses timetabling problem. In order to evaluate our proposal, A series of are carried out on Ghardaia University Timetabling data, the performance of the proposed approach are evaluated and compared with other algorithms developed to solve the same problem. The results show a clear superiority of our proposal against the other in terms of execution time and result quality.

Keywords: University course timetabling · Artificial Orca Algorithm (AOA) · Metaheuristics

1 Introduction

Scheduling and planning issues arise frequently in a variety of real-world contexts. These issues can manifest in a wide range of ways, and the activities or processes that must be carried out are frequently constrained by a number of variables, where resources have to be assigned or allocated within a set of periods including the optimization process. Additionally, where resources like space, and time are critical, creating useful and timetables is not simple task due to the significant impact of scheduling over socio-economic fields (transport, manufacturing, sports and education), many of approach and methods have been arisen aiming to automate these processes.

The role of Timetabling is the organization of a set of events (surgeries, sport competition, exams, courses) allocating human resources (medical doctors, sport officials, examination proctors, instructors, athletes, nurses) to space resources (operating theatres, exam halls, sport fields, classrooms) at a set of periods

M. Salem et al. (Eds.): ICAITA 2022, CCIS 1769, pp. 160–172, 2023.
https://doi.org/10.1007/978-3-031-28540-0_13

[1]. It is a well-liked subject in operations research, thus, it attracts a lot of specialists to participate at ITC (International Competition Timetabling), due to its importance in many daily life areas, such as healthcare, education, sports, transportation, private businesses, work and many more [2] This combinatorial optimization topic is a common challenge for a many Academic institutions like universities, colleges, or schools [3,4]. Thus, several of researchers interest in to accomplish a particular set of goals via finding the best way to arrange a set of variables [5]. In the real world, where resources like people, space, and time are scarce, creating useful and appealing timetables is not an easy feat. Even for seasoned designers, it becomes a difficult task when resources are restricted. The people who use these schedules are significantly impacted by them. The construction of these schedules should be as good as possible. Such timetables are frequently changed or entirely rescheduled. This essay addresses on the issue with university course scheduling.

At first, the timetabling problem was dealt with classical methods due to the simplicity of data, namely integer programming model [6] which belong to the exact methods, however, these methods cannot produce optimal results particularly for complex problem with fast computation rates. Furthermore, the fixed charge transportation problem was the base to solve teacher assignment after formulation as mixed integer programming [7]. The study's focus has shifted to heuristic-based as a result of the limitations of classical methods, which make it inadequate. Genetic algorithm (GA) in [8] have been combined with local search, Tabu search algorithm is proposed in [9] to solve timetabling problem in order to compete at ITC2007. Genetic algorithm hybridized with local search in [10] to tackle UCTP. Simulated Annealing (SA) algorithm with their multiple variant were proposed to solve this issues kind, furthermore, Great deluge and Ant colony algorithms are proposed as solver.

In this paper, an artificial orca algorithm is based to solve UCTP, this approach is subject to two-stage, the first one ensure feasibility of the solution found in the second one, the task is switched to the optimization mission aiming to enhance solution. In fact, in order to avoid violating any hard constraint, initial solution generated deterministically checking resource availability before any assignment or allocation. Subsequently, the algorithm try to optimize solution through minimizing the score of penalized soft constraints violations.

The remainder of this present the problem in Sect. 2. Section 3 reveal a literature review. Section 4, is dedicated to Artificial Orca Algorithm for Timetabling issues. Section 5 contains experiments and results with some comparisons, whereas Sect. 6, conclude the work giving a set of remarks.

2   Problem Description

UCTP belongs to the classical challenge, which make it a famous issue in the range of optimization problems, since it reflects an important side of real life. In this work on focus in UCTP for the university of Ghardaia- Algeria-, the required task is scheduling courses to attendees, instructors, rooms and timeslots, where satisfying a set of constraints. A problem scenario includes:

1. A set of N events must be assigned into 36 timeslots (6 days of 6 periods each).
2. A set of I instructors with their availabilities and preferences.
3. A set of R rooms with their type, capacity, availabilities and preferences.
4. A set of S attendees (sections, groups, subgroup) with their availabilities and preferences.

The required solutions have to be feasible, needs to fulfilling all the hard constraints including:

1. No event that not be assigned to an instructor.
2. No event that not be assigned to a room.
3. No instructor teaches more than class at the same time.
4. No room assigned to more than class at the same time.
5. No attendees (section, group or subgroup) attend to more than event at the same time.

In addition, a set of soft constraints should be met as possible as including:

1. No instructors are assigned to their no preferred timeslots.
2. No events are assigned to the no preferred rooms.
3. No attendees are assigned to their no preferred timeslots.
4. No practical meeting is assigned to the first week day, due some far students might absent where attend is obligatory.
5. No lectures meeting are assigned at evening.
6. No events are assigned to the latest timeslots of days.

3   Related Works

In the recent decades, many methods addressing the university courses timetabling problem. These approaches might divide as follows: Single solution-based meta-heuristics (SSBMH), population-based meta-heuristics (PBMH), operational research (OR), hybrid approaches (HA), multi criteria/objective and hyper-heuristic (HH).

For SSBMH, algorithms look in the area around a single solution at the beginning to find the optimal one, thus, it qualifies as a local search methods. [11] Divided SSBMH into one-stage category, two-stage category, and relaxation-allowing category algorithm optimization, the first category fulfil all constraints types, hard and soft simultaneously, in other hand, for second category, making feasible solution consider the hard constraints in the started stage, while the latest category, it permits a relaxation to the violation constraints via putting infeasible events apart, or adapting artificially the events by creating a dummy period [12]. Furthermore, involving to modify of neighbourhood size randomly for the tabou search algorithm proposed in [13] generated a competitive result. Simulated Annealing algorithm avoid being trapped in the local optima via accepting of bad moves accordingly to probabilistic criteria [14,15].

For PBMH, the idea of approach is generating a set of solutions using some operators and rules, among these types are ant colony optimization(ACO), genetic algorithms (GA), the later one based on biological evolution, its strategy is choosing the optimal parents in order to generate children solutions [16,17]. To solve UCTP instance involving Pheromone signal used in foraging behaviour of ant, [18,19] provided an ACO approach, the given results proved its efficiency. Groups of animals like bees, birds, elephants and other, motivated researcher to inspire approach from their behaviour that called particle swarm optimization, each individual have its own fitness and represent a candidate solution which ensures an extra intensification and diversification of search process. [20] Tackled the UCTP using PSO technique, where less parameter and fast convergence criteria, compose the main advantages.

Toggling to RO kind, many early work have based on the resemblance between timetabling and graph colouring problem where vertices can denote events, further each pair of vertices can't share the same colour reducing the number of colours which represent periods [21,22]. A clique-based algorithm presented in [23] to create feasible solutions where clique denotes some courses that might be scheduled at the same time period. Resizing of clique was obtained through perturbation and recombination operations, this strategy leads to considerable results. Relaxation of the integer programming approach was proposed in [24], pattern formulation model leads to assign course to a set of time slots on the same day. Hyper-Heuristic method was developed in [25] combining tabou search and Variable Neighbourhood Search (VNS), this method outperformed the traditional one. Regarding hybrid approach, In [26] clustering hybridized with colour mapping to generate UCTP enabling enrolment of students to courses without conflicts. Real datasets were involved to automate UCTP using clustering within two stage heuristic in [27] predicting course enrolment and increasing rooms occupancy. Hybrid Genetic Algorithm presented in [28] involving four opera-tors to treat UCTP where optimization teaching workload is considered. Regarding multi objective approaches, in [29] lectures disruption which degrade the timetable quality, was addressed defining a good Pareto front identified by solution quality and robustness measure, solver was based on simulated annealing approach with multi-objective trend, outperforming the genetic algorithm.

4   Artificial Orca Algorithm (AOA)

Artificial Orca Algorithm (AOA) inspired from orca behaviour in terms of hunting, travelling and structure, Orca swarm structured as clans which are divided into some pods that include individuals orca, each clan, pod or the whole population guided by its matriarch which represent the fittest member of the structure, degree closeness can distinguish between hierarchical levels. Echolocation and Wave Washing are the interesting methods used by this specie for communication and foraging applying discrete calls, clicks and whistles, in addition, the orcas use their heads to generate waves cooperatively that submerge their prey after the checking of environment. Exploitation and exploration phases are described in [30] as follows:

1. Empirical parameters definition, solutions creation after initialization of orcas population reserving closeness degree between pods members or clans members.
2. Sorting pods by their evaluated fitness.
3. Update the orca individuals during exploitation phase following either Eqs. 1 to 3 which simulate Echolocation search, or Eqs. 4 to 6 which represent Cooperation Update modelling the teamwork when generating waves.

$$f_{group} = f_{min} + \alpha \times (f_{max} - f_{min}) \tag{1}$$

$$v_{pi}^t = v_{pi}^{t-1} + f_{pi} \times D_p + f_c \times D_c + f_{pop} \times D_{pop} \tag{2}$$

$$x_{pi}^t = x_{pi}^{t-1} + v_{pi}^t \tag{3}$$

$$x_{temp,pi}^t = A \times \sin(\frac{2 \times \pi}{L})x_{pi}^{t-1} \times \cos(\frac{2 \times \pi}{T})t_x \tag{4}$$

$$x_{m,p}^t = \frac{\sum_{j=1}^n x_{temp,pi}}{n} \tag{5}$$

$$x_{pi}^t = x_p^* - \beta \times x_{m,p}^t \tag{6}$$

4. Update orcas individuals who have the worst fitness of all pods applying Eq. 7.

$$x_{new,pi}^t = \frac{\gamma \times x_{pop_{rand1}}^t + \omega \times x_{crand2}^t}{2} \tag{7}$$

5. Evaluation of orcas individuals, calculating fitness for each.
6. If stopping criteria is met, return the fittest individual of the population; otherwise return to step3. For more detail, please refer to [30].

5   Artificial Orca Algorithm for Timetabling Issues

The solution is based only on the mathematical model inspired from echolocation technique mentioned in Eq. 1 to Eq. 3. where group represent a random frequency corresponding to the clan, pod or population, generated after adjusting the empirical parameters f_{min}, f_{max} into 0 and 1 respectively, knowing that α is a random value in range [1,12] considering the week composed of 36 (6 days x 6 slots) time slots, distances between the current individual and the matriarch of population, clan and pod are denoted by D_{pop}, D_c, D_p respectively. v_{pi}^t represent velocity of the orca member within the pod p at position i, in addition, x_{pi}^t denote the new orca state at position i in the pod p at iteration t. divide and rule policy is exploited, only the first year students of the science and technology discipline are considered, they are the most populous, 567 students sectioned into two sections (A, B), each section divided into 8 groups, the number of meeting required for each of the students led to creating an events space composed from 116 events, it contains a set of lectures, directed works (TDs) and practical works (TPs) meeting, each event require a feasible assignment into instructors, rooms and time slots respecting the availability of these resources that probably incoming from the output of other timetable of other discipline considering students who have debts, or instructor is occupied on Monday, ... The objective of the optimization process is minimizing the number of violated soft constraints for each event. Only move operator is used, moving into a free and feasible time slots. This move obtained via the above EQs defining the values of each EQ and its components as follows:

$$f_{min} = 0, f_{max} = 1 \tag{8}$$

$$v_{pi}^{t-1} = 0 \tag{9}$$

$$v_{pi}^t = v_{pi}^{t-1} + (f_{pi} + f_c + f_{pop}) \times Event_{i,ViolatedConstraintsCount}^{t-1} \tag{10}$$

$$x_{pi}^{t-1} = (Event_{i,DayIndex}^{t-1} \times 6) + Event_{i,SlotIndex}^{t-1} \tag{11}$$

$$x_{pi}^t = x_{pi}^{t-1} + v_{pi}^t \tag{12}$$

If we project the components of algorithm in the real components, we find events match orca individuals, pods match a group, sections match clans and the whole events which related to entire of students match population, on the other hand, each event cooperate with others to improve fitness by reducing the number of violated constraints via the random walk following Eq. 12 through computing the new position based on the old one plus the velocity which is calculated by Eq. 11 assuming that all matriarch have the same distance value which equal to zero where distance simulates the number of violated constraints and the matriarch have the optimal value corresponding surely to an event somewhere. In Eq. 10 $Event_{i,DayIndex}^{t-1}$ means the value of day index that the current event is assigned, in addition $Event_{i,SlotIndex}^{t-1}$ means the value of the time slot index which the current event is assigned. Algorithm 1 shows Artificial Orca Algorithm for timetabling, whereas all actors belonging to the input of algorithm have the own table which contain availability, preference and unassignment columns, its values are 0 or 1, where 0 mean yes and 1 mean no, so we can count NPSI like this:
select count(NonPrefferedSlotInstr) from event where unprefferedSlotInstr=1. or like this:
select sum(NonPrefferedSlotInstr) from Event.

Since orca depend on the soft constraints satisfaction, so the objective function is composed of a set of Sub objective function described in Table 1 as mentioned in Eq. 13

$$Objective\ Function = \sum^{n} Sub\ Objective\ Function_n \qquad (13)$$

Table 1. Modeling of Sub-objective functions that will be minimized where their values represent the counts of constraints violation committed by each event penalized by score equal to 1, with weight=1.

Sub-Objective function	Computational model	Description
Non Preferred Slots of Instructors (NPSI)	$Sub_{OF1} = \sum_{i=1}^{116} Event_{i,NPSI}$	Count NPSI.
Non Preferred Slots of Rooms (NPSR)	$Sub_{OF2} = \sum_{i=1}^{116} Event_{i,NPSR}$	Count NPSR.
Non Preferred Slots of Attendees (NPSA)	$Sub_{OF3} = \sum_{i=1}^{116} Event_{i,NPSA}$	Count NPSA.
TD tP at First Week Day (TDPFWD)	$Sub_{OF4} = \sum_{i=1}^{116} Event_{i,TDPFWD}$	Count TDPFWD.
Lectures at Evening (LE)	$Sub_{OF5} = \sum_{i=1}^{116} Event_{i,LE}$	Count LE.
Assignment at Last time Slot (ALS)	$Sub_{OF6} = \sum_{i=1}^{116} Event_{i,ALS}$	Count ALS

6 Experiments and Results

6.1 Setup and Dataset Representation

Since the mathematical model of algorithm is not enough complicated, and dataset is not very large, so the chosen technique is the simplest and the mastered one, therefore, the program have developed in Delphi7 and Ms Access database, run on a machine featured as: LENOVO Laptop, Processer Intel(R) Core™ i3-2310M CPU @ 2.10 GHz, 2100 MHz, 4 GB of RAM under Microsoft Windows 7 operating system.

The timetable to be obtained relates only to the students of the first year of science and technology, which contains 567 students, divided into two sections, each section is divided into 8 groups, and each group attends 10 courses related to 7 different courses. That means that each group attends weekly 16 meetings (If only the Group unit is adopted for sectioning and the subgroup unit is eliminated, the solution will remain feasible). These data engender 116 events which will be assigned to the set of rooms illustrated in Table 2.

Algorithm 1: Artificial Orca Algorithm for timetabling
Input: set of M meeting, set of A attendees, set of I instructors, set of R rooms, all cited actors with their availability and preferences contained in tables.
Output: optimized timetable
Begin
1: Building events table from M, A
2: **For each** event //make a feasible assignment
3: Assign instructor according to availability priority with checking feasibility.
4: Assign room and time slots with checking availability and feasibility.
5: **End For**
6: **For each** Event //optimize the fitness using orca model with checking feasibility
7:     **while** (EventFitness not enough optimal) **and** (t ≤ Max Iteration) **do**
8:         search an optimal assignment in a feasible space eliminating the visited one, using Eq.8 to Eq.12
9:     **End while**
10: **End For**
11: Return optimized timetable
12: **End.**

Table 2. List of rooms.

Room type	Number	Capacity
Amphitheatre	1	350
Classrooms	5	35
Physics LAB	1	20
Chemistry LAB	1	20
Computer Science LAB	1	35

6.2   Results and Discussions

In order to prove Artificial Orca Algorithm efficiency in solving timetabling issue, the average of five executions on our dataset shows the best convergence after 96 iterations, whereas the worst execution solve the problem after 138 iterations. Also, we note that our tests converge toward a same score each time.

In order to prove our proposition efficiency, we compared with similar approach proposed by [31], the author participate with this algorithm International Timetabling Competition 2021, in which simulated annealing Algorithm (SA) is chosen to solve timetabling problem. SA consist of a random move, this move is accepted if it improves fitness, else it is accepted only if $Random(0,1) < EXP(-\Delta F/T)$ where ΔF denote the difference between Fitness before and after move, whereas T represents the current temperature which is updated following certain cooling scheme involving a cooling rate, and the process stop when coldest temperature or max iterations is meted. The updated position of the current event is obtained through Eq. 14 and Eq. 15, where x_i^t

Table 3. Parameters used by SA and AOA.

Algorithm	Parameter name	Value	Comment
SA	MaxIter	150	Max iterations for each event
	InitTemp	100	Initial temperature
	FinTemp	5	Final temperature
	CoolRat	0.95	Cooling rate
AOA	MaxIter	150	Max iterations for each event
	Population	116	Population size
	f_{min}	0	Minimum frequencies
	f_{max}	1	Maximum frequencies

represent the position of $Event_i$ at the iteration t, and x_i^{t-1} denote the oldest position of the same event

$$x_{pi}^{t-1} = (Event_{i,DayIndex}^{t-1} \times 6) + Event_{i,SlotIndex}^{t-1} \qquad (14)$$

$$x_{pi}^{t} = x_{pi}^{t-1} + Random(1,300) \qquad (15)$$

With the aim to make the comparison totally fair, we have chosen the best parameters for each algorithm. As known, the parameters selection process is crucial task in metaheuristic algorithms, The selected parameters are presented in 3.

Within this instance dataset, we have 116 events which have scheduled after ensuring a compromise between available time slots of instructors, rooms and attendees, which ensure the feasibility. However, this scheduling is penalized by score of 84 in terms of violated soft constraints, after five runs for each algorithm, all runs converge toward 3, all runs converge toward 3, because LE constraint is violated while the number of lectures (20) is larger than morning timeslots space(6 days 3 time slot), in addition the majority of instructors are not totally available at morning considering external occupancy, but the difference reside in the number of required iterations and execution time. Results, illustrated in Fig. 1 shows the global fitness evolution of the best and worst case corresponding to a set of runs for both Orca and SA algorithms. The outcomes prove that Artificial Orca Algorithm outperform Simulated annealing algorithm in both worst and best cases, consequently. Also, we noted, during our tests, the superiority of AOA in all our runs. This impressive results can be explained by the good trading between exploitation and exploration in the Artificial Orca Algorithm, which lead the population to the best optimal solutions. Moreover, AOA accept only the fittest moves, and all visited positions are eliminated immediately, both good or bad. Whereas, SA accepts some bad positions aiming to be a starting base towards promising areas, but it remains only a hope subject to a certain probability. Therefore, this trick give an opportunity to reduce the feasible space more than SA, so AOA increase its probability to outperform SA.

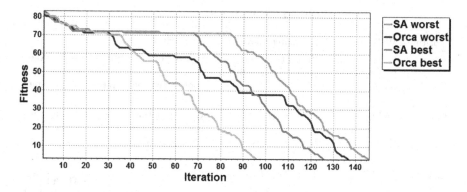

Fig. 1. Fitness function convergence in terms of iteration number.

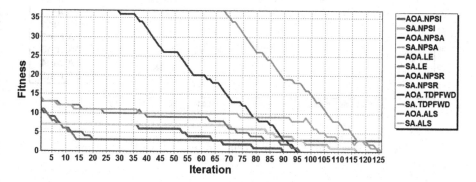

Fig. 2. Evolution of constraints satisfaction in terms of iteration number.

Figure 2 reveal that some constraints don't maintain their outperformance during the satisfaction process, especially at starting phase, such as AOA.ALS vs SA.ALS and AOA.LE vs SA.LE. whereas AOA.TDPFWD maintain their outperformance against SA.TDPFWD after getting stuck until 30th iteration, where SA.TDPFWD have got stuck until 68th iteration. Certainly, it comes down to the characteristics of each algorithm illustrated in the Table 4.

Results in Fig. 3 present the global fitness evolution in terms of CPU time corresponding to five runs of AOA and SA matching red and blue colour gradient, respectively. Again, results prove that our method outperform SA in execution time thanks to the low complexity of Artificial Orca Algorithm. In order to discover the robustness of AOA via its promised walking, In this context, the convergence speed is considered as performance metric in terms of time or number of iterations which match the number of moves.The speed criteria allow AOA to find the fittest solution at a limited time against SA when the time is a critical requirement for some issues, with aiming to tackle other benchmark especially those presented at ITC,surely basing on other metric.

Table 4. AOA and SA features.

Algorithms	Walking model	Accepted moves	Stop criteria
SA	Old position plus a random number	The fittest and probably non fittest	Getting optimal fitness, reaching the max iteration(max=30), expiration of cooling scheme, expiration of feasible space.
AOA	Old position plus velocity	The fittest only	Getting optimal fitness, reaching the max iteration(max=30), expiration of feasible space

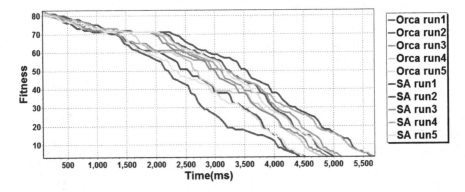

Fig. 3. Fitness function convergence in terms of CPU time (ms).

7   Conclusion

In this paper, the university course timetabling issue have been tackled using a novel metaheuristic called Artificial Orca Algorithm. Our proposed algorithm containts two stages, the first one construct a feasible timetable, whereas, the other ensure optimization process and constrains validations. Our algorithm have compared with a simulated annealing algorithm which is used until now as strong approach to solve this issue. The impressive results obtained through the carried out experiments, which outperform SA at a significant time. Thus, this results motivate us to complete with the leading algorithms through comparisons and analysis in the near future. Knowing that, we used just the simple move operator, and therefore we intend to involve other operators such as mutation, and other techniques like Kempe chains in order to improve efficiency.

References

1. Zhang, D., Liu, Y., M'Hallah, R., Leung, S.C.: Discrete optimization a simulated annealing with a new neighborhood structure based algorithm for high school timetabling problems. Eur. J. Oper. Res. **203**(3), 550–558 (2010)
2. Bettinelli, A., Cacchiani, V., Roberti, R., Toth, P.: An overview of curriculum-based course timetabling. TOP **23**(2), 313–349 (2015). https://doi.org/10.1007/s11750-015-0366-z

3. Tan, J.S., Goh, S.L., Kendall, G., Sabar, N.R.: A survey of the state-of-the-art of optimisation methodologies in school timetabling problems. Exp. Syst. Appl. **165**, 113943 (2021)
4. Tan, J.S., Goh, S.L., Sura, S., Kendall, G., Sabar, N.R.: Hybrid particle swarm optimization with particle elimination for the high school timetabling problem. Evol. Intell. 1–16 (2020)
5. Habashi, S.S., Salama, C., Yousef, A.H., Fahmy, H.M.: Adaptive diversifying hyper-heuristic based approach for timetabling problems. In: 2018 IEEE 9th Annual Information Technology, Electronics and Mobile Communication Conference (IEM-CON), pp. 259–266 (2018)
6. Breslaw, J.A.: A linear programming solution to the faculty assignment problem. Soc.-Econ. Plan. Sci. **10**, 227–230 (1976)
7. Hultberg, T.H., Cardoso, D.M.: The teacher assignment problem: a special case of the fixed charge transportation problem. Eur. J. Oper. Res. **101**, 463–473 (1997)
8. Abdullah, S., Turabieh, H., McCollum, B., Burke, E.K.: An investigation of a genetic algorithm and sequential local search approach for curriculum-based course timetabling problems. In: Proceedings of the Multidisciplinary International Conference on Scheduling: Theory and Applications (MISTA 2009), pp. 10–12, Dublin, Ireland (2009)
9. Lu, Z., Hao, J.K.: Adaptive Tabu search for course timetabling. Eur. J. Oper. Res. **200**, 235–244 (2010)
10. Rezaeipanah, A., Matoori, S.S., Ahmadi, G.: A hybrid algorithm for the university course timetabling problem using the improved parallel genetic algorithm and local search. Appl. Intell. **51**(1), 467–492 (2020). https://doi.org/10.1007/s10489-020-01833-x
11. Lewis, R.: A survey of metaheuristic-based techniques for university timetabling problems. OR Spectr. **30**(1), 167–190 (2008). https://doi.org/10.1007/s00291-007-0097-0
12. Kiefer, A., Hartl, R.F., Schnell, A.: Adaptive large neighborhood search for the curriculum-based course timetabling problem. Ann. Oper. Res. **252**(2), 255–282 (2016). https://doi.org/10.1007/s10479-016-2151-2
13. Nagata, Y.: Random partial neighborhood search for the post enrollment course timetabling problem. Comput. Oper. Res. **90**, 84–96 (2018)
14. Ceschia, S., Di Gaspero, L., Schaerf, A.: Design, engineering, and experimental analysis of a simulated annealing approach to the post-enrolment course timetabling problem. Comput. Oper. Res. **39**(7), 1615–1624 (2012)
15. Lewis, R.: A time-dependent metaheuristic algorithm for post enrolment based course timetabling. Ann. Oper. Res. **194**(1), 273–289 (2010)
16. Assi, M., Halawi, B., Haraty, R.A.: Genetic algorithm analysis using the graph coloring method for solving the university timetable problem. Proc. Comput. Sci. **126**, 899–906 (2018)
17. Harada, T., Alba., E.: Parallel genetic algorithms: a useful survey. ACM Comput. Surv. (CSUR) **53**(4), 1–39 (2020)
18. Badoni, R.P., Gupta, D.K.: A new algorithm based on students groupings for university course timetabling problem. In: 2015 2nd International Conference on Recent Advances in Engineering & Computational Sciences (RAECS), pp. 1–5. IEEE (2015)
19. Nothegger, C., Mayer, A., Chwatal, A., Raidl, G.R.: Solving the post enrolment course timetabling problem by ant colony optimization. Ann. Oper. Res. **194**(1), 325–339 (2012)

20. Chen, R.M., Shih, H.F.: Solving university course timetabling problems using constriction particle swarm optimization with local search. Algorithms **6**(2), 227–244 (2013)
21. Werra, D.: Graphs, hypergraphs and timetabling. Methods Oper. Res. **49**, 201–213 (1985)
22. Burke, E.K., Kingston, J., De Werra, D.: 5.6: Applications to timetabling. Handb. Graph Theory **445**, 4 (2004)
23. Liu, Y., Zhang, D., Chin, F.Y.: A clique-based algorithm for constructing feasible timetables. Optim. Methods Softw. **26**(2), 281–294 (2011)
24. Bagger, N.C.F., Desaulniers, G., Desrosiers, J.: Daily course pattern formulation and valid inequalities for the curriculum based course timetabling problem. J. Sched. **22**(2), 155–172 (2019)
25. Muklason, A., Irianti, R.G., Marom, A.: Automated course timeta-bling optimization using tabu-variable neighborhood search based hyper-heuristic algorithm. Proc. Comput. Sci. **161**, 656–664 (2019)
26. Shatnawi, S.M., Albalooshi, F., Rababa'h, K.: Generating timetable and students schedule based on data mining techniques. Int. J. Eng. Res. Appl. **2**(4), 1638–1644 (2012)
27. Sze, S.N., Bong, C.L., Chiew, K.L., Tiong, W.K., Bolhassan, N.A.: Case study: university lecture timetabling without pre-registration data. In: Proceedings of the 2017 IEEE International Conference on Applied System Innovation: Applied System Innovation for Modern Technology, ICASI 2017, pp. 732–735 (2017)
28. Matias, J.B., Fajardo, A.C., Medina, R.P.: A hybrid genetic algorithm for course scheduling and teaching workload management. In: 2018 IEEE 10th International Conference on Humanoid, Nanotechnology, Information Technology, Communication and Control, Environment and Management, HNICEM 2018, pp. 1–6 (2019)
29. Gülcü, A., Akkan, C.: Robust university course timetabling problem subject to single and multiple disruptions. Eur. J. Oper. Res. **283**(2), 630–646 (2020)
30. Bendimerad L.S., Drias, H.: An artificial orca algorithm for continuous problems. In: Abraham, A., Hanne, T., Castillo, O., Gandhi, N., Nogueira Rios, T., Hong, T.-P. (eds.) HIS 2020. AISC, vol. 1375, pp. 700–709. Springer, Cham (2021).https://doi.org/10.1007/978-3-030-73050-568
31. Sylejmani, K., Gashi, E., Ymeri, A.: Simulated annealing with penalization for university course timetabling. J. Sched. 1-21 (2022)

Global Automatic Tuning of Fuzzy Sliding Mode Controller for an Inverted Pendulum: A Genetic Solution

Soumia Mohammed Djaouti[1]([✉]), Mohamed Fayçal Khelfi[2,3], and Mimoun Malki[1]

[1] High School of Computer Science of Sidi Bel-Abbes, Sidi Bel Abbès, Algeria
soumia.mohammed.djaouti@univ-mascara.dz
[2] Faculty of Applied and Exact Sciences, University of Oran, 1 Ahmed BEN BELLA, Es Senia, Algeria
[3] ESGEE Oran, Oran, Algeria

Abstract. In this paper, we proposed a real-coded genetic solution for a Fuzzy Sliding Mode Controller (FSMC) of a nonlinear inverted pendulum system to stabilize the pole angle and avoid the chattering phenomenon. To get the best performances of the system, we need to tune simultaneously a large number of controller parameters to satisfy a cost function. A manual tuning design would be time-consuming. The genetic algorithm based on the Darwinian principle of evolution is used to overcome these difficulties providing an automatic tuning scheme for all the parameters of the FSMC where a fitness function is developed reflecting a minimum steady-state error with fast rise time and low overshoot. Parameters tuned are the switching gain for sliding mode, the membership functions for inputs/output, and the rules base of the fuzzy logic controller (FLC). The efficiency of the proposed genetic tuning method is tested and compared with the conventional method with free and additional disturbances. Simulation results have shown the advantages of automatic tuning of the FSMC's parameters to achieve the desired results. The superiority of the tuned controller has been proved to control the pole angle of the inverted pendulum despite the presence of additional disturbances.

Keywords: Sliding mode control · Fuzzy logic · Genetic algorithm · Optimization · Nonlinear systems · Inverted pendulum

1 Introduction

Several control techniques are used to improve the performance of industrial processes against external or internal disturbances. Stability is a very important characteristic of any system because almost every control system is designed to be stable.

The inverted pendulum has become a hot topic in the control field. It is a well-known example of a nonlinear, unstable control problem. It is a simple pendulum whose mass is located in the air. The system presents an unstable equilibrium in a vertical position; this position is maintained by the control of a movable cart. To achieve that, a variety of

M. Salem et al. (Eds.): ICAITA 2022, CCIS 1769, pp. 173–185, 2023.
https://doi.org/10.1007/978-3-031-28540-0_14

methods for inverted pendulum control are presented in the literature to control the cart position, to stabilize the pole, or for both of them.

Sliding mode control (SMC) is easy to tune and implement special nonlinear control featuring remarkable properties of accuracy and robustness. SMC has proved its effectiveness through several theoretical studies. The advantages provided by such control are quick response, insensitive to parameters variation, and robustness against disturbances and uncertainties of the model. However, the appearance of the chattering phenomena, caused by the discontinuous control, is a severe problem when the state of the system is close to the sliding surface.

To relieve the chattering phenomenon, some works [10, 17] used a modified sliding controller based on the boundary layer to improve it. Whereas, other researchers added fuzzy logic (FLC) to obtain a soft, robust, and smooth control (FSMC) that can reduce or eliminate chattering [5, 15]. This control structure combines the advantages of the robustness of the sliding mode control with those of the speed and the excellent tracking of the fuzzy control. This makes it possible to overcome the problem of the chattering of the SMC control but requires an expert to design the fuzzy controller.

The drawback of the former controller could be solved using an optimization technique like the genetic algorithms to tune the fuzzy parameters. Indeed, the genetic algorithm has been widely used to address difficulties in nonlinear control to solve complex optimization problems.

In [8], the authors proposed a new adaptive fuzzy sliding mode control (AFSMC) strategy where the fuzzy terms are introduced in the sliding mode controller and optimized by using an improved genetic algorithm to control a permanent magnet linear synchronous motors (PMLSM) direct-driven servo control system. While, authors of [4] use a robust SMC with the fuzzy evolutionary procedure on a remotely operated vehicle (ROV) subjected to model perturbation and external uncertainties. The SMC was next integrated into GA to develop a real-time code of GA control systems with good position responses with less chattering. In [3], the genetic algorithm is used for self-learning for deciding on the fuzzy control rules and the initial parameter vector for an adaptive fuzzy sliding mode controller (AFSMC). In the study of [7], researchers proposed three methods among them GAs to select the appropriate gain in the discontinuous part of the sliding mode controller to achieve good overall performance and reducing the effect of chattering.

In the work of [14], a genetic-based fuzzy sliding mode control design is proposed to track the problem of a two degree of freedom rigid robot manipulator. A genetic algorithm is used for automating the parameters of the fuzzy logic controller (rule base and membership functions).

The work presented in this paper is dedicated to the implementation of a Fuzzy Sliding Mode Controller (FSMC) for a nonlinear inverted pendulum. We propose a tuning scheme of an FSMC based on genetic algorithm optimization with the objectives of reducing the chattering phenomena and achieving precise tracking performance for an inverted pendulum.

Indeed, the proposed fuzzy controller is added to avoid high frequencies that cause chattering but the design of its parameters is done usually by a domain expert or by a trial error which decreases its performance. This drawback is solved in this paper by real-coded genetic algorithms where we aim to find the optimal parameters set for the controller. Unlike other related works, the proposed fuzzy-genetics sliding mode approach tunes both the parameters of the sliding mode controller and the fuzzy controller which are the membership functions parameters and base rules where nine rules are initially considered. This base could be reduced during the simulations.

The remainder of this paper is organized as follows. Fuzzy sliding mode control design steps are presented in Sect. 2. Section 3 describes the transformation of the designed FSMC problem to an optimization problem solved by GAs. Simulation results and discussion for nonlinear inverted pendulum are given in Sect. 4. Section 5 concludes this work.

2 Sliding Mode for Inverted Pendulum

2.1 Modeling the Inverted Pendulum

A classical inverted pendulum is composed of a pendulum attached to a cart (Fig. 1) [1]. It consists of a nonlinear second-order system given by Eq. (1):

$$\dot{\theta} = f(x) + g(x).u(t) + d(t) \tag{1}$$

where $x(t) = \begin{bmatrix} \theta & \dot{\theta} \end{bmatrix}$ is the state vector, $d(t)$ is the external disturbances and $u(t)$ is the control vector [13]. f and g are two nonlinear functions describing the dynamical system.

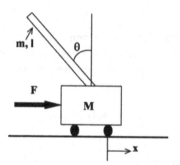

Fig. 1. An inverted pendulum

By applying a horizontal force F, the cart moves in the horizontal direction and provokes a deviation of the pole of θ radians.

$$\dot{\theta} = \frac{gsin\left(\dot{\theta}\right) - ml\dot{\theta}^2cos(\theta)sin(\theta)/(M+m)}{l\left(4/3 - mcos^2(\theta)/(M+m)\right)} + \frac{cos(\theta)/(M+m)}{l\left(4/3 - mcos^2(\theta)/(M+m)\right)}u(t) + d(t) \tag{2}$$

With θ and $\dot{\theta}$ are respectively angular position and the velocity of the pole, M and m are the mass of the cart and the pendulum respectively. l is the half-length of the pole [19].

2.2 Sliding Mode Control for Inverted Pendulum

Sliding Mode Control is a control strategy that provides high precision, good stability, invariance, and robustness which allows it to be suitable for systems with an imprecise model. The idea behind this control is to bring the system to a stable switching surface then it slides on to the desired balance point [17].

Considering a nonlinear system described by Eq. (1), referring to a given desired trajectory $\theta_d(t)$, the tracking error is given by:

$$e(t) = \theta(t) - \theta_d(t) \tag{3}$$

The objective is to find a control law such as the tracking error $e(t)$ tends to zero even with the presence of disturbances.

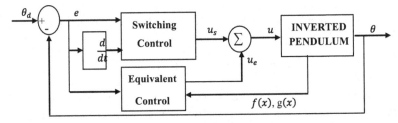

Fig. 2. Sliding mode control bloc

The control input consists of two components (Fig. 2): a discontinuous component $u_e(t)$ to drive the system states to the sliding surface and a continuous component $u_s(t)$ which is responsible for keeping the system on the surface $s = \dot{e} + \lambda e$ to force the error variable to the origin:

$$u(t) = u_e(t) + u_s(t) \tag{4}$$

with

$$u_e(t) = \frac{1}{g(x)}[-f(x) + \ddot{x}_d(t) - \lambda x(t) + \lambda \dot{x}_d(t) - d(t)] \tag{5}$$

and

$$u_s(t) = -k.\text{sign}(s) \tag{6}$$

where, λ is the slope of the sliding line and k is the switching gain chosen large enough to compensate disturbances. Sign is a signum function defined by:

$$\text{sign}(s) = \begin{cases} 1 & \text{if } s > 0 \\ -1 & \text{if } s < 0 \end{cases} \tag{7}$$

The choice of the gain k is very influential: a large gain provides faster convergence to the desired point but on the other hand, an undesirable problem occurs in the control of the sliding: "chattering phenomenon". A small switching gain gets slower convergence with small chattering. In literature, various approaches have been suggested to attenuate this undesirable effect like using a boundary layer [16] but none of them led to a complete chattering avoidance or canceling.

To overcome to this problem, we propose to introduce a new fuzzy controller to obtain a variable sliding gain which decreases near the sliding surface and increases when it moves far away from it [12].

3   Genetic Fuzzy Sliding Mode Control

3.1   Fuzzy Sliding Mode Control

The performance of a fuzzy logic controller FLC depends on its control rules and membership functions. Hence, adjusting these parameters to the controlled process may be very important [6, 11].

The FLC's feature of producing a smooth control action can be used to overcome the disadvantages of the SMC systems; this is achieved by introducing the fuzzy logic controller into the sliding mode structure to form a fuzzy sliding mode controller (FSMC) [9, 18]. The structure of the proposed fuzzy sliding mode controller is shown in Fig. 3.

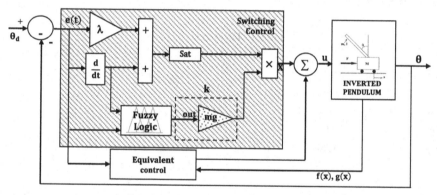

Fig. 3. Fuzzy sliding mode control bloc

The description of the fuzzy logic bloc is represented in Fig. 4. The error e, which measures the difference between desired and actual input, and the change of error $\dot{e}$ are the inputs of the fuzzy system and *out* denotes the output. The switching gain k is calculated by multiplying the fuzzy output *out* from the fuzzy logic controller by a gain mg.

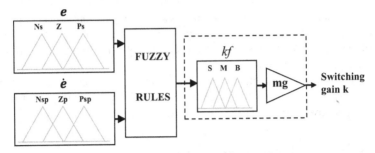

Fig. 4. Fuzzy logic bloc

This controller is based on a trial-and-error tuning procedure where the appropriate membership functions parameters should be selected.

The Mamdani's method is considered. The triangular membership function shape is chosen. The fuzzy subsets of inputs/output variables are expressed as follows: Negative small (*Ns*), Zero (*Z*), and positive small (*Ps*) for error *e*. Negative small (*Nsp*), zero (*Zp*), and positive small (*Psp*) for the derivative of error $\dot{e}$. . Small (*S*), Middle (*M*), and Big (*B*) for the fuzzy output.

The If-Then rules base of FSMC is illustrated in Table 1 where 9 rules, indexed from 1 to 9, are considered.

Table 1. Fuzzy rules table

e	$\dot{e}$		
	Ns	Z	Ps
Nsp	B (1)	M (2)	B (3)
Zp	M (4)	S (5)	M (6)
Psp	B (7)	M (8)	B (9)

The tuning of the FSMC controller via trial-and-error is extremely difficult and requires considerable experience on the designer part. Employing automatic tuning techniques can be more convenient. It is sufficient to perform manual tuning to get acceptable bounds of the tuning parameters, followed by automatic tuning via an off-line optimization.

This optimization process could be solved using genetic algorithms (GAs) which have been demonstrated to be a powerful tool for automating the fuzzy control rule base and membership functions.

3.2 Genetic Optimization for Fuzzy Sliding Mode

Recently, Genetic Algorithm (GA) plays an important role in the area of control. GA is a stochastic general search method that can effectively explore large search spaces where we start by generating an initial population composed of N individuals (chromosomes) randomly [2]. Each chromosome is evaluated based on its fitness calculated by an objective function. Best chromosomes with higher fitness have the chance to be selected for crossover and mutation operators giving a new set of solutions (offspring).

In this work, GA is used to tune and optimize parameters of the FSMC controller where the main scheme is depicted in Fig. 5.

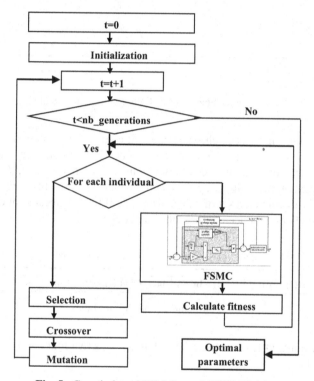

Fig. 5. Genetic-based FSM Control (GFSMC) bloc

The idea is to optimize the rules, the mg factor, and fuzzy inputs /outputs parameters. Thus, as described in Fig. 6, each membership function is defined in the range [a,e] by three parameters.

Real encoding is considered. This method has been wielded for many problems and to represent the chromosomes in simulation programs.

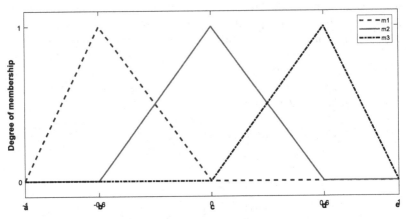

Fig. 6. The membership functions coding of a variable

As described in Fig. 7, each chromosome will contain the definition of 9 variables representing 9 membership functions for the inputs (e and ė) and the output (*out*) of the FLC controller (for each membership function, the parameters {*b, c, d*} are selected inside the range]*a, e*[), a multiplying gain, and the indexes of the fuzzy rules (Table 1). The rules base will contain a set of values from 1 to 9 for the selected rules. This base could be adjusted or be reduced during the simulations.

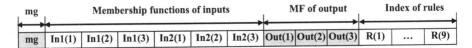

Fig. 7. Structure of the chromosome.

The Mean Squared Error (MSE) is used as an objective function to evaluate each chromosome (Eq. 8)

$$MSE = \frac{1}{N} \sum_{t=1}^{N} e(t)^2 \tag{8}$$

With *e(t)* is the control error given by Eq. (3).

4   Simulation Results

To evaluate the proposed approach in the former section, we will carry out simulations under MATLAB environment to stabilize an inverted pendulum with the following parameters:

$$g = 9.8\,^m/_s,\, M = 0.57\,kg,\quad m = 0.23\,kg,\quad l = 0.3302\,m.$$

First of all, we will test the proposed genetic-based FSM controller tuning with additional external disturbances d(t) = 20 sin(2πt) and next, a comparison is carried out with manual tuning of the fuzzy sliding controller. If control parameters are set suitably, the FSM controller can provide the properties of insensitivity and robustness to uncertainties and external disturbances.

The proposed method is used to obtain the best real-coded parameters of switching gain and the FLC controller for the FSM controller. We aim to have minimal rise time, overshoot, settling time, and steady error. Automatic tuning is performed by a genetic algorithm having as parameters: crossover probability = 0.8; mutation probability = 0.05; number of generations = 100; population size = 20 and the roulette wheel as selection type.

When implementing GA for optimization, one will need to set the upper and lower boundaries of the membership functions for the inputs/output, and the range of rules of the FSMC and the multiplying gain mg. In the present work, we considered a range of [-1,1] for error and change of error, [0,1] for fuzzy output, [1,100] for multiplying gain, and [1, 9] for indexing rules.

4.1 Evaluation of the Genetic-Based FSM Controller

To evaluate genetic-based FSMC (GFSMC), the first simulations are carried out with a step input with and without additional disturbances.

For comparison, we used the same conditions including initial population, population size, and the number of generations.

After running, we measured the minimal fitness in the presence/absence of disturbances using Eq. (7). The two minimum fitness costs are depicted in Fig. 8.

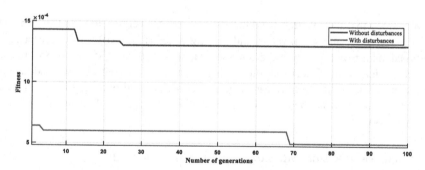

Fig. 8. Evolution of fitness function for a step input.

The angular errors and real/desired angles of the pole obtained using the optimal parameters obtained using the GFSMC are in Fig. 9.

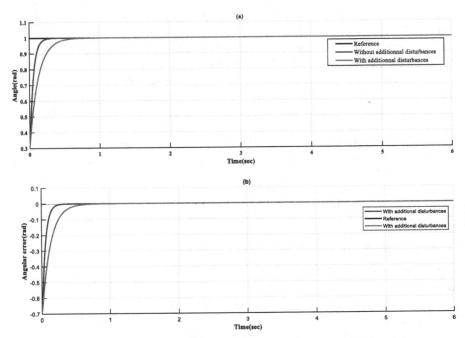

Fig. 9. Simulation results for a step input. (a) Real and desired Position, (b) Angular error using GFSMC

We notice that the inverted pendulum angle start converging to zero from $(\pi/10)$ with no overshoot and small rising and settling times (Fig. 9(a)). In the presence of additional disturbances, we notice a short delay in the rise time (0.35 s and 0.85 s for the other case) and settling time, this is due to the adaptation of the controller with additional disturbances when trying to converge to the desired set-point. From Fig. 9(b), the optimal parameters resulted from the use of GA allowed the convergence of the angular error to zero in a short time.

4.2 Comparison Results with Fuzzy Sliding Mode Controller

To test the effectiveness of the genetic-based fuzzy sliding mode controller, we compared it with manual tuned FSMC in the presence and absence of additional disturbances to control a nonlinear inverted pendulum system. To get the best results for the FSM controller, we need to tune simultaneously the system parameters which would be time-consuming.

To test the influence of disturbances for both FSMC and GFSMC, the desired/real angular positions and the angular errors for a sinusoidal input are presented in Fig. 10.

A zero steady-state error is achieved. The two controllers start converging to the set-point but the FSMC has taken more time than GFSMC. The genetic-based FSMC

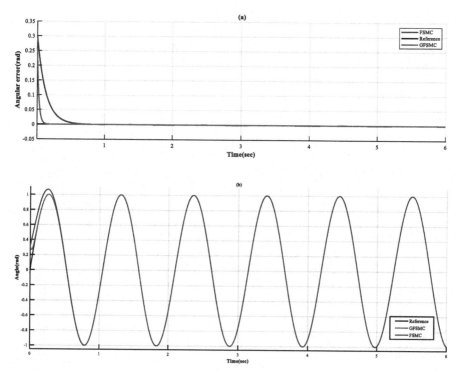

Fig. 10. Response of the inverted pendulum to a sinusoidal input for GFSMC and FSMC (a) Angular error (b) Real and desired Positions

with automatic tuning of parameters gives good results than FSMC in terms of steady error and rising time (Fig. 10(a)).

The optimal combination of parameters obtained by genetic FSMC brings the pole angle to its origin which enhances the system performances. Indeed, the method gets good results in his first generations with no overshoot. This is due to the ability of GA to avoid local optima and continuing searching in the search space which improves the performance of the algorithm.

The genetic optimization has allowed the pruning of the rule base, so it discarded three non-influent rules keeping only the rules 1, 2, 4, 6, 7, and 8. The resulted membership functions are presented in Fig. 11.

The simulation results showed the robustness of the proposed method to external disturbances. High performance of control was obtained with a reduced number of rules.

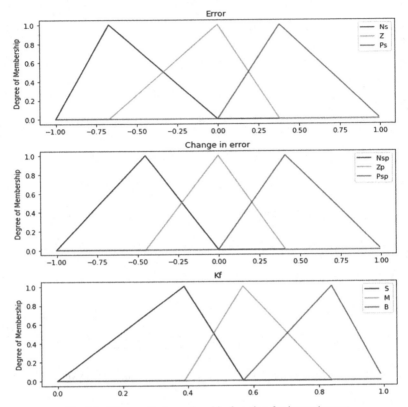

Fig. 11. The resulted membership function for inputs/output

5   Conclusion

In this paper, a robust control system with a fuzzy sliding mode controller for an Inverted Pendulum is presented. The issue of the improvement of the robustness and effectiveness of the sliding mode control design is investigated. Our objective is to propose an efficient method using a genetic algorithm to tune appropriate parameter sets of FSMC to attenuate the chattering phenomenon, where the parameters to tune are: the switching gain for sliding mode controller and membership functions and rules of the fuzzy controller. The advantage of GA is that they do not need any tuning just a suitable fitness function as a guide to find the best set of parameters. It is defined in such a way that the selected parameters can drive then keep the state slide along the surface with less chattering.

In this work, the effectiveness of a genetic-based fuzzy sliding mode was validated through numerical simulation with Simulink/MATLAB. A comparison is illustrated with free and additional disturbances injected in empirical FSMC. We show that automatic tuning by the genetic algorithms gives better control of FSMC for the considered system despite the presence of disturbances. The rule base has been reduced obtaining a new optimal FLC. Results showed a good adaptation of FSMC with additional disturbances and good performance where the chattering is small. Regardless of the good results given by the GA tuning but, this algorithm expends a long computation time in his searching for optimal parameters. Increasing the population size ensures and increases its diversity and reduces the probability to converge prematurely to a local optimum.

The modified controller can provide the properties of insensitivity and robustness to uncertainties and external disturbances, and the response of the Inverted Pendulum.

References

1. Banerjee, R., Pal, A.: Stabilization of inverted pendulum on cart based on pole placement and LQR. In: International Conference on Circuits and Systems in Digital Enterprise Technology (ICCSDET), pp. 1–5. Kottayam, India (2018)
2. Chen, P., Chen, C., Chiang, W.: GA-based fuzzy sliding mode controller for nonlinear systems. Mathematical Problems in Engineering **2008**, 1–16 (2008)
3. Chen, P., Chen, C., Chiang, W.: GA-based modified adaptive fuzzy sliding mode controller for nonlinear systems. Expert Systems with Applications **36**, 5872–5879 (2009)
4. Chin, C., Lin, W.: Robust genetic algorithm and fuzzy inference mechanism embedded in a sliding-mode controller for an uncertain underwater robot. IEEE/ASME Transactions on Mechatronics **23**(2), 655–666 (2018)
5. Dotoli, M., Maione, B., Naso, D. Evolutionary Techniques for Tuning Fuzzy Sliding Mode Controllers. In: Masulli, F., Parenti, R., Pasi, G. (eds.) Advances in Fuzzy Systems and Intelligent Technologies. Shaker Publishing (2000)
6. Dotoli, M.: Fuzzy sliding mode control with piecewise linear switching manifold. Asian Journal of Control **5**(4), 528–542 (2003)
7. Farzdaq, R., Mina, Q.: Chattering attenuation of sliding mode controller using genetic algorithm and fuzzy logic techniques. Eng. & Tech. Journal **27**(14), 2595–2610 (2009)
8. Guo, L., Zheng, C.: Optimization of fuzzy sliding mode controller with improved genetic algorithm. In: 22nd International Conference on Electrical Machines and Systems (ICEMS). Harbin, China (2019)
9. Ji-Chang, L., Ya-Hu, K.: Decoupled fuzzy sliding-mode control. In: IEEE Trans on Fuzzy Systems **6**(3) (1998)
10. Kapoor, N., Ohri, J.: Integrating a few actions for chattering reduction and error convergence in sliding mode controller in robotic manipulator. IJERT **2**(5), 466–472 (2013)
11. Katsuhiko, O.: Modern Control Engineering. Pearson (2009)
12. Kawamura, A., Itoh, H., Sakamoto, K.: Chattering reduction of disturbance observer-based sliding mode control. IEEE Transactions on Industry Applications **30**(2), 456–461 (1994)
13. Khalil, S.: Inverted pendulum analysis, design and implementation. The Instrumentation and Control Lab at the Institute of Industrial Electronics Engineering, Pakistan (2003)
14. Moshiri, B., Jalili-Kharaajoo, M., Besharati, F.: Application of fuzzy sliding mode based on genetic algorithms to control of robotic manipulators. In: IEEE Conference on Emerging Technologies and Factory Automation (2003)
15. Palm, R., Driankov, D., Hellendoorn, H.: Model Based Fuzzy Control. Springer-Verlag (1997)
16. Slotine, J., Sastry, S.: Tracking control of nonlinear systems using sliding surfaces with application to robot manipulator. International Journal of Control **38**, 931–938 (1983)
17. Slotine, J., Li, W.: Applied Nonlinear Control. Prentice-Hall, New Jersy (1991)
18. Spyros, G., Gerosimos, G.: Design and stability analysis of a new Sliding mode Fuzzy logic Controller of reduced Complexity. Machine Intelligence & Robotic Control **1**(1) (1999)
19. Wang, W.: Adaptive fuzzy sliding mode control for inverted pendulum. In: Proceedings of the Second Symposium International Computer Science and Computational Technology (ISCSCT'09), vol. 26, pp. 231–234 (2009)

Intelligent ITSC Fault Detection in PMSG Using the Machine Learning Technique

Issam Bahloul[✉] [iD], Monia Bouzid, and Sejir Khojet El Khil

LR11ES15 Laboratoire des Systèmes Electriques, Ecole Nationale d'Ingénieurs de Tunis, Tunis, Tunisia
Bah.issam@gmail.com

Abstract. In a wind turbine system, the permanent magnet synchronous generator (PMSG) is the key element that converts wind energy into electrical energy. According to global statistics, Inter-turns short circuit (ITSC) faults are the frequent electrical stator faults and the major cause of failures.

This paper presents an automatic and intelligent method to detect an ITSC fault in the PMSG based on the machine learning technique. For this aim, this work presents first the two used relevant indicators of ITSC fault extracted from the negative sequence voltage to detect the ITSC fault, then, it describes the database useful to train four machine learning algorithms which are the K-Nearest Neighbors "K-NN", Naïve Bayes "NB", Support Vector Machines "SVM", and Decision tree "DT". Finally, a comparative study is elaborated to evaluate the effectiveness of each of them and choice the best algorithm that gives the best classification performances in terms of turnaround time and the following metrics: Accuracy, Precision, Recall, and F1 score.

The outcomes of the comparison study show that the SVM algorithm is the best classification machine-learning algorithm for automatic ITSC fault detection since it has the highest performances.

Keywords: Inter-Turn Short-Circuit fault detection · Machine learning algorithms · Permanent magnet synchronous generator · Negative sequence voltage · KNN algorithm · NB algorithm · DT algorithm · SVM algorithm · Metrics score

1 Introduction

Nowadays, permanent magnet synchronous generators (PMSG) have gained great popularity due to their high efficiency, high torque, and high power density. It is used in many applications: in the field of electric vehicles [1, 2], metro rail [3], aeronautical transport and aviation [4, 5], industrial applications, and even in industry 4.0 [6] and especially in wind turbine systems [7]. A failure of the PMSG system can result in high repair costs and production losses.

Several electrical faults can affect the generator such as short circuit winding fault [8], open circuit fault [9], inverter pole fault [10], and demagnetization [11]. However, stator electrical faults and especially inter-turn short-circuit (ITSC) faults remain the

M. Salem et al. (Eds.): ICAITA 2022, CCIS 1769, pp. 186–201, 2023.
https://doi.org/10.1007/978-3-031-28540-0_15

most common and the most frequent in PMSG. They account for up to 38% of all faults [12]. Three main approaches have been used to diagnose an ITSC fault. The three classic approaches are the signal-based method, the model-based method, and the knowledge-based method. The first approach is based on frequency analysis and time-based current and voltage analysis [13–16], the second approach uses electrical diagrams, magnetic equivalent circuits, and mathematical models [17–20] and the third approach utilizes the artificial intelligence (AI). Machine learning (ML) is one of the various branches of artificial intelligence (AI) technique that can perform numerical analysis and can depend on real data records for classification and estimation [21].

Currently, with the massive availability of data and with the evolution of computer and information processing systems, data-based fault diagnosis methods have developed rapidly and are based on ML algorithms. The ML algorithms are categorized into supervised and unsupervised algorithms [22]. The supervised learning techniques used in fault diagnosis are classification algorithms [23]. Many classification algorithms are available in scientific literature such as K-Nearest Neighbors (K-NN), Naïve Bayes (NB), Support Vector Machines (SVM), and Decision tree (DT) and perform well in classification accuracy [24]. SVM is used as a good classifier in diagnosing faults in PMSG like bearing and pump [25–29]. Similarly, KNN is an efficient classification algorithm, in [30], the authors used KNN to realize an online detection and classification of PMSM stator winding faults. In [31], a decision tree algorithm is well used to diagnose broken rotor bar and in [32] they also used this algorithm in temperature fault detection of PMSM. In [33] the NB algorithm is used to classify bearing faults in PMSG.

Therefore, in this paper, a negative sequence voltage (NSV)-based method associated with an ML algorithm is proposed to detect an ITSC fault in the PMSG automatically. However, to choose the adequate algorithm that gives us the best fault detection performances, this paper elaborates a thorough comparison between four ML algorithms which are "K-NN", "NB", "SVM" and "DT". For each ML algorithm, to predict the class of each type of operating mode of the PMSG which are the healthy mode, ITSC fault in phase "a", ITSC fault in phase "b" and ITSC fault in phase "c", different metric tools are used [34]. These metric tools are useful to compare the different algorithm results and to identify the best algorithm based on performance metrics for ITSC prediction.

This paper is arranged as follows: a description of the proposed ITSC fault detection method is described in Sect. 2. A brief overview of ML techniques and evaluation is introduced in Sect. 3. Diagnostic and comparison using ML techniques: simulation and discussion in Sects. 4 and 5 respectively. A conclusion and future work are given in Sect. 6.

2 Description of the Proposed Method to Detect an ITSC Fault in the PMSG

To detect and locate an ITSC fault in the PMSG, relevant fault indicators were thoroughly investigated and specifically chosen. These ITSC fault indicators are the amplitude V_2 and the phase shift φ_{V2} of the NSV. Therefore, to detect ITSC fault automatically the proposed method is based on the ML technique to achieve intelligent fault detection. Thus, a large database is mandatory to carry out the training and test procedures. This database is constituted of different values of the amplitude V_2 and the phase shift φ_{V2} of the NSV under different operating conditions. Note that to extract these indicators an efficient faulty PMSG model, validated and published in [35] is used.

The used faulty PMSG model can simulate different stator faults. The three output voltages of the PMSG $V_a(t)$, $V_b(t)$, and $V_c(t)$ are the input of the NSV block. This block permits the calculation of the NSV $\bar{V}_2$ and the extraction of its amplitude V_2 and its phase angle φ_{V2} using the complex Fortescue's transformer expressed by (1).

$$
\begin{bmatrix} \bar{V}_1 \\ \bar{V}_2 \\ \bar{V}_0 \end{bmatrix} = \frac{1}{3} \begin{bmatrix} 1 & \bar{a} & \bar{a}^2 \\ 1 & \bar{a}^2 & \bar{a} \\ 1 & 1 & 1 \end{bmatrix} \cdot \begin{bmatrix} \bar{V}_a \\ \bar{V}_b \\ \bar{V}_c \end{bmatrix}
\tag{1}
$$

where $\bar{a} = e^{j\frac{2\pi}{3}}$ and $\bar{a}^2 = e^{j\frac{4\pi}{3}}$

The flowchart in Fig. 1 clearly demonstrates the procedure of the ITSC fault detection based on the NSV. According to this flowchart, the proposed method consists of monitoring the value of V_2. If V_2 is equal to zero, the PMSG is healthy. However, if not, there is an ITSC fault in one phase of the machine. In this case, the value of the phase angle φ_{V2} leads to locating the faulty phase. If φ_{V2} is around $-120°$ the fault is in phase "a". If φ_{V2} is around $0°$ the fault is in phase "b" and if, φ_{V2} is around $120°$ the fault is in phase "c". More information about these results can be found in [36].

In this paper, the ML technique is used to automatically detect and locate the faulty phase "a", or "b" or "c" of the PMSG in addition to the healthy operating mode. The basic principle of the proposed method to detect and locate an ITSC fault automatically in the PMSG is represented in Fig. 2.

The system proposed in Fig. 2 generates two indicators of fault (V_2 and φ_{V2}). These indicators are the items of the database that will be used in ML, named "features".

The label is what one would predict or expect. In our case, four labels or classes are defined, which correspond to the different operating states of the PMSG. As it is shown in Fig. 2, these four labels are Healthy PMSG, fault in phase "a", fault in phase "b" and fault in phase "c".

3   Machine Learning Techniques and Evaluation

In practice, it is difficult to program the flowchart obtained in Fig. 1 using classic programming because the input variables can change if, for example, the faulty resistance changes. To solve this problem, machine learning makes it possible to overcome this limitation.

3.1   Machine Learning Approaches

The most popular machine learning approaches used in classification and fault diagnosis are the following:

Support Vector Machines (SVMs): Support Vector Machines are a type of supervised machine learning algorithm. SVM is mostly used for classification. It is a frontier method

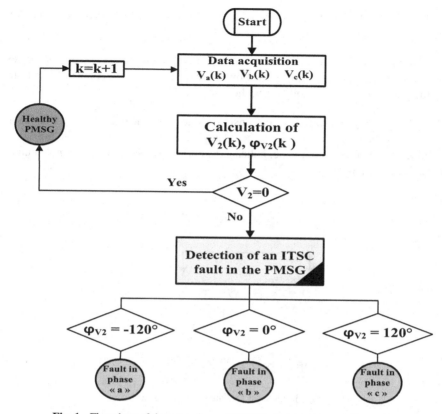

Fig. 1. Flowchart of the procedure of ITSC fault detection and diagnosis

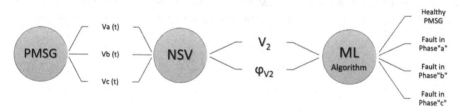

Fig. 2. Diagram of the principle of the proposed method to detect and locate an ITSC fault in the PMSG

for segregating classes. Their principle is to separate data into classes using a boundary that is as simple as possible, in such a way that the distance between the different groups of data and the border which separates them is maximal [23].

K-NN (K-Nearest Neighbour): This algorithm is used for solving classification, regression, or anomaly detection problems, its principle is to calculate the distance between the test data and all the training points, then select the number K of points that are closest to the test data and finally it chooses the class with the highest probability [23].

Decision Trees (DT): A decision tree is a diagram representing the possible outcomes of a series of interconnected choices. It allows the evaluation of different possible actions according to their cost, probability, and their benefits [23].

As shown in Fig. 3, a decision tree consists of one root node from where the decision tree starts, decision nodes are used to make any decision and have multiple branches, leaf nodes are the final output node, indicate the class to be assigned and do not contain any further branches.

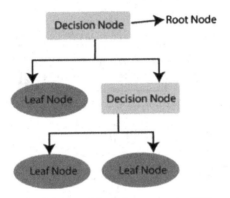

Fig. 3. General Decision tree nodes [23]

Naïve Bayes (NB): is a supervised learning algorithm used for classification. It is a probabilistic algorithm [23] where its principle is based on the Bayes Theorem as it is expressed by (2)

$$P(A|B) = \frac{P(B|A).P(A)}{P(B)} \qquad (2)$$

With:
P(A|B): how often does A happen given that B happens
P(B|A): how often does B happen given that A happens
P(A): how likely A is on its own
P(B): how likely B is on its own

3.2 Classification Metrics

To evaluate the performance of any algorithm model, a confusion matrix is illustrated in Fig. 4. It is a table that is used to describe the performance of a classification model on a set of test data for which the true values are known. All metrics measures "Accuracy,

	Predicted class	
	Negative	Positive
Actual class Negative	TN	FP
Actual class Positive	FN	TP

Fig. 4. The confusion matrix [37]

Precision, Recall & F1 Score metrics" can be calculated by using the four parameters: TP, FP, FN, and TN.

With:

TP: (True Positives) which means that the value of the actual class is yes and the value of the predicted class is also yes.

TN: (True Negatives) which means that the value of the actual class is no and value of the predicted class is also no.

FP: (False Positives) when actual class is no and predicted class is yes.

FN: (False Negatives) when actual class is yes but predicted class is no.

The Metrics measures used to evaluate the considered algorithms are the following:

Accuracy: is a ratio of correctly predicted observations to the total observations (3). The value of Accuracy varies between 0% and 100%. If we have high accuracy, the model is then the best.

$$Accuracy = (TP + TN)/(TP + FP + FN + TN) \tag{3}$$

Precision: is the ratio of correctly predicted positive observations to the total predicted positive observations (4). Like accuracy, the value of precision varies between 0% and 100%. A high precision relates to the low false positive rate.

$$Precision = TP/TP + FP \tag{4}$$

Recall: is the ratio of correctly predicted positive observations to all observations in actual class "yes" (5). A good recall is equal to 100%.

$$Recall = TP/TP + FN \tag{5}$$

F1 score: is a weighted average of accuracy and recall (6). Therefore, this result takes into account both false positives and false negatives. F1 score varies between 0% and 100%

$$F1\ Score = 2 * (Recall * Precision)/(Recall + Precision) \tag{6}$$

4 Fault Diagnosis Using Machine Learning Techniques: Simulation and Results

The four algorithms are implemented in Jupiter 6.4.5 with python programming language. The scikit-learn libraries are used for implementation and evaluation. The different algorithms were trained and tested by a large dataset with the same method and dataset. The dataset is distributed as 80% reserved for the training procedure and 20% for the test procedure.

In this study, the total dataset is composed of successive samples of 64 examples of healthy operating and different cases of ITSC fault operating of the PMSG in phase "a", "b" and "c". Each sample is composed of a vertical vector of two values $[V_2, \varphi_{V2}]^T$. The algorithm is trained by 80% of the dataset, which corresponds to 51 samples, and tested by the remaining 20% of the dataset which are the 13 samples.

The 51 input training samples, are composed of:

- 3 examples of healthy PMSG with no load
- 3 examples of healthy PMSG with different loads
- 20 examples of ITSC fault in phase "a"
- 20 examples of ITSC fault in phase "b"
- 18 examples of ITSC fault in phase "c"

For each case of operating mode corresponds to specific desired class. Therefore, four classes are defined as follows:

- Class 0: for a health PMSG
- Class 1: for an ITSC fault in phase "a"
- Class 2: for an ITSC fault in phase "b"
- Class 3: for an ITSC fault in phase "c"

To evaluate the performance of the four ML algorithms, five metrics score are considered which are *Accuracy Precision, Recall, F1 Score, and Program running time (ms)*. After a comparison study, the chosen algorithm is the best one that has the best metrics score. The obtained comparison results are presented in detail in the following sub-subsections.

4.1 Simulation Results Using the Support Vector Machines (SVM) Algorithm

Support Vector Machines algorithms use a set of mathematical functions that are defined as the kernel. These functions can be of different types. The best known and usable are linear, polynomial, and radial basis functions (RBF) [38].

To apply and evaluate SVM classifiers, we apply in the first step, the libraries NumPy, pandas and the scikit-learn libraries of Python. The second step in the implementation of SVM is the use of our ITSC fault dataset with the three chosen kernels. The last step is the generation of metrics score that we will use to evaluate our algorithm.

Table 1 shows the simulation results of the different SVM algorithms and their metrics evaluation. It can be seen that the SVM algorithm with the linear kernel (*SVM_Lin*) has the shortest execution time equal to 0.34ms, and the best accuracy, precision, recall, and F1 scores.

To clarify and highlighted the performances of each SVM algorithm, a comparison of four different metrics score of the different SVM kernels is depicted in Fig. 5. Note that the *SVM_Lin*, has the best scores, all are above 80%. It can be therefore concluded that the best algorithm among the different SVM algorithms is the *SVM_Lin*.

Table 1. SVM simulation results

	SVM_Lin	*SVM_Pol*	*SVM_Rbf*
Accuracy	*0,85*	*0,77*	*0,77*
Precision	*0,9*	*0,62*	*0,62*
Recall	*0,85*	*0,77*	*0,77*
F1 Score	*0,82*	*0,68*	*0,68*
Program running time(ms)	*0,34*	*0,37*	*0,36*

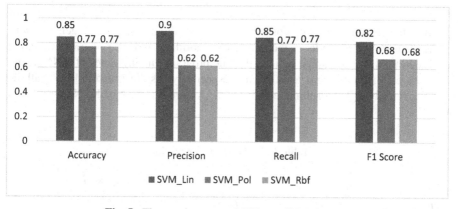

Fig. 5. The metric scores of different SVMs kernels

4.2 Simulation Results Using K-Nearest Neighbors (KNN) Algorithm

K-nearest neighbors is one of the most basic classification algorithms used in ML. In this algorithm, an object is classified by a plurality vote of its k nearest neighbors. The best choice of parameter k depends upon the dataset; a good k can be selected with a dynamic programming technique.

After calling Python's libraries, the second step is the implementation of the KNN algorithm and the selection of the best parameter k by using a script based on specific programming. The classification results are shown in Fig. 6 where the accuracy score is represented as a function of the variation of k from 1 to 9. Note that, if k ϵ [1, 2] the accuracy is equal to 85%, and for k ∈ [3, 9] the accuracy decreases to 77%. Therefore, to analyze the performance of this algorithm two cases are considered, which are k = 2 (KNN_2) and k = 9 (KNN_9).

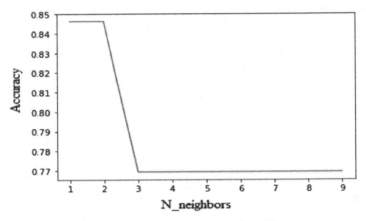

Fig. 6. k-number accuracy score of neighbors

After the choice of the two possible values of k, the last step is the generation of metric measures. The results of the two KNN algorithms and their metrics evaluation are listed in Table 2. It can be noticed that the program running times of KNN_2 and KNN_9 are equal to 0.36 ms and 0.39 ms respectively. As illustrated in Fig. 7, all the scores of the KNN_2 algorithm are above 80%. As a result, the best KNN algorithm is the KNN_2.

Table 2. KNN simulation results

	KNN_2	KNN_9
Accuracy	0,85	0,77
Precision	0,92	0,62
Recall	0,83	0,77
F1 Score	0,82	0,68
Program running time (ms)	0,36	0,39

4.3 Simulation Results Using Naïve Bayes (NB) Algorithm

Naïve Bayes is a classification and probabilistic algorithm used in ML based on the Bayes Theorem (2) $P\ (A|B)$, in our case the variable A is the class variable (Faulty/Healthy) and the variable B represents the attributes /features, B is given as $B = (V_2\ (V), \varphi_{V2}\ (°))$.

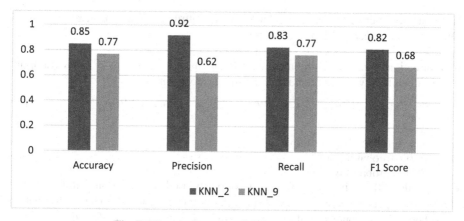

Fig. 7. The metrics score of different k_neighbors

There are several types of naive bayesian classifiers: Multinomial Naïve Bayes, Complement, Bernoulli and Gaussian Naïve Bayes Classifier. Bernoulli Naïve Bayes is the popular naive bayesian classifiers [39].

One of the disadvantages of Naïve-Bayes is that certain attribute values are negative like some values of the ITSC data set, as in the case of Multinomial and Complement Naïve Bayes. Thus, this aspect leads to the use of the Bernoulli Naïve Bayes algorithm (NB_B) to study the metrics score which are illustrated in Fig. 8. The "NB_B" algorithm was run during 0.38 ms. It can be noted that only the Recall percentage takes a good score equal to 75% while the three other metrics are less than 70%.

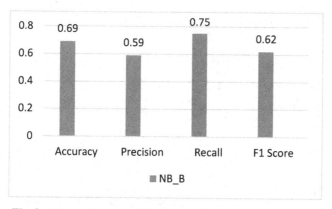

Fig. 8. The metrics score of the Bernoulli Naïve Bayes algorithm

4.4 Simulation Results Using Decision Trees (DT) Algorithm

The decision tree is a tree-structured classifier used in ML where decision nodes, shown in Fig. 9, represent the decision rules and each leaf node represents the outcome. To solve such problems, DT uses a technique called attribute selection measure (ASM). The two popular techniques for ASM are:

- Information gain (IG): is the measurement of changes in entropy (7). According to the value of IG, we split the node and build the decision tree.

$$Entropy(T) = -P(y)log2\,P(y) - P(n)log2\,P(n) \qquad (7)$$

where,
T = Total number of samples
P(y) = probability of yes
P(n) = probability of no

- Gini index (8) is a measure of impurity or purity used as a function that determines how well a decision tree was split. Gini impurity ranges values from 0 to 1. It is better to choose the attribute with the lowest possible Gini index, it is preferred to be equal to zero.

$$\text{Gini index} = 1 - \sum_{i=1}^{n} (Pi)^2 \qquad (8)$$

After importing the libraries needed to build a decision tree in Python and load the ITSC dataset, the structure of the decision tree is obtained by using the "tree.plot_tree" command from the library "sklearn. Tree" is represented in Fig. 9.

It can be seen in Fig. 9 that there is one root node from where the decision tree starts with a Gini index (GI) equal to 0.69. Two decision nodes to make any decision with GI 0.56 and 0.29 and finally four final output leaf nodes their GI value equal to zero these are the four classes state Healthy, fault in phase "a", fault in phase "b" and fault in phase "c".

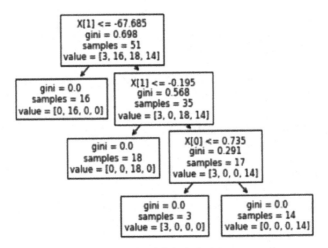

Fig. 9. Structure of ITSC fault decision tree nodes

Figure 10 displays DT metrics for an execution time of 0.37 ms, where it can be observed that all the metrics score of the *DT* are above 80%.

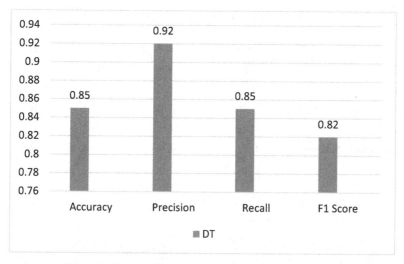

Fig. 10. The metrics score of the Decision tree algorithm

5 Comparison and Discussion of the Performances of the four Machine Learning Algorithms

This section aims to evaluate the performance of the four used classifiers. Table 4 summarizes the comparison of the different algorithms applied in ITSC fault diagnosis.

It can be noticed from Table 3 that SVM_Lin has the shortest time running equal to 0.34 ms, unlike KNN_2, which takes about 36 ms followed by DT and NB_B with 0.37 ms and 0.38 ms respectively.

This can be explained by the fact that the distribution of the attributes of the ITSC dataset is linear ($\varphi_{v2} = -120°$ if the fault is in phase "a", $\varphi_{v2} = 0°$ if the fault is in phase "b" and $\varphi_{v2} = 120°$ if the fault is in phase "c").

According to Fig. 11, the accuracy of SVM, KNN, and DT algorithms to classify the different operating modes is equal to (85%) is better than the one obtained by Naïve Bayes which has an accuracy of 69%. On the other hand, the precision obtained by KNN_2 and DT is equal to 92% is better than that obtained by SVM_Lin and DT which successively have a precision of 90% and 59% respectively.

Figure 11 shows that DT and SVM_lin achieved the highest value of recall 85% followed by 82% for KNN_2 and 75% NB_B.

It can also be easily seen from Fig. 11 that NB_B has the lowest value of F1 Score with 62% while the other algorithms have a score of 82%.

Based on this comparison, it can be conclude that the DT algorithm slightly outperformed other classifiers a bit since it has the best performances. However, the SVM has the shortest time running. As a conclusion, the chosen algorithm to detect ITSC fault automatically is the SVM algorithm.

Table 3. Classifiers performances

	SVM_Lin	KNN_2	NB_B	DT
Accuracy	0,85	0,85	0,69	0,85
Precision	0,9	0,92	0,59	0,92
Recall	0,85	0,83	0,75	0,85
F1 Score	0,82	0,82	0,62	0,82
program running time (ms)	0,34	0,36	0,38	0,37

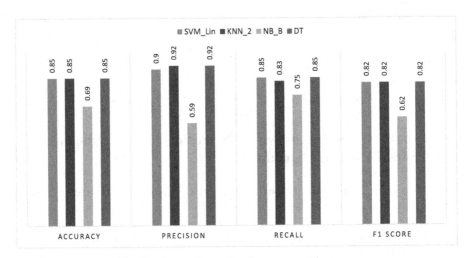

Fig. 11. Comparison of performance metrics score

6 Conclusion and Future Work

This paper proposes a fault diagnosis method that can accurately and automatically detect ITSC faults in a PMSG based on ML technique. Four machine learning algorithms are trained and tested with an ITSC fault diagnosis dataset obtained from several simulations. The four used machine learning algorithms used in this work,are SVM, Decision tree, KNN, and Naive Bayes. In this study, a comparison of the performances of the four algorithms according to many metrics: Accuracy, Precision, Recall, F1 score, and turnaround time is elaborated.

The decision tree and SVM have proven good performances on several metrics in front of the other algorithms,

However, since the SVM has the shortest time running, this paper selects the SVM algorithm for automatic detection of ITSC faults in PMSG.

In future work, we intend to conduct a study on the prediction of the number of faulty turns using regression algorithms.

References

1. Gupta, U., Yadav, D.K.: Inbuilt charging system of electric vehicles through generator installed on the rear shaft of the vehicle. In: Global Conference for Advancement in Technology (GCAT). IEEE (2019)
2. Akrad, A., Hilairet, M., Diallo, D.: Design of a fault-tolerant controller based on observers for a PMSM drive. IEEE Trans. Industr. Electron. **58**(4), 1416–1427 (2011)
3. Wang, W., Cheng, M., Zhang, B.F., Zhu, Y., Ding, S.C.: A fault-tolerant permanent-magnet traction module for subway applications. IEEE Transactions on Power Electronics **29**(4), 1646–1658 (2014)
4. Haylock, J.A., Mecrow, B.C., Jack, A.G., Atkinson, D.J.: Operation of a fault tolerant PM drive for an aerospace fuel pump application. IEE Proceedings – Electric Power Applications **145**(5), 441–448 (1998)
5. Yan, L., Dong, C.: Flux weakening control technology of multi-phase PMSG for aeronautical high voltage DC power supply system. In: 2019 22nd International Conference on Electrical Machines and Systems (ICEMS). IEEE (2019)
6. Tian, B., Mirzaeva, G., An, Q.T., Sun, L., Semenov, D.: Fault-tolerant control of a five-phase permanent magnet synchronous motor for industry applications. IEEE Trans. Ind. Appl. **54**(4), 3943–3952 (2018)
7. Freire, N.M.A., Cardoso, A.J.M.: Fault-tolerant PMSG drive with reduced DC-link ratings for wind turbine applications. IEEE J. Emerg. Selec. Topics in Power Electro. **2**(1), 26–34 (2014)
8. Sun, Z.G., Wang, J.B., Howe, D., Jewell, G.: Analytical prediction of the short-circuit current in fault-tolerant permanent-magnet machines. IEEE Trans. Industr. Electron. **55**(12), 4210–4217 (2008)
9. Arafat, A., Choi, S., Baek, J.: Open-phase fault detection of a five-phase permanent magnet assisted synchronous reluctance motor based on symmetrical components theory. IEEE Trans. Industr. Electron. **64**(8), 6465–6474 (2017)
10. Zhang, W.P., Xu, D.H., Enjeti, P.N., Li, H.J., Hawke, J.T., Krishnamoorthy, H.S.: Survey on fault-tolerant techniques for power electronic converters. IEEE Trans. Power Electron. **29**(12), 6319–6331 (2014)
11. Mínaz, M.R., Akcan, E.: An effective method for detection of demagnetization fault in axial flux coreless PMSG with texture-based analysis. IEEE Access **9**, 17438–17449 (2021)
12. El Sayed, W., Abd El Geliel, M., Lotfy, A.: Fault diagnosis of PMSG stator inter-turn fault using extended kalman filter and unscented kalman filter. Energies **13**(2972), 1–24 (2020)
13. Khalaf, R., Watson, I.-S.: Stator winding fault diagnosis in synchronous generators for wind turbine applications. In: 5th IET International Conference on Renewable Power Generation (RPG), January (2017)
14. Mazzoletti, M.A., Bossio, G.-R., De Angelo, C.H.: Interturn short-circuit fault diagnosis in PMSM with partitioned stator windings. IET Electr. Power Appl. **14**(12), 2301–2311 (2020)
15. Yong, C., Xu, Z., Liu, X.: Sensorless control at low speed based on HF signal injection and a new signal processing method. In: Proceedings of the Chinese Automation Congress (CAC", Jinan, China, pp. 3041–304520–22 October (2017)
16. Obeid, H., Battiston, N., Boileau, A., Nahid-Mobarakeh, T., Early, B.: Intermittent interturn fault detection and localization for a permanent magnet synchronous motor of electrical vehicles UsingWavelet transfor. IEEE Trans. Transp. Electrif. **3**, 694–702 (2017)
17. Usman, A., Joshi, B.M., Rajpurohit, B.S.: Review of fault modeling methods for permanent magnet synchronous motors and their comparison. In: IEEE 11th International Symposium on Diagnostics for Electrical Machines, Power Electronics and Drives (SDEMPED), pp. 141–146. Tinos, Greece, 29 August–1 September (2017)

18. Yassa, N., Rachek, M., Djerdir, A., Becherif, M.: Detecting of multi phase inter turn short circuit in the five permanent magnet synchronous motor. Int. J. Emerg. Electr. Power Syst. **17**, 583–595 (2016)

19. Faiz, J., Nejadi-Koti, H., Exiri, A.H.: Inductance-based inter-turn fault detection in permanent magnet synchronous machine using magnetic equivalent circuit model. Electr. Power Compon. Syst. **45**, 1016–1030 (2017)

20. Khan, M.S., Okonkwo, U.V., Usman, A., Rajpurohit, B.S.: Finite element modeling of demagnetization fault in permanent magnet direct current motors. IEEE Power & Energy Society General Meeting (PESGM), Portland, OR, USA, pp. 1–5 and 5–9 (2018 August)

21. Khoshaba, F., Kareem, S., Awla, H., et al.: Machine learning algorithms in Bigdata analysis and its applications: A Review. In: 2022 International Congress on Human-Computer Interaction, Optimization and Robotic Applications (HORA), pp. 1–8. IEEE (2022)

22. Kareem, S., Okur, M.C.: Bayesian Network Structure Learning Using Hybrid Bee Optimization and Greedy Search. Çukurova University, Adana, Turkey (2018)

23. Sen, P.C., Hajra, M., Ghosh, M.: Supervised classification algorithms in machine learning: A survey and review. Emerging technology in modelling and graphics, pp. 99–111. Springer, Singapore (2020)

24. Christobel, Y., Sivaprakasam, A.: An empirical comparison of data mining classification methods. Int. J. Comput. Inf. Syst. **3**(2), 24–28 (2011)

25. Kou, Z., et al.: Application of ICEEMDAN energy entropy and AFSA-SVM for fault diagnosis of hoist sheave bearing. Entropy **22**(12), 1347 (2020)

26. Wang, M., et al.: Roller bearing fault diagnosis based on integrated fault feature and SVM. J. Vibr. Eng. Technol. **10**(3), 853–862 (2022)

27. Zhang, X., et al.: A novel fault diagnosis procedure based on improved symplectic geometry mode decomposition and optimized SVM. Measurement **173**, 108644 (2021)

28. Akpudo, U.E., Hur, J.-W.: Intelligent solenoid pump fault detection based on MFCC features, LLE and SVM. In: 2020 International Conference on Artificial Intelligence in Information and Communication (ICAIIC). IEEE (2020)

29. Yao, G., et al.: VPSO-SVM-based open-circuit faults diagnosis of five-phase marine current generator sets. Energies **13**(22), 6004 (2020)

30. Pietrzak, P., Wolkiewicz, M.: On-line detection and classification of PMSM stator winding faults based on stator current symmetrical components analysis and the KNN algorithm. Electronics **10**(15), 1786 (2021)

31. Quiroz, J.C., Mariun, N., Mehrjou, M.R., Izadi, M., Misron, N., Radzi, M.A.M.: Fault detection of broken rotor bar in LS-PMSM using random forests. Measurement **116**, 273–280 (2018)

32. Gupta, A.: Prediction of electric motor temperature (pmsm) motor using decision tree. IJSRD-Int. J. Sci. Res. Develop. 197–198 (2021)

33. Senanayaka, J.S.L., Robbersmyr, K.G.: A robust method for detection and classification of permanent magnet synchronous motor faults: Deep autoencoders and data fusion approach. Journal of Physics: Conference Series. vol. 1037. No. 3. IOP Publishing (2018)

34. Saranya, T., Sridevi, S., Deisy, C., Chung, T.D., Khan, M.A.: Performance analysis of machine learning algorithms in intrusion detection system: a review. Procedia Computer Science **171**, 1251–1260 (2020)

35. Ben Khader Bouzid, M., Gerard, C., Ahmed, M., Slim, T.: Efficient simplified physical faulty model of a permanent magnet synchronous generator dedicated to the stator fault diagnosis part i: faulty model conception. IEEE transactions on industry applications **53**(3) (2017 may/june)

36. Bouzid, M.B.-K., Champenois, G.: An efficient simplified physical faulty model of permanent magnet synchronous generator dedicated to the stator fault diagnosis - part ii: automatic

stator fault diagnosis. IEEE Transactions on Industry Applications **53**(3), 2762–2771 (2017 Mai-Juin)

37. Choudhary, R., Gianey, H.K.: Comprehensive review on supervised machine learning algorithms. International Conference on Machine Learning and Data Science (MLDS), pp. 37–43. IEEE (2017)

38. Scikit-learn Homepage: https://scikit-learn.org/stable/modules/svm.html. Last accessed 15 Augusr 2022

39. Scikit-learn Homepage: https://scikit-learn.org/stable/modules/naive_bayes.html. Last accessed 15 August 2022

Series Elastic Actuator Cascade PID Controller Design Using Genetic Algorithm Method

Amira Sersar$^{(\boxtimes)}$ and Mohammed B. Debbat🆔

Departement of Electrical Engineering, University of Mascara, 29000 Mascara, Algeria
sersaramira@gmail.com, mohamed.debbat@univ-mascara.dz

Abstract. Series Elastic Actuator (SEA) has many advantages over traditional stiff actuator, such as intrinsic safety and reduction of energy consumption. SEA has been successfully applied in the robotic field as exploration vehicles or actuation mechanisms on bipedal robots and intelligent prostheses. However, a good performance of this electromechanical system depends a lot on the design of its control system. This paper deals with the design of a cascade PID controller based on genetic algorithm (GA) method for SEA position control. The purpose of this controller makes the system settle faster, the steady-state error tends to zero, the overshoot lower, and makes the system less sensitive to disturbances. After built of mathematical model and compute of the system transfer functions, GA method is used to tune, simultaneously, the inner (torque control) and outer (impedance control) loops of cascade PID controller. The sum of integral absolute error (IAE) values of two controller inputs is used as the objective function. The performance of the designed controller is evaluated by simulation under the MATLAB/Simulink software. The same controller is re-designed using the Ziegler-Nichols method in order to compare both methods in terms of response performance and robustness. The comparison shows that the GA method is more effective than conventional method.

Keywords: Series elastic actuator · PID controller · Tuning rules · Genetic algorithm · Optimization

1   Introduction

First introduced by G. A. Pratt and M. M. Williamson [1], SEA has many advantages over rigid actuators. It has been widely used in physical human-robot interaction systems such as rehabilitation exoskeletons [2], humanoids [3] and bionic prostheses [7]. SEA connects a traditional actuator such as an electrical motor to the output through an intentional, elastic element which can allow for high fidelity force control, impact tolerance and energy storage to a spring (Fig. 1). A good performance of SEA depends a lot on the design of its control

© The Author(s), under exclusive license to Springer Nature Switzerland AG 2023
M. Salem et al. (Eds.): ICAITA 2022, CCIS 1769, pp. 202–211, 2023.
https://doi.org/10.1007/978-3-031-28540-0_16

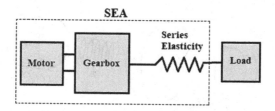

Fig. 1. Series elastic actuator schematic

system. Due to complex design of SEA and the presence of nonlinearities (friction, backlash, hysteresis, etc.), its position control design is a challenge. Many researchers have developed different control strategies for SEA [2–8]. Among the most used one is the cascade PID controller because it is widely used in industries process, simple and can improve the performance in terms of robustness. Difficulty of using this controller is tuning of the its parameters especially when the system is complex and high order. There are various methods which are used for the tuning the cascade PID controller parameters. The most used ones are the classical methods such as *Ziegler − Nichols* and trial error [9]. However, it is very difficult to tune the optimal parameters with these conventional methods. In order to overcome this problem, the genetic algorithm has appeared as an ideal method to tune controller gains [10].

Genetic algorithms (GA's) are search procedures inspired by the laws of a natural selection and genetics, they are computational algorithms which gives better result after every iteration. They can be viewed as a general-purpose optimization method and have been successfully applied to search, optimization and machine learning tasks. Based on the previous research, genetic algorithm have better result than another algorithm [11–13].

The main of this work is to use the GA method for tuning the gains of Cascade PID controller for SEA and for comparison of these processes with the ones occurring in cascade systems with PID controllers tuned by the classical methods.

2   Modeling of Series Elastic Actuator

The SEA type considered is this work is represented by the schematic diagram in (Fig. 2) [6].

By applying the rotational equivalent of Newton's second law, the dynamic behavior of SEA can be described by the following linear differential equations:

$$J_m\ddot{\theta}_m + b_m\dot{\theta}_m + \frac{1}{N}K(\frac{1}{N}\theta_m - \theta_l) = K_t i_m \tag{1}$$

$$J_l\ddot{\theta}_l + b_l\dot{\theta}_l + K(\frac{1}{N}\theta_m - \theta_l) = 0 \tag{2}$$

$$\tau_k = \frac{1}{N}K(\frac{1}{N}\theta_m - \theta_l)) \tag{3}$$

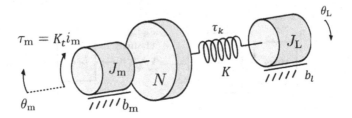

Fig. 2. The schematic diagram of SEA.

where

J_m is the equivalent inertia of motor,

J_l is the load inertia,

K is the stiffness coefficient of spring,

b_m and b_l are the damping coefficients of the motor and load, respectively,

N is the reduction ratio of the gear box,

i_m is the motor current,

θ_m, θ_l are motor and output position, respectively,

$\ddot{\theta}_m, \ddot{\theta}_l$ are motor and load angular acceleration respectively,

$\dot{\theta}_m, \dot{\theta}_l$ are motor and load velocity respectively and

τ_k is the actuator's output torque.

2.1 Transfer Function

To obtain the transfer functions of the linear system equations, we must first take the Laplace transform of the system equations (Eq. (1) and (2)) assuming zero initial conditions. The resulting Laplace transforms are presented below:

$$(J_m s^2 + b - ms + \frac{K}{N^2})\theta_m(s) - \frac{K}{N}\theta_l(s) = K_t i_m(s) \tag{4}$$

$$(J_l s^2 + b - ls + K)\theta_m(s) = \frac{K}{N}\theta_m(s) \tag{5}$$

$$\tau_k(s) = \frac{1}{N}K(\frac{1}{N}\theta_m(s) - \theta_l(s)) \tag{6}$$

A transfer function represents the relationship between one input and one output. The system is described by three transfer functions:

$$\frac{\theta_m(s)}{i_m(s)} = \frac{N^2 K_t(J_l s^2 + b_l s + K)}{D(s)} \tag{7}$$

$$\frac{\theta_l(s)}{i_m(s)} = \frac{N^2 K_t K}{D(s)} \tag{8}$$

$$\frac{\tau_k(s)}{i_m(s)} = \frac{N K_t K s^2 + N K_t K b_l s}{D(s)} \tag{9}$$

where: $D(s) = J_m J_l N^2 s^4 + (J_l N^2 b_m + J_m N^2 b_l)s^3 + (J_l K + J_m K N^2 + N^2 b_m b_l)s^2 + (K b_m N^2 + K b_l)s$.

3   Cascade PID Controller Design

The cascade control loop is purposely to improve the overall process performance particularly in dealing with the external disturbance [13]. The cascade loop consists of an outer loop (impedance controller), which is nested to the inner loop (torque controller).

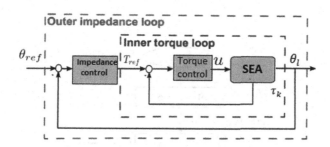

Fig. 3. Cascade PID control system

The concept of impedance control is developed by Hogan [4]. Impedance control is the relationship between the position and force (or torque). The impedance control law resembles a proportional-derivative (PD) motion controller with stiffness (propotional) and damping (derivative) coefficients as gains [6]. So we can write as:

$$G_{PD}(s) = k_{pi} + k_{di}s \tag{10}$$

where k_{pi} and k_{di} are the stiffness and damping respectively of the desired model. The main function of the SEA is to achieve a desired actuator force or torque control. The torque control is presented here by a PID controller and Its transfer function is given by:

$$G_{PID}(s) = k_{pt} + \frac{k_{it}}{s} + k_{dt}s \tag{11}$$

where k_{pt}, k_{it} and k_{dt} are propotional, integral and derivative gains respectively.

Tuning of cascade PID controller refers to the tuning of its parameters (k_{pi}, k_{di}, k_{pt}, k_{it} and k_{dt}) to achieve an optimized value of the desired response. The basic requirements of the output will be the stability, desired rise time, peak time and overshoot. There are different methods to tune this parameters such as classical methods and genetic algorithm method.

3.1   Genetic Algorithm Method

GA is a stochastic global adaptive search optimization technique based on the mechanisms of natural selection. GA starts with an initial population containing a number of chromosomes where each one represents a solution of the problem in which its performance is evaluated on a fitness function. Basically, GA consists of three main stages: Selection, Crossover and Mutation. The application of these three basic operations allows the creation of new individuals. The fittest chromosomes are taken as parents further they are reproduced, crossed over and mutated. The offspring is checked for the value of the fitness depending on the value it is either taken or neglected from the population. This algorithm is repeated for many generations and finally stops when reaching individuals that represent the optimum solution to the problem. The GA architecture is shown in (Fig. 4).

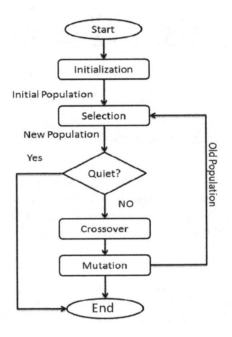

Fig. 4. Basic flowchart of genetic algorithm

The Genetic Algorithm as described consists of the following steps:

- **Initialization:** An initial population comprising of individuals are generated in this phase at the genotype level by filling the bit strings randomly by 1 or 0 values
- **Evaluation:** It is made to each of the chromosomes(k_{pi} k_{di}, k_{pt}, k_{it} and k_{dt}) that make up a population. At the end of the cycle through all the generations of the algorithm the solution will be given by the aptitude chromosome is the fittest, being this individual the optimal solution to the problem.
- **Selection:** In the selection process, individuals are chosen from the current population to enter a mating pool devoted to the creation of new individuals for the next generation such that the chance of given individual to be selected to mate is proportional to its relative fitness.

- **Crossover:** Crossover provides the means by which valuable information is shared among the population. It combines the feature of two parent individuals to form two children individuals that may have new and possibly better phenotype structures compared to those of their parents and play a central role in the GA optimization process.
- **Mutation:** The purpose of mutation is to introduce occasional perturbations to the variables to maintain the diversity in the population.
- **Termination:** While an acceptable solution is found

The design control problem is to find simultaneously the optimal gains for both controllers, namely (k_{pi}, k_{di}) and (k_{pt}, k_{it}, k_{dt}) that satisfy the objective function. In this work, the objective function J is taken as the sum of the integral absolute error (IAE) values of both controller inputs (Fig. 5) and it is defined as

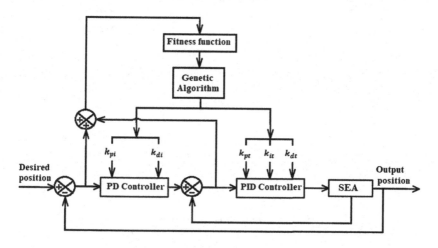

Fig. 5. Block diagram of GA based cascade PID controller

$$J = \int |e(t)| dt \tag{12}$$

where $e(t) = e_1(t) + e_2(t)$.

The error $e_1 = \theta_{ref} - \theta_l$ is the error signal that is given as input of PD controller and the error $e2 = \tau_{ref} - \tau_k$ is the error signal that is given as input of PID controller. where θ_{ref} and τ_{ref} are the reference or desired position and reference actuator torque (generated by PD and PID controllers) respectively.

4 Simulation Results

The SEA parameters taken from [5] are as follow: $J_m = 5.72^-5 Kg.m^2$; $b_m = 0.0015 N.m.s/rad$; $J_l = 0.198 Kg.m^2$; $b_l = 1.008 N.m.s/rad$; $K = 8.0769^3 N.m/rad$; $K_t = 0.19 N.m/A$; $N = 100$.

Tuning gains of the inner and outer loop controllers are determined using a genetic algorithm with following parameters: Generation = 50; Population size = 25. After several simulation tests in MATLAB, the optimal gains of cascade PID controller. Figure 6 shows the variation of the fitness function over the generations. It can be seen how the value of the IAE index decreases faster from the fifth generation, remaining almost constant in the rest of the generations, reaching a minimum value of 12.5434. This value corresponds to the optimal gains $k_{pi} = 198.4796$, $k_{di} = 0.1489$, $k_{pt} = 80.0036$, $k_{it} = 0.001$ and $k_{dt} = 0.0905$. Using these gain values, the response of SEA under a pulse input $\theta_{ref} = 0.05$ is illustrated in Fig. 7. It shows that the system reaches the reference position with a less overshoot and a shorter settling time $(0.5s)$.

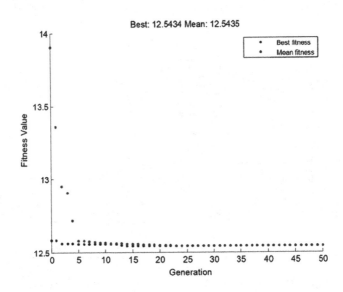

Fig. 6. Variation of fitness function over the generation

A comparison between IAE used in this work and others fitness function ISE $(J = \int e(t)^2 dt)$ and $ITAE$ $(J = \int t|e(t)|dt)$ is presented in Fig. 8. It is observed that IAE better in term of settling time than other performance indices. In order to compare the performances of the GA method with a conventional method, an other cascade PID is designed based on $Ziegler - Nichols$ $(Z - N)$ method ([9]). The optimal gains obtained with this classical method are: $k_{pi} = 160$, $k_{di} = 0.5$, $k_{pt} = 72$, $k_{it} = 0.02$ and $k_{dt} = 0.03$. The comparison results are shown in Fig. 9 and 10. It can see that the designed controller based on GA method is very effective (robust) in rejecting the change effects of θ_{ref} and faster than the designed one based on $Z - N$ method.

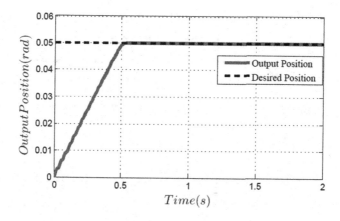

Fig. 7. Nominal response system

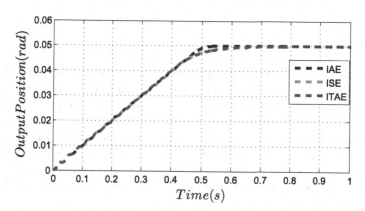

Fig. 8. Comparative analyses of IAE, $ITAE$ and ISE performance indexes

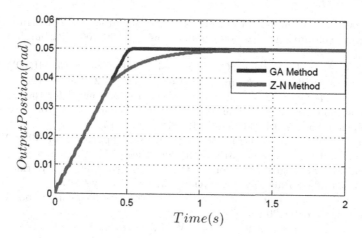

Fig. 9. Comparative analyse of both methods GA and $Z - N$

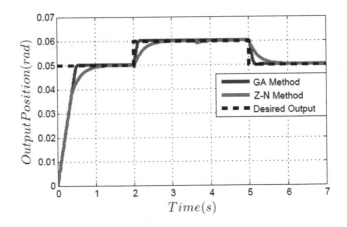

Fig. 10. Robustness test of both methods GA and $Z - N$

5   Conclusions

In this paper, Cascade PID controller based on genetic algorithm is designed in order to control series elastic actuator position. The mathematical model is built and the transfer functions of the system are computed. The gains of controller $(PD + PID)$ are tuned using genetic algorithm method. Using MAT-LAB/Simulink software, the performance of the designed controllers is evaluated a Comparative study with the controllers based on Ziegler-Nichols method is carried out. It has been shown that the GA based controller is better in terms of the settling time and robustness than the Ziegler-Nichols based controller.

References

1. Pratt, G.A., Williamson, M.M.: Series elastic actuators. In: IEEE International Conference on Intelligent Robots and Systems, pp. 399–406. Pittsburgh, PA, USA (1995)
2. Wang, S., Wang, L., Meijneke, C., et al.: Design and control of the MINDWALKER exoskeleton. IEEE Trans. Neural Syst. Rehab. Eng. **232**, 277–286 (2015)
3. Paine, N., Mehling, J.S., Holley, J., et al.: Actuator control for the NASA-JSC Valkyrie humanoid robot: a decoupled dynamics approach for torque control of series elastic robots. J Field Robot. **32**, 378–396 (2015)
4. Hogan, N.: Impedance control: an approach to manipulation. In: 1984 American Control Conference, pp. 304–313. IEEE, San Diego, CA, USA (1984)
5. Focchi, M.: Strategies to improve the impedance control performance of a quadruped robot. PhD thesis, University of Genoa, Italy (2013)
6. Mehling, J.S.: Impedance control approaches for series elastic actuators. PhD thesis, Rice University (2015)
7. Al-Shuka, H.F., Leonhardt, S., Zhu, W.H., Song, R., Ding, C., Li, Y.: Active impedance control of bioinspired motion robotic manipulators: an overview. Appl. Bionics Biomech. (2018). Article ID 8203054, 19 pages (2018)

8. Zhaoyao, S., Pan, Z., Jiachun, L.: High performance control method of electro-mechanical actuator based on active disturbance rejection control. Adv. Mech. Eng. **147**, 1–15 (2022)
9. Ogata, K.: Modern Control Engineering, 4th edn. Prentice Hall, Upper Saddle River, NJ (2003)
10. Holland, J.: Adaptation in Natural and Artificial Systems. University of Michigan Press, Ann Arbor (1975)
11. Ahmed, A., Gupta, R., Parmar, G.: GWO/PID approach for optimal control of DC motor. In: 2018 5th International Conference on Signal Processing and Integrated Networks (SPIN), pp. 181–186. IEEE, Noida, India (2018)
12. Gani, M.M., Islam, M.S., Ullah, M.A.: Optimal PID tuning for controlling the temperature of electric furnace by genetic algorithm. SN Applied Sciences 1880 (2019). Maddi, D., Sheta, A., Davineni, D., Al-Hiary, H.: Optimization of PID controller gain using evolutionary algorithm and swarm intelligence. In: 2019 10th International Conference on Information and Communication Systems, pp. 199–204. IEEE, Irbid, Jordan (2019)
13. Chew, I.M., Wong, F., Bono, A., Nandong, J., Wong, K.I.: Genetic algorithm optimization analysis for temperature control system using cascade control loop model. Int. J. Comput. Digit. Syst. **91**, 119–128 (2020)

Artificial Intelligence in Big Data
and Natural Language Processing

Quantum Natural Language Processing: A New and Promising Way to Solve NLP Problems

Yousra Bouakba[(✉)] and Hacene Belhadef

LISIA Laboratory, University of Abdelhamid MEHRI, Constantine, Algeria
{yousra.bouakba,hacene.belhadef}@univ-constantine2.dz

Abstract. Natural Language Processing (NLP) has known important interest and growth in recent years as may witness the increasing number of publications in different NLP tasks over the world. The primary focus of recent research has been to develop algorithms that process natural language in quantum computers, hence the emergence of new subdomain called QNLP were proposed. Hence the objective of this paper is to provide to NLP researchers a new vision and way to deal with the NLP problems basing on Quantum computing techniques. The present paper aims to provide a list of existing alternatives and classify them by; implementation on classical or quantum hardware, theoretical or experimental work and representation type of sentence. Our study focuses on the Distributional Compositional Categorical model (DisCoCat), from its mathematical and theoretical demonstration into its experimental results.

Keywords: Quantum computing · Natural language processing · Quantum machine learning · Quantum natural language processing

1 Introduction

Natural Language Processing is at the heart of most information processing tasks. It has a long history in computer science and has always been a central discipline in artificial intelligence. With the availability of data and the advancement of processing technology (such as GPUs, TPUs,...), we have recently witnessed a rising success of machine algorithms and deep learning, that marked the renewal of AI and the beginning of its new age. But a new vision to solve NLP problems is to use quantum theory in order to accelerate the process of building an AI model. As a result, quantum machine learning algorithms has involved the natural language processing (NLP) field. This has created the so-called Quantum Natural Language Processing (QNLP). Following that, many works are proposed which implement QNLP on either classical or quantum computers.

The purpose of this paper would provide a general overview of quantum approaches to solve natural language processing tasks with a strong attention

M. Salem et al. (Eds.): ICAITA 2022, CCIS 1769, pp. 215–227, 2023.
https://doi.org/10.1007/978-3-031-28540-0_17

on DisCoCat Model. Firstly, we presenting the most major limitations of current NLP models !. After a summary introduction of quantum computing fundamentals, the connection between quantum and NLP is described. Subsequently, quantum algorithms are listed by their type: implemented in quantum or classical computers and distinguishing between their sentence representation. Focusing on DisCoCat model, we present detailed experimental results of three QNLP application: question-answering, text classification and machine translation. Finally, important libraries for QNLP implementation are discussed.

This paper is structured as follows: Sect. 2, present limits of current NLP models. Then in Sect. 3, introduce the required foundation for understanding the intersection between NLP and quantum computing. First in Sect. 3.1, quantum computing fundamentals are described. And Sect. 3.2, Similarity points between NLP and Quantum theory are specified. Moreover, in Sect. 4 quantum algorithms per type are listed, divided into quantum-inspired algorithms (Sect. 4.1): full theoretical or algorithms with classical implementation, and quantum algorithms (Sect. 4.2): "bag-of-word" and DisCoCat models. Section 5, provides the two main QNLP toolkit: Lambeq and TensorFlow quantum.

2   Limits of Current NLP Models

Transfer learning and the application of Pre-Trained Language Models to varied NLP tasks have identified as the primary directions in current NLP research works. It is well known that more dataset and more parameters are essential for training deep transformers from scratch such as, GPT-3 which has used 175 billion parameters, 96 attention layers, and a Common Crawl data-set that is 45 TB in size [7]. However, it is costly in terms of time, resources, and processing power.

While some IA specialists, such as Anna Rogers, believe that gaining achievements by just exploiting additional data and computer power is not new research, it is a SOTA (State-Of-The-Art): it refers to the best models that can be used for achieving the results in an AI-specific task only [17].

Furthermore, because of their impractical resource requirements, these models are difficult to adapt to real-world business challenges. Several studies, on the other hand, have been carried out to understand if neural language models effectively encode linguistic information! or just replicate patterns observed in written texts! [9] [3]. With both of these reservations about current models, the lookout for new methodologies has taken priority in the field.

3   NLP and Quantum Computing

A number of recent academic contributions investigate the notion of using quantum computing advantages to improve machine learning methods. It has resulted an increasing number of strong applications in fields including NLP, cryptography [22].etc. QNLP is a branch of quantum computing that Implementing quantum algorithms for NLP problems.

3.1   Quantum Computing Fundamentals

This section introduces basic quantum computing fundamentals for a better understanding of QNLP.

Quantum Computer and Qubits. In quantum computing, qubits are the basic unit of information. A qubit can be 0, 1 or a superposition of both and represented using the Dirac notation: $|0>$ and $|1>$.

Superposition. A qubit can be represented by the linear combination of states:

$$|\psi>= a|0> +b|1> \tag{1}$$

where a and b are complex numbers and

$$|a|^2 + |b|^2 = 1 \tag{2}$$

When we measure a qubit we obtain either 0, with probability $|a|^2$, or 1, with probability $|b|^2$.

Entanglement. Bell states are specific quantum states of two qubits representing the simplest and maximal examples of quantum entanglement.

$$\frac{|00> +|11>}{\sqrt{2}} \tag{3}$$

Due to the entanglement property, the second qubit must now obtain exactly the same measurement as the first qubit because the measurement results of these two entangled qubits are correlated.

Measurement. When a qubit is in a superposition state, once we measure it collapse the superposition state and takes either 1 or 0.

QRAM: Quantum Random Access Memory. Is the quantum equivalent of classical random access memory (RAM). QRAM architecture has been proposed by [8] using n qubits to address any quantum superposition of N memory cells where a classical RAM uses n bits to randomly address $N = 2^n$ distinct memory cells. However, it is still unachievable at the implementation level.

NISQ Device: Noise Intermediate-Scale Quantum. Is a quantum hardware to run quantum algorithms with a maximum memory size of 100 qubits. It's "Noisy" because there aren't enough qubits for error correction. And "Intermediate-Scale" because the quantity of quantum bits is insufficient to calculate advanced quantum algorithms but sufficient to demonstrate quantum advantage [16]

3.2   Why NLP Is Quantum-Native?: Similarity Points

The term quantum-native appears often in several literature, showing the intuitive relationship between quantum computing and NLP; we outline this connection below:

1. A word multiple meaning is a superposition state; where each meaning represent a quantum state. for example the word "apple" can have different meanings, it can indicate Fruit or Enterprise depending the context. Using the Dirac notation it can be represented as a superposition state:

$$|apple >= a|fruit > +b|enterprise > \tag{4}$$

   the representation above means, the probability of being a fruit is $|a|^2$ and the probability of being an enterprise is $|b|^2$.
2. Context of sentence act like measurement; once a word is observed in context, one of the available meanings is typically selected. for example, "I received a phone from Apple". When it appears in the context of purchasing, it collapses to a fixed meaning. where Apple refers to Enterprise.
3. The grammatical structure entangles words: grammar is what connects the meanings of words and words are encoded as quantum state, then the grammatical structure is to entangle these states. [4]
4. Vector spaces and linear algebra are very used in NLP and quantum mechanics.

4   Quantum Algorithms Types

There are two types of quantum algorithms: those that can be implemented with classical computers and those that can only be executed with quantum computers.

4.1   Quantum-inspired/Quantum-like Algorithms

Several types of work are concerted quantum-inspired, such as full theoretical work that has never had a real implementation and work based on quantum physics but executing on classical hardware.

Full Theoretical Quantum Approaches: In terms of theory, [23] provides a method for implementing distributional compositional models, such as the Distributional Compositional Categorical model (DisCoCat), on quantum computers that is based on the usage of unavailable QRAM. This study demonstrates theoretically improved performance in sentence similarity. which will be applicable once QRAM is released.

To overcome for this unavailable QRAM limitation, [5] [13] papers proposes a theoretical full-stack NLP pipeline using a NISQ device that makes use of the classical ansätz parameters without the requirement of QRAM. Because they are followed by experimental proofs (Sect. 4.2), these theoretical approaches are considered as being the most fundamental works in the QNLP field.

Quantum-Inspired Algorithms: Quantum-inspired or quantum-like algorithms adopt mathematical foundations from quantum theory, although they are built to run on classical computers rather than quantum computers. A general method of information retrieval is proposed by [18] which models term dependencies using density matrix for more general representation of text. Paper [11] is another quantum learning model to information retrieval that has been presented in includes a query expansion framework to overcome limits due to limited vocabulary. based on [18], reference [2] implement the same general method for a speech recognition, in addition to using the evolution of the state.

Question Answering (QA) is another task where the use of quantum-inspired approaches have addressed the task in very different ways. In [24], A Neural Network-based Quantum-like Language Model (NNQLM) has been developed and utilizes word embedding vectors as the state vectors, from which a density matrix is extracted and integrated into an end-to-end Neural Network (NN) structure. An connection between the quantum many-body system and language modeling proposed by [25]. Text classification, and sentiment analysis [27] [26], is another NLP applications that has benefited from various quantum-based approaches.

4.2   Quantum Algorithms

The categorization of algorithms is not limited to their ability to be implemented on a quantum computer but also in terms of how they represent a sentence!There are various approaches to achieving this in literature including, sentence is a "bag-of-word", map words to vector [11] [4] [20] and map words to matrix [18]. This section concentrates on two different methods: "bag-of-word model" and "DisCoCat model".

Bag-of-word Model. The term'bag-of-words' refers to the use of "just" individual words as classifier features: The position of words, or which combination of words exist, is ignored. Reference [15] proposed a"bag-of-word" classifier with accuracy 100. The following is a full explanation of the bag-of-word classifier: training and classification phases.

Training phase:
1. Each (word, topic) pair is assigned to a specific qubit
2. Every time a given word occurs with a given topic, the (word, topic) weights increment the rotation for the corresponding qubit.
3. An additional qubit is declared for each topic in order to keep the summation for each topic.

Classification phase:
1. Pre-processing step, identifies the most common words for all topics.
2. Each detected word in the sentence has its topic weights associated with the sum qubit for that topic.
3. Each of the topic qubits is measured, and the winner is the topic that measures the most $|1>$ states over a number of shots.

A circuit implementing this process is shown in Fig. 1.

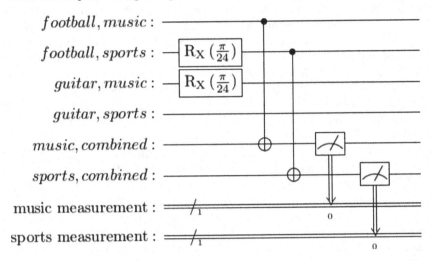

Fig. 1. Classification circuit of two words and two topics [21]

As a result, this method is only used to demonstrate extremely small vocabularies because the number of qubits required is (number words + 1) * number topics [21]. This model treating sentence as a structure-less "bag" containing the meanings of individual words.

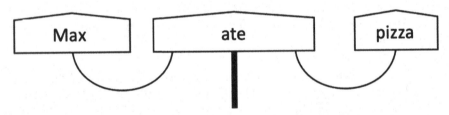

Fig. 2. A sample sentence using string diagram [6]

DisCoCat Model. A sentence in the DisCoCat model is not just a "bag-of-words" but also taking into account types and grammatical structure. In this model, using a string diagram to represent the meaning of words by tensors whose order is determined by the types of words, expressed in a pregroup grammar [6]. This graphical framework use boxes to represents the meaning of words that are transmitted via wires. Figure 2 shows that, the subject noun "Max" and the object noun "pizza" are both connected to the verb "ate" and the combination of these words contributes to the overall meaning of the sentence. In quantum terms, [5] provide a diagrammatic notation in which sentence meaning is independent to grammatical structure (see Fig. 3).

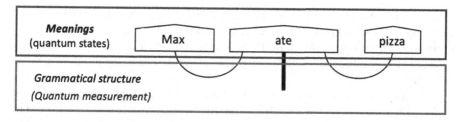

Fig. 3. Diagrammatic notation: word meaning is quantum states and grammatical structure is quantum measurements [5].

The DisCoCat model real origin is the categorical quantum mechanics (CQM) formalism [1]. As a result, it is natural to assume that it is suitable for quantum hardware.

Following to [13], The first NLP Questions-Answering task implementation on NISQ hardware has been proposed by [14] using a labeled dataset of 16 randomly generated phrases with 6 words vocabulary then the circuit runs on two different IBM quantum computers names respectively, *ibmq_ montreal* and *ibmq_ toronto*. The experimental results are, a train error of 12.5% and a test error of 37.5% on *ibmq_ toronto* and, a train error of 0% while the test error is 37.5% on *ibmq_ montreal*.

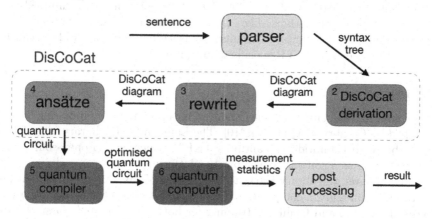

Fig. 4. The general quantum pipeline [12]

Therefore, to explore the challenges and limitations of training and running an NLP model on a NISQ device at a medium-scale, [12] propose the first quantum pipeline equivalent to the classical one (see Fig. 4).

This pipeline has been tested on two different tasks, meaning classification task and prediction task. The first task, Meaning classification(MC) uses a dataset of 130 sentences plain-syntax generated from a 17 words vocabulary using a simple CFG that can refer to one of two possible topics, food or IT. MC running on *ibmq_ bogota* has achieving a train error of 0% and a test error of 16.78%.

The second task, is to predict whether a noun phrase contains a subject-based or an object-based relative clause. This task rise the challenge by using 105 noun phrases are extracted from the RelPron dataset with 115 words vocabulary. The experimental results are, a train error of 0% and a test error of 32.3 % on *ibmq_bogota*.

Machine translation is an another task that has benefited from the DisCoCat model. [19] work trying to use DisCoCat for Spanish-English translation. The work achieved a high percentage (95 %) of sentence similarity between the two languages and good result even when the complexity and length of the sentence increase.

5   QNLP Implementation Tools

The objective of quantum language processing toolkit and libraries is to enable real-world quantum natural language processing applications. Currently there are two ways to implement QNLP tasks: TensorFlow quantum or Lambeq libraries.

5.1   TensorFlow Quantum

Is a framework that uses the NISQ Processors for development of hybrid quantum-classical ML models. This Quantum Machine Learning library provides updated algorithms like QNN (Quantum Neural Networks) & PQM(Parameterized Quantum Circuits). TensorFlow Quantum (TFQ) can be implemented to design QNLP specific quantum circuits.

5.2   Lambeq

It is the first and the only available Quantum natural language processing library developed by Cambridge Quantum [10]. The Lambeq QNLP library is perfectly compatible with Cambridge Quantum's TKET, which is also open-source and frequently used as a software development platform.

At a high level, the library allows the conversion of any sentence to a quantum circuit, based on a given compositional model and certain parameterisation and choices of ansätze, and facilitates training for both quantum and classical NLP experiments. The process of using Lambeq is divided into four steps [10]:

Step 1: Sentence Input is to convert a given sentence into a string diagram using a given compositional models. In fact, any compositional schema that presents sentences as string diagrams or as tensor networks can be added to the toolbox. Currently, the toolkit includes a variety of compositional models that use different levels of syntactic information:

1. Bag-of-words models, do not use any syntactic information instead it renders simple monoidal structures consisting of boxes and wires. For example Spiders reader which composing the words using a spider (see Fig. 5).

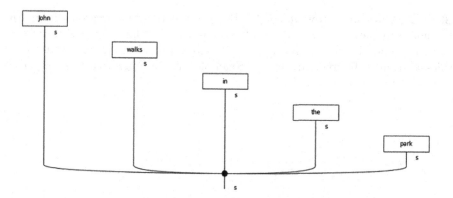

Fig. 5. Spiders reader diagram [10]

2. Word-sequence models, respect word order. There are two models, Cups reader generates a tensor train (see Fig. 6) and Stairs reader combines consecutive words using a box ("cell") in a recurrent fashion, similarly to a recurrent neural network.

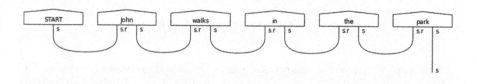

Fig. 6. Cups reader diagram [10]

3. fully syntax-based models, based on grammatical derivations given by a parser. Two cases of syntax-based models in lambeq are: BobCatParser in order to obtain a DisCoCat-like output (see Fig. 7), and tree readers which can be directly interpreted as a series of compositions without any explicit conversion into a pregroup form.

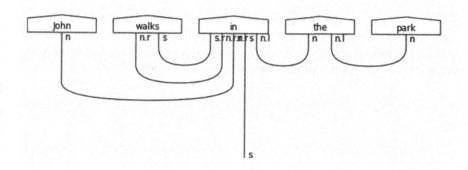

Fig. 7. BobCatParser diagram [10]

Step 2: Diagram Rewriting Syntactic derivations in pregroup form can become extremely complicated, resulting in unnecessary hardware resource usage and unacceptably long training times. Lambeq provide a predefined rewriter rules that simplify string diagrams. Example in (see Fig. 8) where eliminating completely the determiner "the".

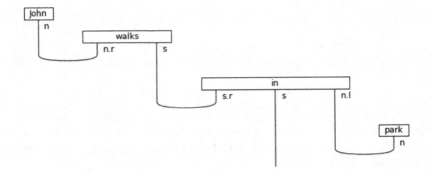

Fig. 8. BobCatParser simplified diagram [10]

This is clear that string diagrams are more flexible than tensor networks. The Toolkit supports auxiliary verbs, connectors, coordinators, adverbs, determiners, relative pronouns, and prepositional phrases as basic rewriting rules.

Step 3: Parameterisation A string diagram can be turned into a concrete quantum circuit (quantum case) or tensor network (classical case) by applying ansätze. An ansatz specifies options like the number of qubits associated with each string diagram wire and the concrete parameterised quantum states that correspond to each word (see Fig. 9).

5.3 Step 4: Training

Using the Lambeq toolkit, provides easy high-level abstractions for all essential supervised learning situations, both classical and quantum. The most significant aspect of Lambeq is that it is a high level tool i.e. we don't need to focus on low level architectural details and can therefore focus on our application.

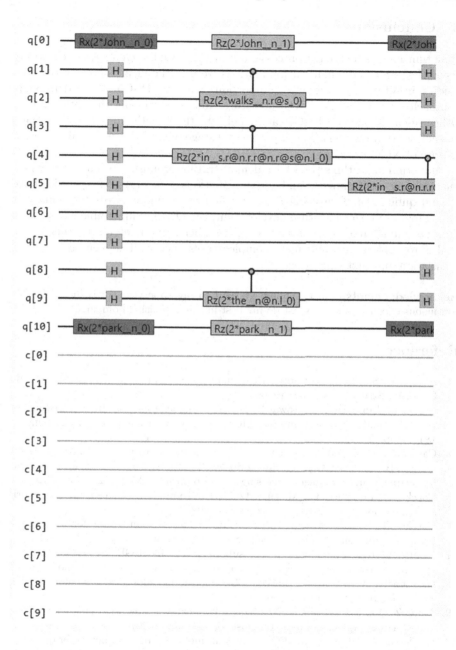

Fig. 9. IQP circuit in pytket form [10]

6 Conclusion

Quantum natural language processing is in the initial stage of research. Many theoretical and mathematical foundations were initiated in the early research studies in literature, the most important was the DisCoCat model. Some approaches have used DisCoCat structure while not utilizing the entire model. Other research, include DisCoCat model into their methodology, while others present alternate models. The experimental stage of QNLP is now realistic, with the use of NISQ devices.

Quantum algorithms [12,14,19] demonstrating promising results even experimenting on small datasets and simple situations. Currently NISQ device have 50–100 qubits therefore using a more sentences or more words vocabulary is impossible!. In terms of future development, the QNLP will be able to use large datasets and complex models with a high number of parameters. Another future goal is to explore new and more complex tasks such as Text generation, Text translation and text summarizing.

Acknowledgements. I'm extremely grateful to my supervisor Pr.Belhadef for his continuous encouragement to just do my best in this promising domain.

References

1. Abramsky, S., Coecke, B.: Categorical quantum mechanics. Handb. Quantum Logic Quantum Struct. **2**, 261–325 (2009)
2. Basile, I., Tamburini, F.: Towards quantum language models. In: Proceedings of the 2017 Conference on Empirical Methods in Natural Language Processing, pp. 1840–1849 (2017)
3. Casals, A.: Medical robotics at UPC. Microprocess. Microsyst. **23**(2), 69–74 (1999)
4. Coecke, B., de Felice, G., Meichanetzidis, K., Toumi, A.: Foundations for near-term quantum natural language processing. arXiv preprint arXiv:2012.03755 (2020)
5. Coecke, B., de Felice, G., Meichanetzidis, K., Toumi, A., Gogioso, S., Chiappori, N.: Quantum natural language processing (2020)
6. Coecke, B., Sadrzadeh, M., Clark, S.: Mathematical foundations for a compositional distributional model of meaning. arXiv preprint arXiv:1003.4394 (2010)
7. FUJII, A.: Reach and limits of the supermassive model GPT-3 (2022). https://medium.com/analytics-vidhya/reach-and-limits-of-the-supermassive-model-gpt-3
8. Giovannetti, V., Lloyd, S., Maccone, L.: Quantum random access memory. Phys. Rev. Lett. **100**(16), 160501 (2008)
9. Jiang, Z., Xu, F.F., Araki, J., Neubig, G.: How can we know what language models know? Trans. Assoc. Comput. Linguist. **8**, 423–438 (2020)
10. Kartsaklis, D., et al.: LAMBEQ: an efficient high-level python library for quantum NLP. arXiv preprint arXiv:2110.04236 (2021)
11. Li, Q., Melucci, M., Tiwari, P.: Quantum language model-based query expansion. In: Proceedings of the 2018 ACM SIGIR International Conference on Theory of Information Retrieval, pp. 183–186 (2018)
12. Lorenz, R., Pearson, A., Meichanetzidis, K., Kartsaklis, D., Coecke, B.: QNLP in practice: running compositional models of meaning on a quantum computer. arXiv preprint arXiv:2102.12846 (2021)

13. Meichanetzidis, K., Gogioso, S., De Felice, G., Chiappori, N., Toumi, A., Coecke, B.: Quantum natural language processing on near-term quantum computers. arXiv preprint arXiv:2005.04147 (2020)

14. Meichanetzidis, K., Toumi, A., de Felice, G., Coecke, B.: Grammar-aware question-answering on quantum computers. arXiv preprint arXiv:2012.03756 (2020)

15. Nielsen, M.A., Chuang, I.: Quantum computation and quantum information (2002)

16. Preskill, J.: Quantum computing in the NISQ era and beyond. Quantum **2**, 79 (2018)

17. Rogers, A.: How the transformers broke NLP leaderboards (2022). https://hackingsemantics.xyz/2019/leaderboards/

18. Sordoni, A., Nie, J.Y., Bengio, Y.: Modeling term dependencies with quantum language models for IR. In: Proceedings of the 36th International ACM SIGIR Conference on Research and Development in Information Retrieval, pp. 653–662 (2013)

19. Vicente Nieto, I.: Towards machine translation with quantum computers (2021)

20. Wang, B., Zhao, D., Lioma, C., Li, Q., Zhang, P., Simonsen, J.G.: Encoding word order in complex embeddings. arXiv preprint arXiv:1912.12333 (2019)

21. Widdows, D., Zhu, D., Zimmerman, C.: Near-term advances in quantum natural language processing. arXiv preprint arXiv:2206.02171 (2022)

22. Xu, F., Ma, X., Zhang, Q., Lo, H.K., Pan, J.W.: Secure quantum key distribution with realistic devices. Rev. Mod. Phys. **92**(2), 025002 (2020)

23. Zeng, W.; Coecke, B.: Quantum algorithms for compositional natural language processing (2016)

24. Zhang, P., Niu, J., Su, Z., Wang, B., Ma, L., Song, D.: End-to-end quantum-like language models with application to question answering. In: Proceedings of the AAAI Conference on Artificial Intelligence, vol. 32 (2018)

25. Zhang, P., Su, Z., Zhang, L., Wang, B., Song, D.: A quantum many-body wave function inspired language modeling approach. In: Proceedings of the 27th ACM International Conference on Information and Knowledge Management, pp. 1303–1312 (2018)

26. Zhang, P., Zhang, J., Ma, X., Rao, S., Tian, G., Wang, J.: TextTN: probabilistic encoding of language on tensor network (2020)

27. Zhang, Y., Li, Q., Song, D., Zhang, P., Wang, P.: Quantum-inspired interactive networks for conversational sentiment analysis (2019)

Learning More with Less Data: Reaching the Power-Law Limits in Steganalysis Using Larger Batch Sizes

Mohamed Benkhettou$^{(\boxtimes)}$ (ID)

Univ Montpellier, LIRMM, CNRS, Montpellier, France
m.benkhettou@outlook.fr

Abstract. Since the new trend of deep learning in steganalysis, practitioners tend to apply state-of-the-art deep learning techniques, good practices, and more generally grope to get the best performances possible. One major way is to increase datasets sizes, as it is likely to follow a power-law decrease of the testing error. However, once datasets get bigger and bigger, training becomes very costly and time-consuming. In this paper, we explore new perspectives of the batch size variation. We show that it can significantly improve models performances by impacting its complexity and helping learn faster by transforming the dataset error power-law curves, enabling models to train in shorter times, while also learning more features on relatively small datasets, without resorting to millions of images.

Keywords: Deep learning · Batch size · Steganalysis · Dataset power-law · Million images

1 Introduction

Steganalysis is the art of detecting the presence of hidden messages, through an apparently innocent medium like images. The hiding process is called steganography, which aims to insert as much information as possible, without leaving clues of any sort of manipulation on the used support to avoid suspicion [9].

The use of deep-learning in steganalysis is no longer revolutionary, it is even a trend that has become established since 2015. Although, its use implies managing several constraints, some of which are independent of models' architectures: datasets samples diversity, hyper-parameters tuning, etc. And others related to the domain, such as the use of models and blocks that best preserve the steganographic signal, controlled datasets' development, etc.

It is well known that deep learning models' performances are highly dependent on the dataset sizes, as larger datasets usually introduce more diversity which often has a very significant impact on the overall model's performance. This led practitioners to use and develop many data augmentation techniques and strategies to artificially mimic the availability of larger datasets, as it is not always abundant [34].

© The Author(s), under exclusive license to Springer Nature Switzerland AG 2023
M. Salem et al. (Eds.): ICAITA 2022, CCIS 1769, pp. 228–240, 2023.
https://doi.org/10.1007/978-3-031-28540-0_18

As far as we can tell, studying the batch size and linking model's performances/behaviors in steganalysis, to more general deep learning phenomena has never been addressed before, thus:

- We reconcile the model's complexity and the concept of generalization gap to assess model's behavior, according to learning regimes, and highlight the main works to whom we relate, in the next section.
- We assess the dataset power-law tendency for EfficientNet in the JPEG steganalysis domain and we show that it also tend to follow a dataset power-law error decrease
- We show that increasing the batch size can affect model's complexity allowing a better training, while also acting like a regularization technique (Sect. 3.1).
- We push the model to the limits till reaching the irreducible error region, showing that hyper-parameters can help switch faster between error regions (Sect. 3.2).
- We vary the batch size over a wide ranger of values and show how it transforms the error power-law curves by affecting its exponent, like previously hinted in [11] (Sect. 3.3).

All of the experimental results are further discussed and analyzed in Sect. 5, after a detailed description of the used setup in Sect. 4. We conclude this paper in Sect. 6.

2   Addressing Performance Goals Through Observed Deep Learning Phenomena

It has long been a goal to achieve the highest accuracy on certain public training datasets like ImageNet [27]. However, models' sizes tend to get larger and larger, which caused a certain amount of constraints, these large models presented a heavy duty for training regarding time, and were not accessible to everyone regarding their sizes. This introduced multiple efficiency goals, hence, the new models' development has widened the objectives from accuracy only to include training efficiency [32], training speed [23] and parameters' efficiency [14,31].

2.1   Scaling and Error Power-Law

The process of elaborating efficient models is hard to achieve for large size models, as it implies training aware neural architecture search (NAS) [30] and scaling techniques after achieving a satisfying base model, often after a wide automatic search [17]. Models can then be scaled according to the width [37], depth [10] or input resolution [15]. Recently, compound scaling has raised in order to maximize the gain of each scaling method, which is the protocol for elaborating the EffiecientNet family [31]. Other optimizations can be done using different blocks, or by performing more adaptive scaling strategies, according to layers' efficiency [32].

Scaling in deep learning concerns both the models and the datasets, both scaling have shown to follow a power-law in many studies [3,11,28].

In [24], they propose a formulation for linking both the model's and the dataset's power-law, enabling them to extrapolate without much error to unknown scales. Having a model size m and a dataset size n, it is given as:

$$\epsilon : \mathbb{R} \times \mathbb{R} \to [0,1]$$

$$\epsilon(m,n) \to \underbrace{a(m)n^{(-\alpha(m))}}_{\text{dataset power-law}} + \underbrace{b(n)m^{(-\beta(n))}}_{\text{model power-law}} + \underbrace{c_\infty}_{\text{irreducible error}} \qquad (1)$$

where $\alpha(m) \geq 0$ and $\beta(n) \geq 0$ (depending respectively on m and n) control the global rate at which error decreases, $a > 0$ and $b > 0$ are a form of unit conversion between data and model sizes and error, all depending on m and n respectively, with $c_\infty > 0$ the asymptotic lower value attainable.

When a, b, α and β are constant, which corresponds to fixed model and dataset, the simplified expression would be:

$$\epsilon(m,n) = a(m)n^{(-\alpha)} + b(n)m^{(-\beta)} + c_\infty \qquad (2)$$

If we fix the model, its power-law term becomes constant and is added to the irreducible error term c_∞ giving c'_∞, changing the expression to become similar to the one in [11]:

$$\tilde{\epsilon}(n) = a'n^{(-\alpha')} + c'_\infty \qquad (3)$$

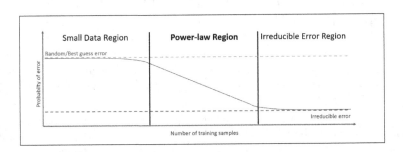

Fig. 1. Generic sketch of the evolution of the test probability of error depending on the number of training samples, for a fixed model and training procedure.

The overall scaling behavior shows that there are three regions, summarized in Fig. 1. The *small data region* is the region where there is not enough data for a model, to escape the random guess error. In the *power-law region*, every increase in the dataset's size would lead to better performance. The *irreducible error region* is considered as the grail as it gives the best performance achievable, where increasing the dataset's size would not make a difference.

It is important to highlight that these observations concern a fixed model architecture and procedure while varying the dataset's size. The same behavior is observed with the models for a fixed dataset.

2.2   Reconciling Model's Behavior and Complexity

The classical statistical learning theory lies on the fundamental bias-variance trade-off [33], stating that models with higher complexity have lower bias and higher variance, and those with a lower complexity have a higher bias and a lower variance, this leads to a certain point that best balances the two notions and considered as the "sweet spot", that gives the best performances for a given dataset, where increasing complexity would lead to over-fitting.

The use of larger models than required, reaching the millions of parameters showed a different behavior, known as the modern or over-parameterized regime [8,39], as models could achieve 0 training error without over-fitting, and still seeing their validation performance increase, even when fitting to random labels [2,38], which raised many questions about the statistical theory.

Both of these regimes are then reconciled with the double descent theory [1,4], that points out the existence of a threshold from which more complex models would interpolate to training data. The interpolation is then linked to model's complexity measures, as it better represents the model's ability to fit the training data.

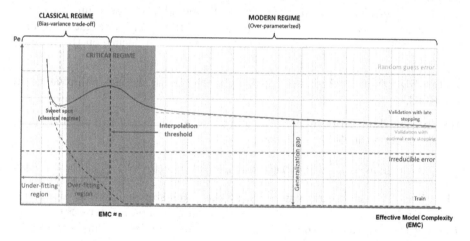

Fig. 2. Generic sketch of the evolution of the model learning curves as a function of the effective complexity of the model (EMC).

Empirical studies [18,29] show the existence of a "critical regime" near the interpolation threshold, where it is harder to predict the model's behavior.

The generalization gap refers to the gap between the train and the validation curves, it is used to give an insight about the model's ability to fit the data in the classical regime, but remains a phenomenon in the modern regime, where many researches try to better understand it, and to a certain extent close it [19,21,38].

A model is defined according to a static structure part, and a dynamic parametric part, that includes all the optimization parameters [13]. This highlights the link between the model and its optimization parameters, that have a direct influence on its performance and on its Effective Model Complexity (EMC). Being able to determine the complexity of a model, permits the precise understanding of the capability and limitation of the model. We will mainly focus on the Effective Model Complexity rather than the Expressive Model Capacity [20], as it is the one that reflects the model's usable capacity in practice. A good definition of the EMC is given in [4] that we reconcile with learning regimes in Fig. 2.

3   Batch Size's Impact on the Addressed Phenomena

We propose to explore 3 main implications of batch size increase that come along with the performance increase that is being observed in the batch size literature [16,22].

3.1   Changing Model's Complexity

In order to better highlight the impact of the batch size on the model's complexity, we consider the definition given in [4], and address it as a function of the batch size bs instead of the training procedure τ. If we consider a dataset n of S samples, we get two cases according to the batch size influence on the training procedure:

Batch Size doesn't Influence τ. Varying the batch size has no influence on the EMC.

Batch Size does Influence τ. We adapt τ as a function of two parameters bs and S, giving a the definition below:

$$EMC_{D,\epsilon}(\tau) := max\{n|\mathbb{E}_{S\sim D^n}[Error_S(\tau(bs,S))] \leq \epsilon\} \tag{4}$$

Therefore, EMC is directly influenced by the batch size change that affects the training procedure τ with respect to distribution D and parameter $\epsilon > 0$. Where $Error_S(M)$ is the mean error of model M on train samples S. Thus, varying the batch size will:

Increase the Effective Model Complexity: This will shift the model's behavior toward the over-parameterized regime direction.

Decrease the Effective Model Complexity: This will shift the model's behavior toward the under-parameterized regime direction.

3.2   Reaching the Model's Irreducible Error

When considering the different error regions related to the dataset's size, if we have bs a random batch size that enables the model to learn, then we consider two cases according to the batch size influence on the model's performance:

Batch Size doesn't Impact the Model's Performance. Varying the batch size have no influence on the model's performance.

Batch Size does Impact the Model's Performance. If we put bs a random batch size that gives a random training performance score with an error of ϵ_{bs}, bs_{max} the batch size that gives the best performance with the lowest error ϵ_{min}, bs_{min} the batch size that gives the lowest performance with the highest error ϵ_{max} while the model is still able to learn some patterns. If we consider that the model learning with bs has an error ϵ_{bs} located in the *power-law region* then:

- bs_{min} will shift ϵ_{max} to the *random guess error region*.
- bs_{max} will shift ϵ_{min} to the *irreducible error region*.

3.3   Transforming the Power-Law Curve

The nature of this shift that we described might act like an artificial augmentation of the number of samples with the same power-law, a translation of the same power-law toward a random direction or eventually transforming the power-law curve. To this end, we must be able to introduce ordering for power-laws. If we consider the same notations as in [11], with a dataset of size n and bs a random batch size, we adjust the equation this way:

$$\epsilon : \mathbb{N} \times \mathbb{R} \to [0, 1]$$

$$\epsilon(bs, n) \approx a(bs)n^{-\alpha(bs)} + c'_{\infty}(bs) \tag{5}$$

So that if we have bs_1, bs_2 two batch sizes, we get these two equations:

$$\epsilon(bs_1, n) \approx a(bs_1)n^{-\alpha(bs1)} + c'_{\infty}(bs_1) \tag{6}$$

$$\epsilon(bs_2, n) \approx a(bs_2)n^{-\alpha(bs2)} + c'_{\infty}(bs_2) \tag{7}$$

If we have bs_1 sufficiently smaller than bs_2, we expect to get:

$$\epsilon(bs_2, n) < \epsilon(bs_1, n) \tag{8}$$

This new formulation implies that the batch size is able to transform the power-law according to its term, but also assumes that the irreducible error won't be a constant, but an asymptotic function that depends on the training procedure, matching the definition given in Eq. 4.

4   Experimental Setup

Models and Hyper-parameters. EfficientNets [31] are a family of neural networks created in 2019 by leveraging NAS to search for the baseline EfficientNet-B0 and scaling it up with the compound scaling strategy to obtain a family of models from B1 to B7. It was by far the most used in Alaska#2 [7] competition due to its relative simplicity and high performance[1], which is surprising for a model that has not been developed for steganalysis purposes. Solutions include many tricks like model ensembling [6] and multiple transfers. Extensive studies on EfficientNet's architecture have been conducted in steganalysis in [35] to improve its performance either by removing stride in the model's stem or by implanting unpooled layers that show improvements in accuracy but with a significant increase in FLOPs and memory consumption.

We use the B0 version that has around 4M parameters pre-trained on the ImageNet dataset[2], as it is mentioned in many papers, training EfficientNet from scratch is not always feasible, regarding its size and the difficulty induced by the low payload of the stego images.

ImageNet pre-trained EfficientNet B0 model backbone[3] being RGB and having our datasets in grayscale, we duplicate the input images across the three input channels to match with the input without losing the ImageNet initialized weights similarly as in [5], and we fine tune it by replacing the fully connected layer, with a 2 nodes layer with a softmax activation and a binary cross-entropy loss function, as it is a binary classification (whether the image is stego or cover). For the optimization process, we used Adamax optimizer with a learning rate set to 0.002 and decreased at the epochs 130 and 230, with a factor equal to 0.1, and a weight decay to 5e–4. We trained for a total of 250 epochs, we shuffled the entire dataset at the beginning of each epoch and pair-constrained the batches.

In order to compare our results with those of the Low Complexity Network (LC-Net) used in [25], we kept the same hyper-parameters except for the batch size that we studied and performed the same images augmentation (random rotations and mirroring).

Dataset's Compilation and Scaling. We performed our experiments following the protocol described in [26] with the same image specifications, namely 256×256 grayscale JPEG images with a quality factor of 75, issued from the development of RAW datasets, known in the steganalysis field, with an ALASKA#2 inspired development script.

It is a compilation of ALASKA#2 (80.005 images from 40 devices), Stego App (24.120 images from 26 devices), BOSS (10.000 images from 7 devices), RAISE (8.156 images from 3 devices), Wesaturate (3.648 images) and Dresden (1.491 images from 73 devices). The largest resulting cover dataset (before inserting hidden messages), that we used is made of 500k images, from which we derive different versions with 100k, 50k, 10k, 7.5k, 5k, and 2500 images. Each of these retains the same proportions of images from the different source datasets and of the devices when the information is available.

---

[1] According to the authors of Alaska#2 [7] paper, this might be induced by getting one of the highest scores few weeks after the competition launch.

[2] Model transfer from ImageNet in steganalysis is explored in [36] and shows the interest of this approach.

[3] Version that scored 0.772 in the top1 accuracy on ImageNet.

To get the stego images (images containing hidden messages), we used the Matlab implementation[4] of the J-UNIWARD steganographic algorithm [12], with a payload of 0.2 bits per non-zero AC coefficient (bpnzacs). The embedding process took almost three days (2 days and 20 h) on an Intel Xeon W2145 (8 cores, 3.74.5 GHz Turbo, 11M cache).

To input MAT files to the models, JPEG images had to be decompressed to the spatial domain without rounding to integers or clipping, this step took us approximately 18 h for all the images.

For all of the datasets, we use both of the cover images as well as the generated stego images during the training, thus, the number of samples doubles, giving datasets of: 1 million, 500k, 200k, 100k, 20k, 15k, 10k and 5000 images.

Training, Validation and Testing Splits. All of the trainings were conducted on an IBM Power System AC922 with 64-bit POWER9 Altivec processors (MCPs) and an Nvidia Tesla V100 SXM2 32 Gb.

We used a split of 90% for the training and 10% for the validation, and kept the testing dataset aside as it remains the same for all of our trainings. It contains a total of 200k (cover and stego) images obtained by developing RAW images which were not present in the training/validation datasets and roughly keeping the same distribution of the sources of the original datasets. Thus, the steganalysis scenario is close to a clairvoyant scenario where the test set and learning sets are statistically very close.

5 Results and Discussion

Closing the Generalization Gap. The first results show how training EfficientNet on 4 different batch sizes (100, 128, 170 and 200) on the 20k version of the dataset affects the learning curves. Figure 3 shows how the generalization gap is being closed while increasing the batch size.

This best illustrates how varying the batch size influences the EMC, and in this particular case, increasing it, reduces the EMC, which is shifting the model's behavior toward the classical regime (see summary in Fig. 2).

This effect can be seen as a strong regularization, and is best understood when assessing the difference between model's capacity to fit data (that is fixed) and it's effective complexity, that depends on the learning procedure, and on batch size in this case.

In practice, this would mean that instead of changing models, and do a wide hyper-parameter search in order to reach the sweet spot, we can get significant performances increase by varying the batch size first with the rest of the hyperparameters fixed, in order to see if it actually does affect the model's behavior.

Learning More and Faster from the Same Data. Table 1 records the average accuracy for different batch sizes, ranging from 32 to 200 and shows how increasing the batch size can boost the accuracy while also shortening the

---

[4] Available at: http://dde.binghamton.edu/download/stego_algorithms/.

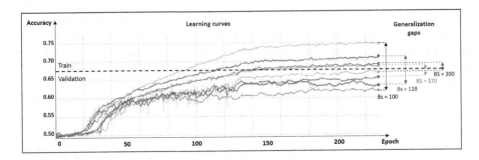

Fig. 3. Learning curves of the ImageNet pre-trained EfficientNet-B0 on a dataset of 20k images, with batch sizes of 100, 128, 170 and 200.

training times. The increase in performance can reach up to 8,9% accuracy, while reducing training time by 34 min for a 4 h train on the same dataset. Table 2 compares our results with those obtained by [25] and shows how we could boost the performances of LC-Net on various datasets.

Table 1. Average accuracy and training time for EfficientNet-B0, on the dataset of 20k images, with batch sizes of 32, 64, 100, 128, 150, 170 and 200.

Batch size	Accuracy	Training time
BS = 32	54.32%	4 h 07 m
BS = 64	59.33%	3 h 47 m
BS = 100	61.46%	3 h 35 m
BS = 128	62.07%	3 h 39 m
BS = 150	62.21%	3 h 36 m
BS = 170	62.29%	3 h 34 m
BS = 200	**63.22%**	**3 h 33 m**

Switching Between Error Regions and Transforming the Power-Law Curve. In order to confirm the shifting propriety between dataset error regions, we first performed training with the best performing batch size (200, which happens to be the largest also) and recorded the accuracy on the largest datasets, to better see the impact. The results show a significant increase in accuracy (See Table 3), and more interestingly that the performances are quite constant for all the datasets from 20k to 1M, which indicates that the model is no longer performing like in the Power-law region (See Fig. 1). Thus, we performed training on smaller datasets of 15k, 10k and 5k images to observe the power-law decrease of the error. Table 4 summarizes the recorded performances using different batch sizes.

This phenomenon can be seen as a long plateau region, ranging from the 20k to the 1M dataset (See Fig. 4), which indicates that the error power-law region is being exited and that the model is actually saturating. Furthermore, we show in Fig. 4 that the decrease is steeper while using larger batch sizes.

Table 2. Average accuracy of LC-Net with a batch size of 100 and 128.

Dataset size	BS = 100	BS = 128
5k	60,43%	59,98%
10k	61,15%	61,68%
15k	62,27%	63,11%
20k	62,33%	63,97%
100k	64,78%	66,97%
200k	65,99%	67,32%

Table 3. Average accuracy for EfficientNet-B0 on different sizes of the LSSD dataset, using a batch size of 200.

	20k	100k	200k	1M
Accuracy	64.16%	65.45%	64.30%	64.42%
Training time	3 h 51 m	17 h 26 m 15	1d 14h 16 m	8d 2 h 59 m 35 s

Table 4. EfficientNet-B0 average accuracy for different batch sizes and different dataset sizes. Random guess errors are underlined, and irreducible errors are highlighted.

Batch size	5k	10k	15k	20k	100k	Average
BS = 32	50,08%	53,11%	52,39%	54,32%	56,78%	53,33%
BS = 64	50,15%	57,32%	58,40%	59,33%	60,66%	57,17%
BS = 100	52,89%	60,24%	61,21%	61,46%	62,11%	59,58%
BS = 200	55,55%	60,79%	62,19%	**64,16%**	**65,45%**	61,63%

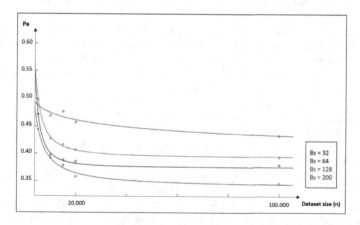

Fig. 4. Dataset power-law error curves for different batch sizes of 32, 64, 100 and 200.

The recorded results and the curves show that the EfficientNet model, could start learning using a dataset of only 2500 cover images, which is impressive for such a large model, that was not designed for steganalysis and without any pre-processing module. Other results show that training with a batch size of 200 on a dataset of 20k images (in less than 4 h), performs better than a training with a batch size of 100 on a dataset of 1M images (in 8d, 4 h and 50 m).

6   Conclusion

In this paper, we discussed different research goals of the deep learning community and proposed new perspectives for the batch size that is often an overlooked hyper-parameter in steganalysis. We showed that its increase leads to get significantly higher performances. We addressed this on a wide range of dataset sizes and showed that it implies faster learning while also learning more from the same datasets. We explained this performance increase using all of model's complexity, dataset error regions and error power-law curves.

Future works might imply exploring other types of learning techniques, like few-shot learning or meta-learning, or more generally, different model initializations.

Acknowledgments. The author would like to thank the LIRMM (Laboratory of Informatics, Robotics and Microelectronics of Montpellier) for providing the computational resources and hosting this research. Special Thanks are also due to the Algerian Ministry of Higher Education and Scientific Research (MESRS), for its funding and scholarship support.

References

1. Adlam, B., Pennington, J.: Understanding double descent requires a fine-grained bias-variance decomposition. Adv. Neural Inf. Process. Syst. **33**, 11022–11032 (2020)
2. Arora, S., Du, S., Hu, W., Li, Z., Wang, R.: Fine-grained analysis of optimization and generalization for overparameterized two-layer neural networks. In: International Conference on Machine Learning, pp. 322–332. PMLR (2019)
3. Bahri, Y., Dyer, E., Kaplan, J., Lee, J., Sharma, U.: Explaining neural scaling laws. arXiv preprint arXiv:2102.06701 (2021)
4. Belkin, M., Hsu, D., Ma, S., Mandal, S.: Reconciling modern machine-learning practice and the classical bias-variance trade-off. Proc. Natl. Acad. Sci. **116**(32), 15849–15854 (2019)
5. Butora, J., Yousfi, Y., Fridrich, J.: How to pretrain for steganalysis. In: Proceedings of the 2021 ACM Workshop on Information Hiding and Multimedia Security, pp. 143–148. IH&MMSec 2021, Association for Computing Machinery, New York, NY, USA (2021). https://doi.org/10.1145/3437880.3460395
6. Chubachi, K.: An ensemble model using CNNs on different domains for alaska2 image steganalysis. In: 2020 IEEE International Workshop on Information Forensics and Security (WIFS), pp. 1–6 (2020). https://doi.org/10.1109/WIFS49906.2020.9360892
7. Cogranne, R., Giboulot, Q., Bas, P.: Alaska# 2: Challenging academic research on steganalysis with realistic images. In: 2020 IEEE International Workshop on Information Forensics and Security (WIFS), pp. 1–5. IEEE (2020)

8. Dar, Y., Muthukumar, V., Baraniuk, R.G.: A farewell to the bias-variance tradeoff? An overview of the theory of overparameterized machine learning. arXiv preprint arXiv:2109.02355 (2021)

9. Fridrich, J.: Steganography in Digital Media. Cambridge University Press, Cambridge (2009). Cambridge Books Online

10. He, K., Zhang, X., Ren, S., Sun, J.: Deep residual learning for image recognition. In: Proceedings of IEEE Conference on Computer Vision and Pattern Recognition, CVPR'2016, pp. 770–778. Las Vegas, Nevada, June 2016

11. Hestness, J., et al.: Deep learning scaling is predictable, empirically. arXiv preprint arXiv:1712.00409 (2017)

12. Holub, V., Fridrich, J., Denemark, T.: Universal distortion function for steganography in an arbitrary domain. EURASIP J. Inf. Secur. 2014(1), 1–13 (2014). https://doi.org/10.1186/1687-417X-2014-1

13. Hu, X., Chu, L., Pei, J., Liu, W., Bian, J.: Model complexity of deep learning: a survey. Knowl. Inf. Syst. 63(10), 2585–2619 (2021)

14. Huang, G., Liu, Z., Van Der Maaten, L., Weinberger, K.Q.: Densely connected convolutional networks. In: Proceedings of the IEEE Conference on Computer Vision and Pattern Recognition, pp. 4700–4708 (2017)

15. Huang, Y., et al.: Gpipe: Efficient training of giant neural networks using pipeline parallelism. Adv. Neural Inf. Process. Syst. 32, 103–112 (2019)

16. Keskar, N.S., Mudigere, D., Nocedal, J., Smelyanskiy, M., Tang, P.T.P.: On large-batch training for deep learning: generalization gap and sharp minima. arXiv preprint arXiv:1609.04836 (2016)

17. Luo, R., Tian, F., Qin, T., Chen, E., Liu, T.Y.: Neural architecture optimization. Adv. Neural Inf. Process. Syst. 31, 7827–7838 (2018)

18. Nakkiran, P., Kaplun, G., Bansal, Y., Yang, T., Barak, B., Sutskever, I.: Deep double descent: where bigger models and more data hurt. J. Stat. Mech. Theory Exp. 2021(12), 124003 (2021)

19. Neyshabur, B., Bhojanapalli, S., Mcallester, D., Srebro, N.: Exploring generalization in deep learning. In: Guyon, I., Luxburg, U.V., Bengio, S., Wallach, H., Fergus, R., Vishwanathan, S., Garnett, R. (eds.) Advances in Neural Information Processing Systems. vol. 30. Curran Associates, Inc. (2017). https://proceedings.neurips.cc/paper/2017/file/10ce03a1ed01077e3e289f3e53c72813-Paper.pdf

20. Poggio, T., et al.: Theory of deep learning iii: explaining the non-overfitting puzzle. arXiv preprint arXiv:1801.00173 (2017)

21. Power, A., Burda, Y., Edwards, H., Babuschkin, I., Misra, V.: Grokking: Generalization beyond overfitting on small algorithmic datasets. ArXiv abs/2201.02177 (2022)

22. Radiuk, P.: Impact of training set batch size on the performance of convolutional neural networks for diverse datasets. Inf. Technol. Manag. Sci. 20, 20–24 (2017). https://doi.org/10.1515/itms-2017-0003

23. Radosavovic, I., Kosaraju, R.P., Girshick, R., He, K., Dollár, P.: Designing network design spaces. In: Proceedings of the IEEE/CVF Conference on Computer Vision and Pattern Recognition, pp. 10428–10436 (2020)

24. Rosenfeld, J.S., Rosenfeld, A., Belinkov, Y., Shavit, N.: A constructive prediction of the generalization error across scales. arXiv preprint arXiv:1909.12673 (2019)

25. Ruiz, H., Chaumont, M., Yedroudj, M., Amara, A.O., Comby, F., Subsol, G.: Analysis of the scalability of a deep-learning network for steganography into the wild. In: Del Bimbo, A., et al. (eds.) ICPR 2021. LNCS, vol. 12666, pp. 439–452. Springer, Cham (2021). https://doi.org/10.1007/978-3-030-68780-9_36

26. Ruiz, H., Yedroudj, M., Chaumont, M., Comby, F., Subsol, G.: LSSD: a controlled large jpeg image database for deep-learning-based steganalysis into the wild. In: Del Bimbo, A., et al. (eds.) ICPR 2021. LNCS, vol. 12666, pp. 470–483. Springer, Cham (2021). https://doi.org/10.1007/978-3-030-68780-9_38

27. Russakovsky, O., et al.: ImageNet large scale visual recognition challenge. arXiv:1409.0575 [cs] (January 2015). http://arxiv.org/abs/1409.0575, arXiv: 1409.0575

28. Sala, V.: Power law scaling of test error versus number of training images for deep convolutional neural networks. In: Multimodal Sensing: Technologies and Applications, vol. 11059, p. 1105914. International Society for Optics and Photonics (2019)

29. Spigler, S., Geiger, M., d'Ascoli, S., Sagun, L., Biroli, G., Wyart, M.: A jamming transition from under-to over-parametrization affects generalization in deep learning. J. Phys. A Math. Theor. 52(47), 474001 (2019)

30. Tan, M., et al.: MnasNet: platform-aware neural architecture search for mobile. In: Proceedings of the IEEE/CVF Conference on Computer Vision and Pattern Recognition, pp. 2820–2828 (2019)

31. Tan, M., Le, Q.: EfficientNet: rethinking model scaling for convolutional neural networks. In: Proceedings of the 36th International Conference on Machine Learning, PMLR'2019, vol. 97, pp. 6105–6114. Long Beach, California, USA, June 2019. http://proceedings.mlr.press/v97/tan19a/tan19a.pdf

32. Tan, M., Le, Q.V.: EfficientNetV2: smaller models and faster training. arXiv:2104.00298 [cs] (June 2021). http://arxiv.org/abs/2104.00298, arXiv: 2104.00298

33. Yang, Z., Yu, Y., You, C., Steinhardt, J., Ma, Y.: Rethinking bias-variance trade-off for generalization of neural networks. In: International Conference on Machine Learning, pp. 10767–10777. PMLR (2020)

34. Yedroudj, M., Chaumont, M., Comby, F., Oulad Amara, A., Bas, P.: Pixels-off: data-augmentation complementary solution for deep-learning steganalysis. In: Proceedings of the 2020 ACM Workshop on Information Hiding and Multimedia Security, pp. 39–48. IHMSec 2020, Virtual Conference due to COVID (Formerly Denver, CO, USA), June 2020

35. Yousfi, Y., Butora, J., Fridrich, J., Fuji Tsang, C.: Improving efficientnet for jpeg steganalysis. In: Proceedings of the 2021 ACM Workshop on Information Hiding and Multimedia Security, pp. 149–157. IH&MMSec 2021, Association for Computing Machinery, New York, NY, USA (2021). https://doi.org/10.1145/3437880.3460397

36. Yousfi, Y., Butora, J., Khvedchenya, E., Fridrich, J.: Imagenet pre-trained CNNs for jpeg steganalysis. In: 2020 IEEE International Workshop on Information Forensics and Security (WIFS), pp. 1–6 (2020). https://doi.org/10.1109/WIFS49906.2020.9360897

37. Zagoruyko, S., Komodakis, N.: Wide residual networks. ArXiv abs/1605.07146 (2016)

38. Zhang, C., Bengio, S., Hardt, M., Recht, B., Vinyals, O.: Understanding deep learning (still) requires rethinking generalization. Commun. ACM 64(3), 107–115 (2021)

39. Zou, D., Gu, Q.: An improved analysis of training over-parameterized deep neural networks. Adv. Neural Inf. Process. Syst. 32, 2053–2062 (2019)

Offline Text-Independent Writer Identification Using Local Black Pattern Histograms

Tayeb Bahram[1]($\boxtimes$) [iD] and Réda Adjoudj[2] [iD]

[1] LGACA Laboratory, University Dr Moulay Tahar of Saida, Saida, Algeria
`tayeb.bahram@univ-saida.dz`
[2] EEDIS Laboratory, University Djillali liabes of Sidi Bel Abbes,
Sidi Bel Abbes, Algeria
`reda.adjoudj@univ-sba.dz`

Abstract. The problem of authenticating a writer from his/her writing samples has been the most important and prevalent one subject of active research in the field of handwriting biometrics for the last decade. In this paper, we have focused mainly on the forensic document analysis, more precisely, the offline automatic writer identification in a truly text-independent mode. A new and simple potential textural descriptor has been analyzed for characterizing the handwriting style of the writers, so as to be used to describe the intra and inter-writer variability by calculating the similarity measurements. In order to extract the textural properties from a scanned handwritten sample, an effective statistical texture descriptor is computed from binary connected-components: Local Black Pattern (LBLP). Classification is performed using k-Nearest Neighbors (k-NN) and the Chi-Square (χ^2) distance in a Holdout strategy. The experimental results obtained on two well-known databases show that the proposed scheme achieves a very satisfactory performance and thus reflecting that our approach is still competitive against the state-of-the-art.

Keywords: Feature extraction · Local black pattern · Writer identification · Texture analysis · Offline handwritten text

1 Introduction

Offline handwriting biometrics is the science of identifying a person from a scanned image of a handwritten document by exploiting his/her unique behavioral traits, which can be used as a complement to other behavioral biometrics (e.g., forensic and historical document analyses) [1]. Moreover, it can be considered as biometrics identifier that is unique to the individual like facial patterns, voice or fingerprints. Since each handwritten document contains the identity of its correct author, the purpose of the automatic writer identification is to identify the author of a specific writing sample (i.e., unknown document) from a set of known authorship in a one-to-many search $(1 : N)$ [2].

In the last few years, text-independent writer identification research in the pattern recognition field has matured gradually and become more competitive

M. Salem et al. (Eds.): ICAITA 2022, CCIS 1769, pp. 241–254, 2023.
https://doi.org/10.1007/978-3-031-28540-0_19

on the challenging public handwritten benchmarks with both script languages: Arabic IFN/ENIT [3] and English IAM [4] databases. A number of recent studies rely on extracting texture features from the handwriting sample (whole document [5,6], blocks [7], connected-components [8–12], writing fragments [13], etc.) using the Local Binary Patterns (LBP) [14], Local Ternary Patterns (LTP) [15], Local Phase Quantization (LPQ) [16], joint distribution of run-lengths and Local Binary Patterns ($LBPruns$) [6], Local gradient full-Scale Transform Pattern ($LSTP$) [12], or Modified Local Binary Patterns and the Ink-trace Width and Shape Letters ($MLBP$-$IWSL$) [17] descriptors, which are the most illustrated handcrafted and widely used features for text-independent writer identification [6,7,11–13]. A Probability Distribution Function (PDF or vector of probabilities) in a given document is computed and employed to characterize the writer's style [5]. In this case, the appearance probability; i.e., histogram bin is a key characteristic for each writer and can be employed to distinguish between inter-writer and intra-writer distances (i.e., dissimilarities).

Additionally, several model-based features; i.e., automatic feature extraction and description (deep learning-based) can be employed as distinctive measure to characterize the writing style of each writer [18,19]. In the recent years, there is a significant increase in the number of approaches proposed and published with good results in the topic of automatic writer identification [20,21]. Moreover, and most importantly, all these techniques require a higher training time and need a large part of the functionalities to achieve the optimal classifier.

It is important to note that, in general, the texture-based techniques are preferred when a minimal amount of text (for training and testing) is available and generally are proven to be efficient in terms of execution time [9,17]. In this paper, we propose a method for offline text-independent writer identification by extracting texture features form black-and-white connected-components of handwriting document images exploiting the Local Black Pattern ($LBLP$) descriptor. The main contributions of this study can be listed in the following:

- An effective solution for text-independent writer identification of handwritten documents has been introduced;
- A fast and discriminative texture descriptor, namely Local BLack Pattern ($LBLP$) has been proposed;
- A comprehensive series of experiments on two challenging public databases with diverse languages (Arabic IFN/ENIT and English IAM). The proposed system is competitive against the state-of-the-art techniques for writer identification.

The rest of this paper is organized as follows. The evaluated benchmark databases with their enrollment and evaluation criteria considered in our study are introduced in Sect. 2. Section 3 elaborates the details of the two main steps of texture-based writer identification methodology; i.e., feature extraction and classification (writer identification). In Sect. 4, we present the experimental results of our study with details and compare our system against state-of-the-art methods. The last section drawn a brief summary and gives an outlook on future works and directions.

2   Databases

In order to validate the effectiveness of F_{LBLP} feature in terms of offline writer identification, two well-known and challenging handwritten benchmark databases are considered in our experimental study (Arabic IFN/ENIT and English IAM). These databases, their enrollment, and evaluation protocols are presented in the following:

2.1   IFN/ENIT Database

The IFN/ENIT [3] is one of the biggest and most widely used Arabic handwriting databases, that is used to evaluate the performance of offline writer recognition (i.e., identification and verification) approaches [9]. It contains 26 459 Arabic name samples of 937 Tunisian village and town provided by 411 different writers. The collect pages are scanned at 300 dpi (8 bits/pixel) resolution and available as binary images. For each writer in our experimental setup, 50-word samples are randomly selected. In order to have a fair comparison, this database setup is arranged as similar to the protocol described in [10,13] where 30 and 20 words are randomly selected for the training and testing sets, respectively.

2.2   IAM Database

The IAM [4] is the well-known used modern English dataset for evaluation and validation of writer and handwriting recognition studies [5]. It consists of handwritten text images from 675 writers, where each writer contributes a variable number of text lines, i.e., between 2 and 11 lines per page (number of pages, i.e., between 1 and 59 page per writer). All writing pages are scanned at 300 dpi and available as gray-scale images. Similar to the [5,11] evaluation protocol, we randomly keep one page for training and one page for testing, where writers provide more than one page. In the case of one page per writer, we also split writing page into testing and training sets. By this modification, our dataset contains a total of 1314 handwriting samples from 657 writers.

3   Proposed Technique

Our proposed system uses textural features for automatic writer identification in a text-independent manner. As illustrated in Fig. 1, the proposed framework involves two main steps: 1- feature extraction and 2- writer identification (classification). Each step of our entire process is detailed in the following sub-sections.

3.1   Feature Extraction

Like most of the automated recognition of patterns in data (e.g., speaker identification, document recognition), the image preprocessing process is always applied before the feature extraction step in order to enhance the performance of the

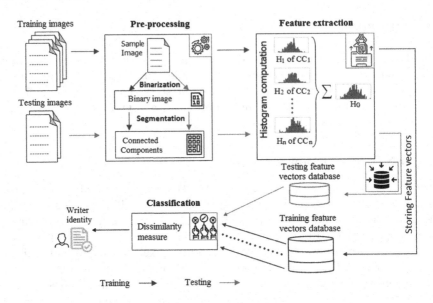

Fig. 1. The framework of our writer identification system.

overall writer identification system. For this purpose, we binarize the handwritten documents (i.e., gray-scale images) by using the Otsu global thresholding method [22]. From the binary image, each black-and-white connected-component (i.e., region of interest) was detected using 8-connected pixels. As stated in [9,10], the black and white pixels corresponding respectively to the ink-trace (foreground) and no-ink (background) in Black-and-White images.

In this step of our system, some non-significant; i.e., common connected-components between writers (e.g., diacritics, scatter noise, periods, dots, commas, and small-components), are removed. Next, for each labeled connected-component in binary image, we compute and extract the Local Black Patterns (*LBLP* descriptors), which are the natural extinctions and modifications of black texture classification and representation.

Local Binary Pattern Code. Most of the textural techniques in the literature employed the basic Local Binary Pattern (*LBP*) descriptor to identify the authorship of the handwriting documents [6,7], which divides the entire image into regions to extract unique and useful texture features. The basic *LBP* code proposed by Ojala et al. [14] in 1996 for texture analysis and description is a fast and computationally efficient texture-based descriptor for grayscale images.

This powerful and most popular model labels each pixel p_c of a 8-bits gray level image I_G by using the value of p_c as a threshold. Figure 2 shows an illustration of the original *LBP* operator, which is determined by comparing the pixel (x_c, y_c) value with its 8 neighbors (x_k, y_k) in a 3×3 neighborhood. A limitation of the simplest *LBP* operator itself is its small 3×3 neighborhood around the

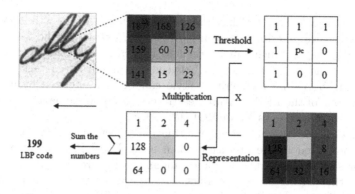

Fig. 2. An example of the basic LBP operator [14].

central pixel: this may leave it unable to detect low frequency patterns. Subsequently, the basic LBP operator was expanded to use different neighborhood sizes, i.e., to capture discriminative features at different scales [14]. For a given pixel p_c, its P neighbors $\{p_k\}$, $_{k=0...P-1}$ are obtained using a bilinear interpolation [17]. Formally, the LBP operator of the center pixel p_c at position (x_c, y_c), can be expressed as:

$$LBP_{P,R}(x_c, y_c) = \sum_{k=0}^{P-1} s(i_k - i_c)\, 2^k \tag{1}$$

where i_c is the 8-bits gray level of the candidate pixel p_c, i_k are the gray values of P surrounding pixels on a circle of radius of R ($R > 0$), and function s is defined as follows:

$$s(i_k - i_c) = \begin{cases} 1, & if \quad i_k \geq i_c \\ 0, & if \quad i_k < i_c \end{cases} \tag{2}$$

Figure 3 illustrates an example of the extended LBP operator using different neighborhood sizes (i.e., different values of P and R).

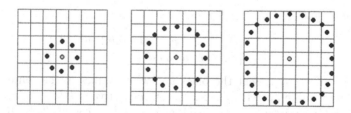

Fig. 3. Examples of the extended LBP operator [14] (three different values of P and R).

The success of the *LBP* has inspired further studies in writer identification. This feature descriptor (texture) is used for extracting the features from handwriting images which can be used to have a measure for the dissimilarity (distance measure, i.e., metric) between two handwritten text images. In addition, an individual handwriting style can be characterized by computing the histograms of the LBP (texture descriptors) from offline handwritten image (Entire image [14], Blocks [7], or ROI [12,13]).

In our view, the use of relevant information present a critical point of many methods presented in the literature when using Local Binary Pattern (LBP) descriptor as a feature extraction and fundamental source information in characterizing the writing style of writers [17]. Specifically, in the approach published by Bahram in 2019 [10], he found that the LBP-based features used for offline writer identification (verification) give an overall idea about the texture information and cannot capture the details in a writing (e.g., character height, orientation, curvature, ink thickness). So, in order to capture dominant descriptors (features) and ignore irrelevant style author information [17], we propose a powerful texture operator named Local Black Pattern (*LBLP*).

Local Black Pattern Code (*LBLP*). Although the gray-scale image could contain additional information like pen pressure on the surface, we chose to work on the pure black-and-white image (binarization of the writing image) which simplifies the representation and feature computation. In black-and-white images of handwriting, the black pixels correspond to the black ink trace (foreground) and the white pixels correspond to the background. Based on this foreground/background separation, the Local Black Pattern (denoted by *LBLP*) is proposed here and it can be computed as follows:

The Local Black Pattern operator labels each pixel p_c of coordinates (x_c, y_c) in a Black-and-White image (i.e., binary connected-components) I_B by examining its P neighbors in a $(2R+1) \times (2R+1)$ neighborhood on the circle of radius, $(n_0, \ldots, n_{P-1})$. The resulting $LBLP$ of a given pixel p_c can be defined as:

$$LBLP_{P,R}(x_c, y_c) = \sum_{k=0}^{P-1} b(i_k - i_c)\, 2^k \qquad (3)$$

The function $b(i_k - i_c)$ is defined as:

$$b(i_k - i_c) = \begin{cases} 1, & if \quad p_k \in BL \ \ and \ \ p_c \in BL \\ 0, & otherwise \end{cases} \qquad (4)$$

where i_c and i_k represent the black-and-white values of the center pixel p_c and its P surrounding pixels $(n_0, \ldots, n_{P-1})$, respectively, BL is the set of black pixels of the binary image I_B, i.e., each pixel in $BL \subseteq I_B$ is called a black pixel and has a value of 1.

For each input handwritten document image, the proposed F_{LBLP} normalized vector is a probability distribution of writer identification feature by the histogram H_{LBLP} and it is defined as follows:

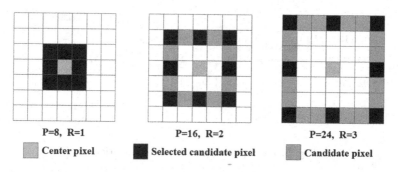

P=8, R=1   **P=16, R=2**   **P=24, R=3**

▨ **Center pixel**   ■ **Selected candidate pixel**   ▨ **Candidate pixel**

Fig. 4. Local information used for $LBLP$ calculation: the circular $(8,1)$, $(16,2)$ and $(24,3)$ neighborhoods.

$$F_{LBLP} = \frac{H_{LBLP}}{||H_{LBLP}|| + \varepsilon} \tag{5}$$

where H_{LBLP} is a histogram computed by using Eq. 6 and ε is a very small value approximately to zero:

$$H_{LBLP}[i] = \sum_{k=1}^{size} \sum_{p_c \in C_k} \delta(i, LBLP_{P,R}), i = 0 \dots P - 1 \tag{6}$$

where $size$ is the number of binary connected components $\{C_k\}_{\varphi(C_k) > \tau}$ in I_B, φ is a function that calculates the number of pixels in a C_k, and τ is a threshold (e.g., $\tau = 120$ pixels for IAM database). Here δ is the Kronecker delta and is given by

$$\delta(i, LBLPP, R) = \begin{cases} 1, & if \quad i = LBLP_{P,R} \\ 0, & otherwise \end{cases} \tag{7}$$

Furthermore, in order to have a more intuitive idea of the proposed $LBLP$ (i.e., necessary information to characterize more intimately the handwriting style of the author of a document), we introduce the use of different neighborhood sizes ($R > 0$), which allows a more accurate description of the Local Black Patterns. For a given neighborhood size R, the $LBLP_{P,R}$ operator produces 2^P different patterns. In this paper, the number of patterns is quantized in 2^8 patterns for every radius R, i.e., from the total number of patterns of a given radius R, we consider only 8 candidate pixels. According to the above proposition, there are two main steps of the proposed candidate pixels detection approach. Firstly, the center pixel p_c is detected. After that, the candidate pixel of each direction in the P neighbors $\{p_i\}$ is obtained. Finally, 8 selected candidate pixels, found on 8 local directions, i.e., horizontal, vertical, and diagonal orientations (right and left), are used to give a sufficiently detailed and robust description of the texture of handwriting. Figure 4 illustrates the candidate pixels detected in an image.

248   T. Bahram and R. Adjoudj

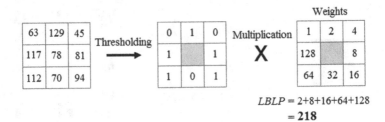

$$LBLP = 2+8+16+64+128$$
$$= 218$$

Fig. 5. Example of *LBLP* code calculation.

The *LBLP* operator is illustrated in Fig. 5. For each input handwritten document image, i.e., training and testing datasets, the proposed F_{LBLP} feature vector is the probability distribution by the normalization of histogram H_{LBLP}. Then, this resulting normalized histogram is computed for each binary image, i.e., representative connected-components $\left(\bigcup_{\substack{N \\ k=1, \varphi(C_k) > \tau}} C_k \right)$. Note that the final F_{LBLP} of black pixels feature vector has a size of 255.

3.2 Writer Identification

As discussed in the previous sections, the offline handwritten samples (training set: $\mathcal{TRAIN} = \{TR_i\}_{1 \leq card(\mathcal{TRAIN})}$, testing set: $\mathcal{TEST} = \{TS_j\}_{1 \leq card(\mathcal{TEST})}$) are represented by their *LBLP* distributions (i.e., *LBLP*-Histograms), where $card(\mathcal{TRAIN})$ and $card(\mathcal{TEST})$ are the number of writing samples in the training and testing data sets, respectively. To precede the identification (classification) step, we need to calculate an appropriate dissimilarity measure (distance) between the *LBLP*-based histograms of compared handwritten samples. Then, the minimum distance $\mathcal{DS}(TS_j, \mathcal{TRAIN})$ between a query writing sample of unknown author TS_j and all of the handwritten images in training set $\mathcal{TRAIN}$ by using a appropriate distance measure and is defined as:

$$\mathcal{DS}(TS_j, \mathcal{TRAIN}) = \min_{1 \leq i \leq card(\mathcal{TRAIN})} \chi^2(F_{LBLP}^j, F_{LBLP}^i) \qquad (8)$$

where F_{LBLP}^j and F_{LBLP}^i represent the normalization histogram feature vectors of $TS_j \in \mathcal{TEST}$ and $TR_i \in \mathcal{TRAIN}$, respectively. In this work, the distance measure used for matching a questioned document TS_j and any other document (TR_i) from the reference base of known writers (training set) for writer identification is Chi-square (χ^2) that can be defined as follows:

$$\chi^2(F_{LBLP}^j, F_{LBLP}^i) = \sum_{x=1}^{N} \frac{(F_{LBLP_x}^j - F_{LBLP_x}^i)^2}{F_{LBLP_x}^j + F_{LBLP_x}^i} \qquad (9)$$

where N is the number of bins in F_{LBLP} (feature histogram dimension). It is worth mentioning that the classification process is performed using the k-Nearest-Neighbor (k-NN) classifier with Chi-Square (χ^2) distance metric.

All distances between the questioned document $TS_j \in \mathcal{TEST}$ and the all writings of training set $TR_i \in \mathcal{TRAIN}$ are sorted and arranged in a ranked hit list. Next, the lowest distance between them is chosen using Eq. 10.

$$Writer(TS_j) = \arg \mathcal{DS}(TS_j, \mathcal{TRAIN}) \tag{10}$$

4 Experiments and Discussion

In this section, we present our empirical results of this study with a detailed evaluation and compared analysis against to the state-of-the-art, i.e., the direct one to one comparison. To show the effectiveness of the $LBLP$-based feature in terms of identification rates, the two benchmark databases (Arabic IFN/ENIT and English IAM) were validated by the "Holdout" strategy and the SoftTop$N$ (or Soft$N$) rates. In this case, the original IFN/ENIT and IAM databases are divided separately into training ($\mathcal{TRAIN}$) and testing ($\mathcal{TS}$) sets (cf. Section 2), which handwritten document is represented by its F_{LBLP} feature vector. Note that the performance measures on the two databases are expressed in the popular SoftTop$N$ identification rate (in percentages %).

4.1 Results and Analysis

The Radius (denoted as R) yielding high-performance is the most important parameter in our writer identification technique. In our experiments, we evaluated

Table 1. Writer identification rates (IR %) of feature vector F_{LBLP} on the Arabic IFN/ENIT (411 writers) database.

Top$k$ accuracy (%) Radius	Top1	Top2	Top3	Top4	Top5	Top6	Top7	Top8	Top9	Top10
1	62.29	77.62	82.00	85.64	89.05	91.24	93.19	94.40	95.13	95.86
2	88.08	95.86	97.08	98.05	98.30	99.27	99.51	99.51	99.51	99.51
3	96.11	98.78	99.76	99.76	100.0	100.0	100.0	100.0	100.0	100.0
4	98.54	99.51	100.0	100.0	100.0	100.0	100.0	100.0	100.0	100.0
5	98.05	99.76	100.0	100.0	100.0	100.0	100.0	100.0	100.0	100.0
6	97.81	100.0	100.0	100.0	100.0	100.0	100.0	100.0	100.0	100.0
7	99.03	99.76	100.0	100.0	100.0	100.0	100.0	100.0	100.0	100.0
8	98.05	99.76	100.0	100.0	100.0	100.0	100.0	100.0	100.0	100.0
9	95.86	99.27	99.27	99.51	99.76	99.76	100.0	100.0	100.0	100.0
10	94.40	98.30	98.78	99/51	99.76	99.76	99.76	99.76	99.76	100.0

the effectiveness of F_{LBLP} feature in the range $\{1, 2, .., 10\}$ for two data sets (training and testing) on IFN/ENIT and IAM databases. The best results

Table 2. Writer identification rates (IR %) of feature vector F_{LBLP} on the English IAM (657 writers) database.

Topk accuracy (%) Radius	Top1	Top2	Top3	Top4	Top5	Top6	Top7	Top8	Top9	Top10
1	74.43	84.78	89.65	92.39	93.61	95.28	95.89	96.35	96.80	96.96
2	86.61	93.00	95.28	96.04	96.65	97.26	97.56	98.02	98.17	98.33
3	92.85	96.19	97.56	98.63	99.24	99.54	99.54	99.70	99.70	99.70
4	96.50	98.33	99.24	99.70	99.70	99.70	99.85	99.85	99.85	99.85
5	97.56	99.39	99.85	99.85	99.85	99.85	99.85	99.85	99.85	99.85
6	97.87	99.85	99.85	99.85	99.85	99.85	99.85	99.85	99.85	99.85
7	97.41	99.09	99.54	99.70	99.70	99.85	99.85	99.85	99.85	99.85
8	97.11	98.78	99.24	99.39	99.70	99.70	99.70	99.85	99.85	99.85
9	95.28	98.02	98.93	99.39	99.54	99.70	99.85	99.85	99.85	99.85
10	94.52	97.41	98.48	98.93	99.09	99.70	99.70	99.70	99.70	99.70

obtained (based on the hyper-parameter R) on the two benchmark databases using the F_{LBLP} feature is tabulated in Tables 1 and 2. Based on these tables, our technique gets the best results (identification rates) for $R = 7$ on the Arabic IFN/ENIT (with scores of Top1 = 99.03%, Top3 = 100.0%) and $R = 6$ on the English IAM (with scores of Top1 = 97.87% and Top6 = 99.85%) databases. The proposed F_{LBLP} achieved encouraging results in terms of correct identification (Top1) on the two databases.

Also, a second analysis is set up to study the influence of the number of writers (size of database) on the identification rate system, when using the textural feature (F_{LBLP}). For this reason, the evaluations are made from 10 to a total number of writers in each benchmark database (411 writer for IFN/ENIT and 657 writers for IAM databases) with a large number of writers. As mentioned previously, the optimum value of the radius R parameter is experimentally kept fixed to 7 and 6 for the Arabic IFN/ENIT and English IAM scripts, respectively. This is illustrated in Fig. 6 where the x- and y-axes represent the number of writers (size database) and the Top1 results, respectively. As this figure shows, there was not a big drop in Top-1 recall rate as the number of writers increased. While the 100.0% recall of the correct writer was obtained for 200 writers of IFN/ENIT and 300 writers of IAM databases by using the F_{LBWP} feature. The 99.03% and 97.87% correct identification rates were obtained on 411 writers of Arabic IFN/ENIT and 657 writers of English IAM databases, respectively.

To confirm the robustness of the proposed system, an analysis of the identification rates as a function of hit list size is presented in Tables 1 and 2. The writer of a query sample is assigned as the author of the document (with the highest similarity, i.e., or the smallest distance) on the Top$N$, $N \in \{1, 10\}$ of the hit list, and Top1,...,Top10 writer identification scores are reported. In view of this, it is interesting to note that, in writer identification searches, an accuracy

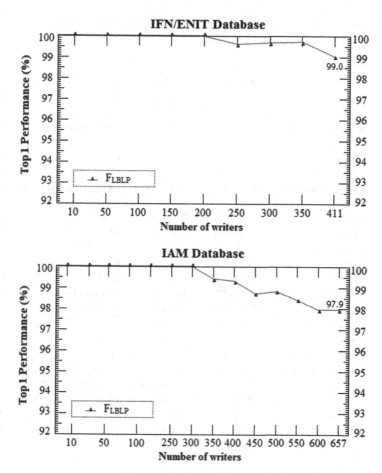

Fig. 6. Writer identification rates (IR %) in function of the number of writers on IFN/ENIT and IAM Databases.

percentage rate of 100.0% on the Top1 hit list, i.e., the correct writer at rank 1 in the hit list is not necessary. The forensic expert can make a final decision based on a correct answer Top5 or Top10 hit list. In our case, our proposed technique provides the highest performance of 100.0% for the three top-ranked writers of IFN/ENIT. Using the same principle, our system realizes an overall identification rates of 99.70% with Top5 and 99.85 with Top6 performances based on the IAM database.

4.2   Comparison with the Literature

In this last decade, a significant number of feature extraction techniques using scanned images of handwriting (typically work with texture and deep learning) have been proposed in the literature. In this section, we compare the performance

Table 3. The comparison of text-independent writer identification performance on the Arabic IFN/ENIT and English IAM databases.

Database	Study	Year	Writers	Feature-based	Feature	IR (%)
IFN/ENIT	Bahram et al. [9]	2016	350	Texture	DL+AL+A/C Co-prob	97.50
	Hannad et al. [13]	2016	411	Texture	LBP	73.48
					LPQ	94.89
	Khan et al. [11]	2019	411	Texture	SIFT+RootSIFT	97.28
	Bahram [10]	2019	411	Texture	LCP-LOC	97.81
	Chahi et al. [12]	2020	411	Texture	LSTP	98.28
	Kumar & Sharma [19]	2020	411	Deep-learning	SEG-WI	98.24
	Semma et al. [21]	2021	411	Deep-learning	FAST+HC+VALD+TE	99.80
	Bahram [17]	2022	411	Texture	MLBP-IWSL	99.27
	Ours	2022	411	Texture	F_{LBLP}	99.03
IAM	Bertolini et al. [7]	2013	650	Texture	LPQ	96.70
	Wu et al. [8]	2014	657	Texture	SDS+SOH	98.50
	Hannad et al. [13]	2016	657	Texture	LBP	65.54
					LPQ	89.54
	He & Schomaker [6]	2017	650	Texture	COLD+BPrun	89.90
	Khan et al. [11]	2019	650	Texture	SIFT+RootSIFT	97.85
	Bahram [10]	2019	657	Texture	LCP-LOC	96.04
	Chahi et al. [12]	2020	657	Texture	LSTP	96.80
	Kumar & Sharma [19]	2020	657	Deep-learning	SEG-WI	97.27
	He & Schomaker [18]	2020	650	Deep-learning	FragNet	96.30
	Lai et al. [20]	2020	657	Deep-learning	PathletPS+bVALD	95.15
	Semma et al. [21]	2021	657	Deep-learning	FAST+HC+VALD+TE	99.50
	Bahram [17]	2022	657	Texture	MLBP-IWSL	98.17
	Ours	2022	657	Texture	F_{LBLP}	97.87

of our new $LBLP$ feature using the Arabic IFN/ENIT and English databases (cf. Section 2) with the recent state of the art approaches (cf. Section 1). Table 3 shows the writer identification rates achieved by using F_{LBLP} feature compared to those obtained by the different state-of-the-art methods. As indicated in Table 3, the recent system proposed by Summa et al. [21] seems interesting and provides the best performance (Top1 identification) (99.80% on 411 Arabic writers of IFN/ENIT and 99.50% on 657 writers of English IAM databases). Note that the authors have evaluated their approach with special evaluation protocols [17] (e.g., for IFN/ENIT 80% and 20% of words are used respectively training and testing sets). On two databases, our F_{LBLP} method surpasses the most successful writer identification works using deep learning introduced in [19] by 0.79% and 0.60% on IFN/ENIT and IAM databases, respectively. On the other hand, as shown in this table, the proposed system clearly surpasses other techniques like those reported in [9,13,18], and [12], where the textural feature was used in characterizing the writing style of an author from his/her scanned document images.

5   Conclusion

The proposed paper investigated the relation between the individual behavioral information using handwriting and the power of feature (i.e., descriptor or measure), where the process is to identify writers from their handwritten documents. We proposed a new feature extraction method: histograms of the Local Black Pattern $(LBLP)$ to capture writer-specific fine details such as: direction, curvature, slant, ink-trace width, interior and exterior character forms, regions confined inside the letters, and the spaces between letters. This texture descriptor is computed from binary writing regions, i.e., black and white connected-components and then normalized to probability distribution F_{LBLP}. Then, the identification of an unknown person on the basis of his/her handwriting images is carried out by using the F_{LBLP} with the nearest neighbor classifier and Chi-square distance measure. The comparisons of our proposed system and the recent existing feature extraction techniques (texture-based and deep learning) prove the robustness of our offline writer identification methodology. In future work, we consider evaluating our technique by using other well-known databases.

Acknowledgements. The authors are grateful to the anonymous referees for their valuable and helpful comments. This research has been carried out within the PRFU project (Grant: C00L07UN220120220001) of the Department of computer science, University Djillali Liabes of Sidi Bel-Abbes. The authors thank the staff of EEDIS and LGACA laboratories for helpful comments and suggestions.

References

1. Franke, K., Köppen, M.: A computer-based system to support forensic studies on handwritten documents. Int. J. Doc. Anal. Recogn. **3**, 218–231 (2001)
2. Srihari, S.N., Cha, S.H., Lee, S., Arora, H.: Individuality of handwriting. J. Forensic Sci. **47**(4), 856–872 (2002)
3. Pechwitz, M., Maddouri, S.S., Märgner, V., Ellouze, N., Amiri, H.: IFN/ENIT database of handwritten Arabic words. In: 7th Colloque International Francophone sur l'Ecrit et le Documentn, CIFED 2002, pp. 129–136. Hammamet, Tunis (2002)
4. Marti, U.V., Bunke, H.: The IAM-database: an English sentence database for offine handwriting recognition. Int. J. Doc. Anal. Recogn. IJDAR **5**, 39–46 (2002). https://doi.org/10.1007/s100320200071
5. Bulacu, M., Schomaker, L.: Text-independent writer identification and verification using textural and allographic features. IEEE Trans. Pattern Anal. Mach. Intell. **29**(4), 701–717 (2007)
6. He, S., Schomaker, L.: Writer identification using curvature-free features. Pattern Recogn. **63**, 451–464 (2017)
7. Bertolini, D., Oliveira, L.S., Justino, E., Sabourin, R.: Texture-based descriptors for writer identification and verification. Expert Syst. Appl. **40**(6), 2069–2080 (2013)
8. Wu, X., Tang, Y., Bu, W.: Offline text-independent writer identification based on scale invariant feature transform. IEEE Trans. Inf. Forensics Secur. **9**(3), 526–536 (2014)
9. Bahram, T., Benyettou, A., Ziadi, D.: A set of features for text-independent writer identification. Int. Rev. Comput. Softw. (I. RE. CO. S) **11**(10), 898–906 (2016)

10. Bahram, T.: A connected component-based approach for text-independent writer identification. In: 2019 6th International Conference on Image and Signal Processing and their Applications (ISPA2019), pp. 1–6. IEEE, Mostaganem, Algeria (2019)
11. Khan, F.A., Khelifi, F., Tahir, M.A.: Dissimilarity Gaussian mixture models for efficient offline handwritten text-independent identification using SIFT and Root-SIFT descriptors. IEEE Trans. Inf. Forensics Secur. **14**(2), 289–303 (2019)
12. Chahi, A., El-merabet, Y., Ruichek, Y., Touahni, R.: Local gradient full-scale transform patterns based off-line text-independent writer identification. Appl. Soft Comput. J. **92**, 106277 (2020)
13. Hannad, Y., Siddiqi, I., El-Kettani, M.E.: Writer identification using texture descriptors of handwritten fragments. Expert Syst. Appl. **47**, 14–22 (2016)
14. Ojala, T., Pietikainen, M., Maenpaa, T.: Multiresolution grayscale and rotation invariant texture classification with local binary patterns. IEEE Trans. Pattern Anal. Mach. Intell. **24**(7), 971–987 (2002)
15. Tan, X., Triggs, B.: Enhanced local texture feature sets for face recognition under difficult lighting conditions. IEEE Trans. Image Process. **19**(6), 1635–1650 (2010)
16. Ojansivu, V., Heikkilä, J.: Blur insensitive texture classification using local phase quantization. In: Elmoataz, A., Lezoray, O., Nouboud, F., Mammass, D. (eds.) ICISP 2008. LNCS, vol. 5099, pp. 236–243. Springer, Heidelberg (2008). https://doi.org/10.1007/978-3-540-69905-7_27
17. Bahram, T.: A texture-based approach for offline writer identification. J. King Saud Univ. Comput. Inf. Sci. (2022). https://doi.org/10.1016/j.jksuci.2022.06.003
18. He, S., Schomaker, L.: FragNet: writer identification using deep fragment networks. IEEE Trans. Inf. Forensics Secur. **15**, 3013–3022 (2020)
19. Kumar, P., Sharma, A.: Segmentation-free writer identification based on convolutional neural network. Comput. Electr. Eng. **85**, 106707 (2020)
20. Lai, S., Zhu, Y., Jin, L.: Encoding Pathlet and SIFT FeaturesWith bagged VLAD for historical writer identification. IEEE Trans. Inf. Forensics Secur. **15**, 3553–3566 (2020)
21. Semma, A., Hannad, Y., Siddiqi, I., Djeddi, C., El-Kettani, M.E.: Writer identification using deep learning with FAST Keypoints and Harris corner detector. Expert Syst. Appl. **184**, 115473 (2021)
22. Otsu, N.: A threshold selection method from gray-level histograms. IEEE Trans. Syst. Man Cybern. **9**(1), 62–66 (1979)

Digital Modulation Classification Based on Automatic K-Nearest Neighbors Classifier

Asmâa Ouessai[1]([✉]) [ID], Abdelkader Tami[2] [ID], and Fatima Abbou[2]

[1] Signals and Images Laboratory, Department of Electronics, University of Sciences and Technology Mohamed Boudiaf, Oran, Algeria
ouessai.as@gmail.com
[2] Technology of Communication Laboratory, Faculty of Technology, University of Saida, Saida, Algeria

Abstract. This paper presents an algorithm for digital modulation classification based on an automatic k nearest neighbors' classifier. The proposed AKNN algorithm attempts to improve the classical kNN classifier accuracy by automatically select the k neighbors for each instance using a model associating the average distances values with the appropriate number of neighbors. The classification procedure was performed over 5 classes of modulation: QPSK, 16PSK, QAM, 16QAM and 4FSK, using cumulants features. In addition to the avoidance of the manual selection of K neighbors, the proposed algorithm outperforms the traditional kNN in terms of accuracy rate mainly in the case of low SNR.

Keywords: Digital modulation · Classification · kNN · Automatic kNN

1 Introduction

Automatic modulation classification systems (AMC) aim to identify the modulation schemes of unknown received signal in order to accurately perform the demodulation task. This process is becoming more and more important for the intelligent communication systems and many civil and military applications.

The AMC system is generally divided into two approaches: maximum likelihood (ML) based and features based (FB) modulation identification. The ML approach estimates the probability density function of the received signal, although it provides optimal solution but it is considered computationally expensive in contrast to the FB method which is less complex [1]. FB approach consists of two steps: features extraction in which some features are extracted from the received signal, and the classification which defines the predicted modulation scheme using a selected machine learning classifier.

K nearest neighbors' classifier is widely used in AMC systems [1–5] due to its simplicity and good accuracy. To improve the performance of this algorithm several works have been proposed in the literature.

In [5] the authors tested different kNN distances to calculate the K nearest neighbors, however, no improvements have been proposed to enhance the efficiency of this algorithm and to mitigate its drawbacks. The authors used in this work a modulated

© The Author(s), under exclusive license to Springer Nature Switzerland AG 2023
M. Salem et al. (Eds.): ICAITA 2022, CCIS 1769, pp. 255–263, 2023.
https://doi.org/10.1007/978-3-031-28540-0_20

signals dataset, they concluded that the Mahalanobis distance slightly outperforms the Minkowski and the Euclidean distances in terms of classification accuracy.

Starting from the assumption that the use of different K values can have a large impact on the predictive accuracy of the kNN, the authors in [6] used different folds of cross validation to find the best value of K, the proposed method was applied to seventeen data sets gathered from the UCI machine learning repository [7]. The obtained results showed that the proposed approach improves the computational performance of the algorithm, unfortunately the classification accuracy was less than the kNN.

Hajizadeh Rassoul et al. [8] proposed two techniques to improve the performance of the kNN: mutual neighbourhood (MN) and modified majority voting (MMV) based decision, on basis of these techniques some new and extended kNN-based classifiers were introduced, the authors used different kinds of databases to investigate the performance of their algorithms.

In [9] the authors presented the DCT-kNN (class contribution and feature weighting) algorithm to improve the kNN classifier, the weights of features were calculated using two accuracies: accuracy of the original dataset using the kNN and the data lack of each dimension feature. Then a weighted distance was calculated in order to obtain the nearest neighbor. Certain degree of improvement in classification accuracy was obtained on the UCI datasets.

Zhibin Pan et al. [10] proposed a new KNN-based classifier, called multi-local means-based k-harmonic nearest neighbor (MLM-KHNN) rule, in where they defined firstly, the number of k-nearest neighbor for each class which was used then to compute k different local mean vectors in order to compute their harmonic mean distance to the query sample. The algorithm was applied on twenty real-world datasets from UCI and KEEL and compared with several kNN algorithms.

In [11] the authors proposed a new kNN algorithm to tackle the problem of the inherent inefficiency of the conventional kNN algorithm. The proposed method combines the use of feature learning techniques, clustering methods, adaptive search parameters per cluster, and the use of pre-calculated K-Dimensional Tree structures. The results show that the proposed algorithm significantly outperforms 16 state-of-the-art efficient search methods while still depicting such an accurate performance as the one by the exhaustive kNN search.

In [12] two improved algorithms, namely KNN^{TS}, and $KNN^{TS-PK}+$ were proposed. These two algorithms used PK-Means++ algorithm which is the essence of KNN^{PK+} algorithm, thus the KNN algorithm is improved on the premise of stabilizing the classification efficiency. KNN^{TS} algorithm is proposed to improve the classification efficiency of KNN algorithm on the premise of the classification accuracy remains unchanged.

Kuan-Chun Huang et al. [13] used a genetic programming to learn the transformation function which develop a differential scalar for measuring the distance between data instances for a dataset, two forms of transformation functions are considered by the authors. The proposed method can effectively enhance the performance of kNN.

In this work we propose to reduce the inefficiency of the kNN algorithm related to the manual selection of the K neighbor's number, thus a simple algorithm is presented to automatically select an appropriate K variable based on a model developed using the linear regression method which connects a previously calculated mean distances and the corresponding k value based on a learning dataset. To our knowledge, none of the previously cited works have proposed a model which connects the distances values with the K neighbor's number.

The proposed algorithm has been evaluated using a dataset including five types of modulation schemes and different value of signal to noise ratio (SNR).

The automatic selection of the k nearest neighbors' task has been evaluated using a modulated signals dataset, knowing that it's important to investigate the efficiency of an improved kNN algorithm over this kind of dataset since it contains several types of information which are usually stained with noise.

The rest of the paper is organised as follow: in Sect. 2 we presented the adopted AMC system and its different components as well as the kNN classifier and the proposed algorithm. In Sect. 3 a detail about the used dataset is firstly presented then we discussed the experimental results. Finally, some conclusions are drawn in Sect. 4.

2   The Proposed Approach

The AMC system involves three stages: signal detection, features extraction and classification as it is presented in Fig. 1.

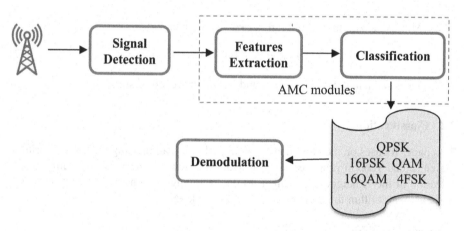

Fig. 1. AMC architecture

The received signal is firstly detected then some features are extracted, in this work we calculated the higher order cumulants (HOC) which are then, applied to the input of the kNN classifier and the automatic kNN (AKNN). We get at the output of the classifiers one class among the 5 classes of modulation considered (QPSK, 16PSK, QAM, 16QAM and 4FSK). Once the modulation type is recognized the signal can be finally demodulated and decoded.

2.1 Features Extraction

In this paper, we used Higher Order cumulants (HOC) which are mathematical tools used to describe the higher-order statistical properties of random process, HOC are broadly utilized in classification of various ASK, PSK, and QAM signals. Cumulants are made up of moments of received signals. For a complex valued stationary signal, the cumulants used in this work are defined as below.

$$C_{20} = |M_{20}| \tag{1}$$

$$C_{21} = M_{21} \tag{2}$$

$$C_{40} = M_{40} - 3M_{20}^2 \tag{3}$$

$$C_{41} = M_{40} - 3M_{20}M_{21} \tag{4}$$

$$C_{42} = M_{42} - |M_{20}|^2 - 2M_{21} \tag{5}$$

$$C_{60} = M_{60} - 15M_{20}M_{40} - 30M_{20}^3 \tag{6}$$

where M is the kth moments of the received signal given by:

$$M_{pq} = E[X(t)^{p-q}X^*(t)^q] \tag{7}$$

E () is a mathematical expectation and X(t) is the received signal.

2.2 Classification

This paper focuses on the classification of digital modulation using the kNN algorithm, the enhanced version proposed aims to avoid the manual intervention relating to the selection of the K neighbor's number. In the following sections, we present firstly the basic kNN algorithm then the proposed automatic kNN.

K Nearest Neighbors

The K nearest neighbors classifier belongs to the family of the supervised classifier, however no learning phase is taking place and no learning model is proposed unlike many other methods (such as neural networks, kernel methods, etc.).

The main idea behind the kNN is based on the selection of one or a number of features which prove their similarity to a feature with unknown class. The number of the selected features is chosen arbitrary as well as the similarity metric.

The kNN algorithm applied to the modulation classification is given bellow.

Algorithm: k-nearest neighbor
Input: features dataset (cumulants), classes of the learning dataset, K
Output: modulation classes of testing dataset (A)
Split the features dataset on 75% for leaning and 25% for testing
For i =1 to length(A)
Calculate the distance between each instance in A and all instances in the learning dataset.
Sort the distances vector
Select the k-nearest neighbors
Select the corresponding class
End

Automatic K Nearest Neighbor

The proposed automatic K nearest neighbors' diagram is presented in Fig. 2.

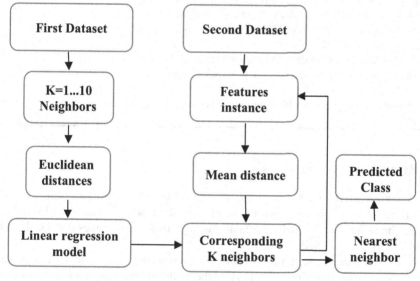

Fig. 2. Flowchart of the proposed automatic kNN classifier

We propose, in this paper, to develop a model connecting K (neighbors number) with the mean distance based on the assumption that the variation of a mean distance depends on the number of neighbours.

In addition to the principal dataset used to evaluate the proposed algorithm (second dataset), we proposed to use a supplementary dataset (first dataset) in order to develop the linear model. In this paper we consider that the K number vary from 1 to 10, for each selected value we calculate the corresponding mean Euclidean distance of the neighbours. The obtained distances and the corresponding K value were used to develop a model using a linear regression approach.

On the other hand, for each instance of the principal dataset (second dataset) we calculate the average distance (using only K = 1 and K = 10) and incorporate it into the previous model to get the number of the nearest neighbors (K).

The proposed AKNN algorithm is presented below.

Algorithm: Automatic k-nearest neighbor
___
First step
Input: first dataset (cumulants, classes)
Output: linear model
Split the features dataset on 75% for leaning and 25% for testing
For k=1 to 10
    Calculate the mean distances
End
Find the linear model that fits the data (k, distances)
Second step
Input: second dataset (features (cumulants), classes)
Output: k, modulation classes of the testing dataset (B)
Split the dataset to 75% for leaning and 25% for testing
For i= 1 to length (B)
    Calculate the mean distance corresponding to k=1 and k=10
    Incorporate the distance in the previous model
    Select the k-neighbors
    Select The nearest neighbor and the corresponding class
End

---

3   Results and Discussions

In this section, we present the evaluations results of the proposed Algorithm.

The dataset used in this work for developing the model contains 1000 simulated signal. The signals were equally modulated using 5 different schemes (QPSK, 16PSK, QAM, 16QAM and 4FSK).

We assume that the K and the distances are connected using a linear relationship, however the linear regression is used to find the model coefficients that fit the variables. The Euclidean distance is calculated in this work between each two instances.

Figures 3 and 4 show five "K-distances" models developed in various environments characterized by different signal to noise ratios (SNR).

In the cases of SNR = 10 dB and SNR = 5 dB the noise is considered lower than the signal intensity, then the mean distances belong to the range [1.2,4], however, in case of SNR = 0 dB the mean distances range was [2, 5]. On other hand, it is noticed that decreasing the SNR increases the mean distance.

High distance value may correspond to the effect of noise on the signal and thus on the values of the calculated cumulants.

It's clearly observed in Figs. 3 and 4 that the developed models accurately fit the data mainly in case of SNR = −10 dB, this mean that the proposed linear relationship is positive as the two variables increase together.

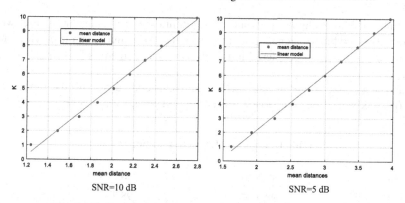

Fig. 3. K-distances models for SNR = 10 dB and SNR = 5 dB

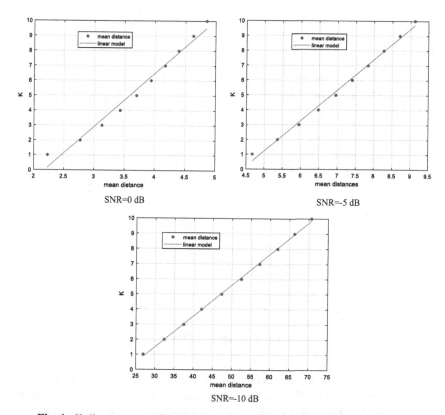

Fig. 4. K-distances model for SNR = 0 dB, SNR = −5 dB and SNR = −10 dB

The proposed models are used in the next step to estimate the k neighbors value corresponding to the average distance obtained.

In the order to evaluate the proposed algorithm we used a second dataset which includes 4000 modulated signals generated by MATLAB simulation, the previous modulation classes are also considered in this dataset, hence, there is 800 signals in each class.

Table 1 and Fig. 5 present the AKNN accuracies compared with the conventional kNN (using k = 1). The comparison was performed using various SNR.

Table 1. KNN vs AKNN classification results

SNR (dB)	kNN (%)	AKNN (%)
10	98.9	98.7
5	98.2	98.2
0	91.1	91.4
−5	75.7	78
−10	69.8	70.7

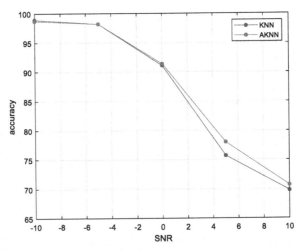

Fig. 5. KNN vs AKNN

The kNN and AKNN accuracies are comparable in cases of high SNR, however the AKNN outperforms the kNN when the SNR decreases (SNR = –5 dB and SNR = 10 dB), this mean that the proposed AKNN algorithm is less sensitive to the noise which makes it suitable for the problem of automatic classification of digital modulated signals.

4 Conclusion

This article discusses a troublesome problem in the kNN algorithm which concern the manual selection of neighbor's number. Automating this task is not trivial, because classical kNN does not rely on a learning model, hence, in this work, we attempted to develop a learning model that connects the average distances and the number of neighbors. The proposed model was able to fit the data accurately, mainly in the case of low SNR. Additionally, it was used for a new dataset to dynamically provide the appropriate k number for each instance based on the average distance. The obtained results are encouraging and open the way for further improvements, particularly in the modeling stage.

References

1. Rao, N.V., Krishna, B.T.: Automatic modulation recognition using machine learning techniques: a review. In: Advances in VLSI, Signal Processing, Power Electronics, IoT, Communication and Embedded Systems, pp. 145–154 (2021)
2. Hazza, A., Shoaib, M., Alsheabili, S.A., et al.: An overview of feature-based methods for digital modulation classification. In: 1st International Conference on Communications, Signal Processing, and their Applications (ICCSPA), pp. 1–6. IEEE (2013)
3. Kharbech, S., Dayoub, I., Zwingelstein-Colin, M., et al.: On classifiers for blind feature-based automatic modulation classification over multiple-input–multiple-output channels. IET Commun. 10(7), 790–795 (2016)
4. Huang, Y., Jin, W., Li, B., et al.: Automatic modulation recognition of radar signals based on Manhattan distance-based features. IEEE Access 7, 41193–41204 (2019)
5. Ghauri, S.A.: KNN based classification of digital modulated signals. IIUM Eng. J. 17(2), 71–82 (2016)
6. Hulett, C, Hall, A., Qu, G.: Dynamic selection of k nearest neighbors in instance-based learning. In: IEEE 13th International Conference on Information Reuse & Integration (IRI). IEEE (2012)
7. Frank, A., Asuncion, A.: UCI machine learning repository (2010)
8. Hajizadeh, R., Aghagolzadeh, A., Ezoji, M.: Mutual neighbourhood and modified majority voting based KNN classifier for multi-categories classification. Pattern Anal. Appl. 1–21 (2022).
9. Huang, J., Wei, Y., Yi, J., et al.: An improved kNN based on class contribution and feature weighting. In: 2018 10th international conference on measuring technology and mechatronics automation (ICMTMA), pp. 313–316. IEEE (2018)
10. Pan, Z., Wang, Y., Ku, W.: A new k-harmonic nearest neighbor classifier based on the multi-local means. Expert Syst. Appl. 67, 115–125 (2017)
11. Gallego, A.J., Rico-Juan, J.R., Valero-Mas, J.J.: Efficient k-nearest neighbor search based on clustering and adaptive k values. Pattern Recognit. 122, 108356 (2022)
12. Wang, H., Xu, P., Zhao, J.: Improved KNN algorithms of spherical regions based on clustering and region division. Alexandria Eng. J. 61(5), 3571–3585 (2022)
13. Huang, K.-C., Wen, Y.-W., Ting, C.-K.: Enhancing k-nearest neighbors through learning transformation functions by genetic programming. In: 2019 IEEE Congress on Evolutionary Computation (CEC), pp. 1891–1897. IEEE (2019)

Improving Multi-class Text Classification Using Balancing Techniques

Laouni Mahmoudi$^{(\boxtimes)}$ (ID) and Mohammed Salem (ID)

Mustapha Stambouli University of Mascara, 29000 Mascara, Algeria
{laouni.mahmoudi,salem}@univ-mascara.dz

Abstract. Social media platforms and micro-blogging websites have grown in popularity in recent years. These platforms are used to express persons' thoughts and feelings regarding items, people, and events. This massive amount of textual data must be exploited. Sentiment analysis is one of the tools used to take advantage of this text data, in which we classify text into different classes such as positive, negative, neutral, or a number of star classes. It has been investigated by many researchers in several languages. Deep Learning approaches such as CNN, RNN, and LSTM applied on balanced datasets have given efficient results compared to classical machine learning approaches such as SVM, NB, and LR. Furthermore, the apparition of BERT has revolutionized the text classification field, even in sentiment analysis tasks. The main problem that the datasets which have been collected from social media platforms, certain classes dominate others, meaning that the datasets are imbalanced. As a result, classifiers lose efficiency. This paper addresses this issue by introducing an ensemble of mathematical balancing techniques to increase the efficiency of sentiment analysis models based on BERT scheme. The obtained results are significant, indicating that our two main metrics, AVG-Recall and F1-PN, are 17% and 19% higher, respectively, when compared to the classifiers' results applied to the imbalanced dataset.

Keywords: Text classification · Sentiment analysis · BERT · araBERT · Imbalanced · Oversampling · Undersampling · Hybrid

1 Introduction

Text classification is a well-known subject in natural language processing [1]. It is one of the most active study topics, especially with the development of social media platforms, where a tremendous amount of text in various domains is created in a matter of seconds. In this study, we focused on sentiment analysis, which is a subfield of text classification.

The goal of sentiment analysis is to forecast and categorize feelings expressed in comments or tweets into categories such as positive, negative, and neutral [2]. In the literature, many strategies have been used to address this issue. Deep learning and related algorithms are becoming increasingly popular [3,4]. The

M. Salem et al. (Eds.): ICAITA 2022, CCIS 1769, pp. 264–275, 2023.
https://doi.org/10.1007/978-3-031-28540-0_21

act of preprocessing is crucial to the creation of models [5]. To feed machine learning algorithms, words and texts must be vectorized.

A suitable data set is required to develop effective deep learning models. A significant exception in the sentiment analysis is that some classes dominate others. For instance, neutral comments are the most common in social media, which means the collected datasets are essentially imbalanced [6,7]. Recent research has used a variety of approaches to overcome this drawback.

Approaches such as oversampling, undersampling, and hybrid balance have been used. Everyone has advantages and disadvantages [7,8]. We compare these strategies in text classification, particularly in sentiment analysis, where we apply a prominent preprocessing methodology and the araBERT classifier variant of BERT algorithm in our study. The F1-PN and AVG-Recall measures are used to evaluate our system.

The rest of the paper is organized as follows: The paper's second section reviews the available literature. Section 3 provides an overview of the baseline algorithm and methodology used in the present study. The suggested balancing techniques are presented in Sect. 4, while the experiments, results, and discussion are presented in Sect. 5. Finally, conclusion is given.

2 Related Work

In recent years, sentiment analysis has grown in popularity as a study area. Several techniques are used in this field. The most common methodologies identified in the literature are: Lexicon-based, Machine Learning, and Deep Learning.

The Lexicon-based technique predicts text feelings based on word polarity [9]. SentiWordNet and SentiStrength 3.0 are used by the authors of [10] and [11], respectively. They did not provide satisfactory results due to the complexity of words with various meanings.

The machine learning approach is primarily based on supervised learning techniques. The basic idea behind machine learning is to divide the dataset into two parts: learning and testing [12,13]. The text features are then manually extracted and fed into the classifiers. The model is tested using the second subset. The authors of [14] employed the LR and SVM algorithms to predict mobile reviews. In [15], authors used several classifiers such as SVM and RBF to classify the polarity of tweets collected from diverse areas.

The deep learning approach is similar to the machine learning approach except that features are not extracted by hand; instead, texts are preprocessed and fed directly to the classifier. [16] proposes detecting hate speech on Twitter using DNN. The dataset for this study was collected from six publically available sources. A mixed model of CNN and GRU are used to classify hate speech on Twitter [17]. Several studies have been conducted to examine the impact of various feature selection methods on classification accuracy. For example, text classification is done using BiLSTM and CNN in [18].

Although the high accuracy of sentiment classification, the research mentioned above has several limitations. For example, the studies often use imbalanced datasets, and consequently, the reported accuracy is higher than the F1-score. An F1 score is preferred as metric on imbalanced datasets, which is very low in the research works discussed. The machine learning algorithms can be overtrained for the majority class on highly imbalanced datasets. Previous studies do not focus on balancing the datasets, and their results may be biased due to the model's overfitting [19]. Predominantly, the proposed approaches follow BERT schemes that are data-intensive [1,20,21], and their accuracy is affected when they use imbalanced datasets. Hence, this study leverages the balancing techniques to perform sentiment analysis and overcomes the issues mentioned above.

3   Deep Learning for Sentiment Analysis

In the literature on sentiment analysis using deep learning approaches, a baseline algorithm is applied to deal with this task. As seen in Fig. 1, this method consists of two sections: processing and model creation.

In our approach a new section named balancing was added and the experiments were carried out using a quantitative research method (see Fig. 2). The first step was preprocessing, which included cleaning and converting text to numerical values. The second step is the primary contribution of our research, which is the use of various balancing methods to obtain appropriate datasets. Finally, we used a deep learning algorithm to create models for each balancing approach and compare the results.

3.1   Preprocessing

Because of the nature of writing on social media platforms, preprocessing is a critical operation. Its goal is to remove noise and words that are not properly written with alphabetical characters, as well as symbols such as Hashtags and URLs, among others. In this step, the use of Natural Language Processing is absolutely necessary.

3.2   Vector Representation

A vector representation is required to use Machine Learning algorithms, including deep learning in order to perform certain mathematical operations. In the literature, a variety of strategies were utilized to attain this purpose, including: Bag of Words (BoW) [22], Term Frequency-Inverse Document Frequency (Tf-IDF), and complicated representations that use semantics and context-named Word Embeddings like FastText [23], GloVe [24], and Word2Vec [25].

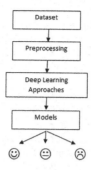

Fig. 1. Baseline methodology

3.3 Deep Learning

Traditional machine learning models, such as support vector machine (SVM), logistic regression (LR), and so forth, require explicit features to be created, making it difficult to select the most important words that predominate the sentences or text. As a result, the accuracy of models varies between datsets. To solve this problem Deep learning approaches are required because they apply automatic techniques to select features.

Numerous methods are employed in text classification, including LSTM, BIL-STM, GRU, CNN, and the evolutionary approach BERT, which stands for (Bidirectional Encoding from Transformer) and offers the greatest accuracy in balanced datasets. Thus, we used this last one in our experiments. However, in imbalanced datasets, all approaches including BERT produce low accuracy due to the overfitting problem caused by text features. This prompted us to include a module called Balancing in the baseline algorithm to address the issue as shown in Fig. 2.

The balancing techniques are discussed in depth in the following section to emphasize the features of each one chosen.

4 Balancing Approaches

In imbalanced datasets, certain classes have far more instances than others. Especially on social media platforms where neutral phrases predominate. As a result, deep learning model creation does not provide good accuracy. There are three types of techniques that can be used to address this issue: Oversampling, Undersampling and Hybrid approach.

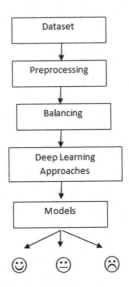

Fig. 2. The new baseline.

4.1   Oversampling

The oversampling is the process of increasing the number of samples from the minority class by replicating some instances or creating new instances from existing ones until a desired balancing ratio R reached which is calculated with the Eq. 1, while $X_{minority}$ is the minority class and $X_{majority}$ is the majority [26,28].

$$R(X) = \frac{X_{minority}}{X_{majority}} \tag{1}$$

The aim is to create new samples from existing ones. There are two techniques for balancing with oversampling: random and synthetic.

The first method is to collect random samples from the minority class and duplicate them until they reach a certain ratio when compared to the majority class. The problem with this method is that it is prone to overfitting due to the use of the same texts in balancing. We can reduce the possibility of overfitting by decreasing the ratio.

The second technique, known as Synthetic Minority Over-Sampling Technique (SMOTE) [27], generates samples through interpolation, which generates new data points within the range of known data points. The minority class is oversampled by creating synthetic samples rather than extracting data at random, which avoid duplication.

4.2 Undersampling

The undersampling method's aim is to reduce the number of samples by eliminating instances from the majority class. This approach have two principles Random, and Boundary [29].

The first is a naive technique based on randomly removing samples until a desired ration R is reached. If R equals one, the class is perfectly balanced. The disadvantage of this method is that we can omit important details that can aid in the creation of a good model.

The second technique extract samples at the boundary of two or more classes. it use an algorithm named Condensed Nearest Neighbours (CNN) [30]. The algorithm tends to select points near the boundary between the classes and transfer them to the new group.

4.3 Hybrid Method

The hybrid method combines the advantages of undersampling and oversampling techniques, allowing us to amplify the minority class while removing any noisy observations.

The next section investigates an ensemble of experiments to compare balancing techniques.

5 Experiments

5.1 Experiment Processes

In this section, we describe the experiments that were conducted and the results that were obtained regarding the accuracy of our models without and with balancing techniques. Python is used to implement all of the codes.
Dataset:
We used an imbalanced dataset known as ASAD (Twitter-based Benchmark Arabic Sentiment Analysis Dataset) in the experiments, which was released for a sentiment analysis challenge sponsored by KAUST. The dataset contains 55000 tweets that have been manually labeled as negative, positive, or neutral, with numeric values of 8820, 8821, and 37359 tweets, respectively. We divided the ASAD dataset into two subsets. The first contains 80% of the data for training, while the second contains the remaining 20% for testing.

All experiments go through the same preprocessing step, which includes cleaning the text of noise, hashtags, URLs, and other symbols such as imogis and non-alphabetical characters. We used fine-tunning BERT, Google's most popular language representation model, for tweet representation vectors and classifications. Experimental results were evaluated using two principle metrics: AVG-Recall and F1-PN.

The four experiments are described in detail in the following subsections. The first uses fine-tunning BERT directly on the imbalanced dataset, while the others use balancing techniques such as oversampling, undersampling, and hybridization.

5.2   Experiment Results

The experimental findings are shown in Tables 1, 2, 3, and 4. They reveal that when compared to the previous tests, the imbalaced dataset (ASAD) fed to BERT has a poor classification accuracy with an AVG-Recall of 0.6 and an F1-PN of 0.54. Furthermore, as shown, each balancing approach works differently, Hybridization clearly outperforms the other two techniques with more than 15% in AVG-Recall and 19% in F1-PN. Furthermore, we noticed that the results of each balancing strategy vary. SMOTE outperforms the random technique in terms of oversampling, but boundary removal outperforms the random method.

Table 1. Without balancing.

	Imbalanced dataset
F1-Positive	0.54
F1-Negative	0.55
F1-Neutral	0.85
Macro-F1	0.65
Positive-Recall	0.46
Negative-Recall	0.45
Neutral-Recall	0.89
AVG-Recall	**0.60**
F1-PN	**0.54**

Table 2. Oversampling balancing.

	Random	SMOTE
F1-Positive	0.66	0.68
F1-Negative	0.66	0.69
F1-Neutral	0.87	0.89
Macro-F1	0.73	0.75
Positive-Recall	0.63	0.67
Negative-Recall	0.64	0.68
Neutral-Recall	0.88	0.89
AVG-Recall	**0.71**	**0.75**
F1-PN	**0.65**	**0.68**

Table 3. Undersampling balancing.

	Random	Boundary
F1-Positive	0.67	0.69
F1-Negative	0.68	0.70
F1-Neutral	0.87	0.89
Macro-F1	0.74	0.76
Positive-Recall	0.64	0.66
Negative-Recall	0.63	0.68
Neutral-Recall	0.87	0.86
AVG-Recall	**0.71**	**0.73**
F1-PN	**0.67**	**0.69**

Table 4. Hybrid balancing.

	Random and Random	SMOTE and Boundary
F1-Positive	0.73	0.75
F1-Negative	0.74	0.74
F1-Neutral	0.88	0.87
Macro-F1	0.78	0.78
Positive-Recall	0.72	0.74
Negative-Recall	0.71	0.73
Neutral-Recall	0.87	0.88
AVG-Recall	**0.76**	**0.78**
F1-PN	**0.73**	**0.73**

5.3   Discussion

The findings presented in the Result section demonstrate how to deal with the imbalanced issue in text classification. A set of initial benchmark results were obtained using the 'Original' data (imbalanced data) (see Table 1). Table 2, 3, and 4 show the AVG-Recall and F1-PN of the different balancing techniques when used on the "Original" ASAD dataset.

From the results, it can be seen that the hybrid balancing technique based on the SMOTE and boundary give better results than other balancing techniques, with more than 18% AVG-Recall and 19% F1-PN.

From Table 4, we can observe that hybrid techniques outperform all other oversampling methods for both AVG-Recall and F1-PN metrics using fine-tuned BERT. The absolute best results are achieved when SMOTE is combined with boundary. It is vital to notice that the accuracy score shows the well-known bias towards the majority class, as discussed in Sect. 2. In a multiclass classification problem with an imbalanced dataset, where the prediction of all the classes is of equal importance, as in many remote sensing applications, accuracy should be of secondary importance compared to more robust metrics, such as F-score and AVG-Recall.

In Table 3, the rankings of the undersamplers are presented and show the superiority of both random and boundary-comparable techniques to the imbalanced dataset. Although undespaling balances the dataset, it is clear from the tables that they produce suboptimal results.

Table 2 compares the performance of the two oversampling methods, SMOTE and random. The findings are greater than in the first experiment and nearly equivalent to the undersampling approach results (see Fig. 3)

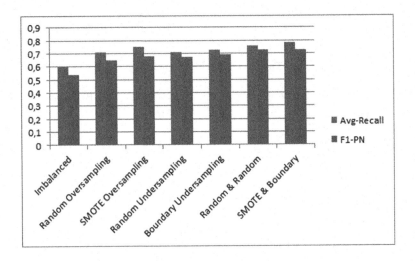

Fig. 3. Results' illustration.

6   Conclusion

This study analyzes the performance of various balancing techniques to perform text classification (i.e., sentiment analysis) using an imbalanced dataset named ASAD, which contains three classes dominated by the neutral class. In addition, we used the efficient text classifier araBERT, a well-known variant of BERT.

Balancing approaches were applied after the embedding process, as compared to text augmentation. Furthermore, we propose balancing after the representation generated by BERT. Moreover, we want to do additional experiments on more datasets for text classification, as well as tests on multi-domain datasets.

References

1. Yu, B., Deng, C., Bu, L.: Policy text classification algorithm based on BERT. In: 2022 11th International Conference of Information and Communication Technology (ICTech), pp. 488–491 (2022). https://doi.org/10.1109/ICTech55460.2022.00103

2. Yang, J., Yang, J.: Aspect based sentiment analysis with self-attention and gated convolutional networks. In: 2020 IEEE 11th International Conference on Software Engineering and Service Science (ICSESS), pp. 146–149 (2020). https://doi.org/ 10.1109/ICSESS49938.2020.9237640

3. Ertam, F.: Deep learning based text classification with Web Scraping methods. In: International Conference on Artificial Intelligence and Data Processing (IDAP), pp. 1–4 (2018). https://doi.org/10.1109/IDAP.2018.8620790

4. Alsukhni, B.: Multi-label Arabic text classification based on deep learning. In: 2021 12th International Conference on Information and Communication Systems (ICICS), pp. 475–477 (2021). https://doi.org/10.1109/ICICS52457.2021.9464538

5. Salur, M.U., Aydin, İ.: The impact of preprocessing on classification performance in convolutional neural networks for Turkish text. In: International Conference on Artificial Intelligence and Data Processing (IDAP), pp. 1–4 (2018). https://doi. org/10.1109/IDAP.2018.8620722

6. Zhang, H., Li, Z., Shahriar, H., Tao, L., Bhattacharya, P., Qian, Y.: Improving prediction accuracy for logistic regression on imbalanced datasets. In: 2019 IEEE 43rd Annual Computer Software and Applications Conference (COMPSAC), pp. 918–919 (2019). https://doi.org/10.1109/COMPSAC.2019.00140

7. Hanskunatai, A.: A new hybrid sampling approach for classification of imbalanced datasets. In: 2018 3rd International Conference on Computer and Communication Systems (ICCCS), pp. 67–71 (2018). https://doi.org/10.1109/CCOMS.2018. 8463228

8. Hanif, A., Azhar, N.: Resolving class imbalance and feature selection in customer churn dataset. In: International Conference on Frontiers of Information Technology (FIT), pp. 82–86 (2017). https://doi.org/10.1109/FIT.2017.00022

9. Raj, R.J.R., Das, P., Sahu, P.: Emotion classification on Twitter data using word embedding and lexicon based approach. In: 2020 IEEE 9th International Conference on Communication Systems and Network Technologies (CSNT), pp. 150–154 (2020). https://doi.org/10.1109/CSNT48778.2020.9115750

10. Agarwal, A., Sharma, V., Sikka, G., Dhir, R.: Opinion mining of news headlines using SentiWordNet. In: Symposium on Colossal Data Analysis and Networking (CDAN), pp. 1–5 (2016). https://doi.org/10.1109/CDAN.2016.7570949

11. Rabab'ah, A.M., Al-Ayyoub, M., Jararweh, Y., Al-Kabi, M.N.: Evaluating SentiStrength for Arabic sentiment analysis. In: 2016 7th International Conference on Computer Science and Information Technology (CSIT), pp. 1–6 (2016). https:// doi.org/10.1109/CSIT.2016.7549458

12. Zheng, Y.: An exploration on text classification with classical machine learning algorithm. In: 2019 International Conference on Machine Learning, Big Data and Business Intelligence (MLBDBI), pp. 81–85 (2019). https://doi.org/10.1109/ MLBDBI48998.2019.00023

13. Venkatesh, Ranjitha, K.V.: Classification and optimization scheme for text data using machine learning Naïve Bayes classifier. In: 2018 IEEE World Symposium on Communication Engineering (WSCE), pp. 33–36 (2018). https://doi.org/10.1109/ WSCE.2018.8690536

14. Pathuri, S.K., Anbazhagan, N., Prakash, G.B.: Feature based sentimental analysis for prediction of mobile reviews using hybrid bag-boost algorithm. In: 2020 7th International Conference on Smart Structures and Systems (ICSSS), pp. 1–5 (2020). https://doi.org/10.1109/ICSSS49621.2020.9201990

15. Dhahi, S.H., Waleed, J.: Emotions polarity of tweets based on semantic similarity and user behavior features. In: 2020 1st Information Technology to Enhance e-

Learning and Other Application (IT-ELA), pp. 1–6 (2020). https://doi.org/10.1109/IT-ELA50150.2020.9253088

16. Putra, B.P., Irawan, B., Setianingsih, C., Rahmadani, A., Imanda, F., Fawwas, I.Z.: Hate speech detection using convolutional neural network algorithm based on image. In: 2021 International Seminar on Machine Learning, Optimization, and Data Science (ISMODE), pp. 207–212 (2022). https://doi.org/10.1109/ISMODE53584.2022.9742810

17. Amrutha, B.R., Bindu, K.R.: Detecting hate speech in tweets using different deep neural network architectures. In: International Conference on Intelligent Computing and Control Systems (ICCS), pp. 923–926 (2019). https://doi.org/10.1109/ICCS45141.2019.9065763

18. Zhou, K., Long, F.: Sentiment analysis of text based on CNN and bi-directional LSTM model. In: 2018 24th International Conference on Automation and Computing (ICAC), pp. 1–5 (2018). https://doi.org/10.23919/IConAC.2018.8749069

19. Santos, M.S., Soares, J.P., Abreu, P.H., Araujo, H., Santos, J.: Cross-validation for imbalanced datasets: avoiding overoptimistic and overfitting approaches [research frontier]. IEEE Comput. Intell. Mag. **13**(4), 59–76 (2018). https://doi.org/10.1109/MCI.2018.2866730

20. Mohammadi, S., Chapon, M.: Investigating the performance of fine-tuned text classification models based-on BERT. In: 2020 IEEE 22nd International Conference on High Performance Computing and Communications; IEEE 18th International Conference on Smart City; IEEE 6th International Conference on Data Science and Systems (HPCC/SmartCity/DSS), pp. 1252–1257 (2020). https://doi.org/10.1109/HPCC-SmartCity-DSS50907.2020.00162

21. Weijie, D., Yunyi, L., Jing, Z., Xuchen, S.: Long text classification based on BERT. In: 2021 IEEE 5th Information Technology, Networking, Electronic and Automation Control Conference (ITNEC), pp. 1147–1151 (2021). https://doi.org/10.1109/ITNEC52019.2021.9587007

22. Shao, Y., Taylor, S., Marshall, N., Morioka, C., Zeng-Treitler, Q.: Clinical text classification with word embedding features vs. bag-of-words features. In: 2018 IEEE International Conference on Big Data (Big Data), pp. 2874–2878 (2018). https://doi.org/10.1109/BigData.2018.8622345

23. Alessa, A., Faezipour, M., Alhassan, Z.: Text classification of flu-related tweets using FastText with sentiment and keyword features. In: IEEE International Conference on Healthcare Informatics (ICHI), pp. 366–367 (2018). https://doi.org/10.1109/ICHI.2018.00058

24. Shrivastava, P., Sharma, D.K.: Fake content identification using pre-trained glove-embedding. In: 2021 5th International Conference on Information Systems and Computer Networks (ISCON), pp. 1–6 (2021). https://doi.org/10.1109/ISCON52037.2021.9702379

25. Yue, W., Li, L.: Sentiment analysis using Word2vec-CNN-BiLSTM classification. In: 2020 Seventh International Conference on Social Networks Analysis, Management and Security (SNAMS), pp. 1–5 (2020). https://doi.org/10.1109/SNAMS52053.2020.9336549

26. Liu, C., et al.: Constrained oversampling: an oversampling approach to reduce noise generation in imbalanced datasets with class overlapping. IEEE Access **10**, 91452–91465 (2020). https://doi.org/10.1109/ACCESS.2020.3018911

27. Srinilta, C., Kanharattanachai, S.: Application of natural neighbor-based algorithm on oversampling SMOTE algorithms. In: 2021 7th International Conference on Engineering, Applied Sciences and Technology (ICEAST), pp. 217–220 (2021). https://doi.org/10.1109/ICEAST52143.2021.9426310

28. Cahyana, N., Khomsah, S., Aribowo, A.S.: Improving imbalanced dataset classification using oversampling and gradient boosting. In: 2019 5th International Conference on Science in Information Technology (ICSITech), pp. 217–222 (2019). https://doi.org/10.1109/ICSITech46713.2019.8987499
29. Veni, C.V.K., Rani, T.S.: Quartiles based undersampling (QUS): a simple and novel method to increase the classification rate of positives in imbalanced datasets. In: Ninth International Conference on Advances in Pattern Recognition (ICAPR), pp. 1–6 (2017). https://doi.org/10.1109/ICAPR.2017.8593202
30. Luqyana, W.A., Ahmadie, B.L., Supianto, A.A.: K-nearest neighbors undersampling as balancing data for cyber troll detection. In: International Conference on Sustainable Information Engineering and Technology (SIET), pp. 322–325 (2019). https://doi.org/10.1109/SIET48054.2019.8986079

Offline Arabic Handwritten Text Recognition for Unsegmented Words Using Convolutional Recurrent Neural Network

Mohamed Amine Chadli[(✉)] [iD], Rochdi Bachir Bouiadjra [iD],
and Abdelkader Fekir [iD]

University of Mustapha Stambouli, 29000 Mascara, Algeria
{mohamed.chadli,r.bachir-bouiadjra,aekfekir}@univ-mascara.dz

Abstract. This paper presents an analytical approach for offline Arabic Handwritten Text Recognition (HTR), based on Convolutional Recurrent Neural Network (CRNN). The suggested method is a three-part end-to-end trainable deep learning system that includes feature extraction, label prediction, and transcription part. The first part is performed by Convolutional Neural Network (CNN) layers, where sequential features are extracted. In the label prediction part, the extracted features are used to generate new sequential contextual features by feeding them to recurrent layers. This set of features for Arabic texts is then used to predict label distributions with fully connected layers. In the third part of the system, the transcription part, the predicted label distributions are translated into actual label sequences, using the Connectionist Temporal Classification (CTC) method. The experiments are carried out and reported on the publicly available IFN/ENIT database. The results of the proposed system are encouraging, and the recognition rates are comparable to those of numerous other systems in the literature.

Keywords: Off-line handwritten recognition · Arabic script ·
IFN/ENIT database · Convolutional neural network · Recurrent neural
network · Connectionist temporal classification

1 Introduction

The Arabic language is the fifth spoken language in the world with more than 300 million native speakers. With this huge number of Arabic speakers, the need for digitalization has never been that important, and the urge of developing recognition systems to recognize Arabic texts arises. The importance of handwriting text recognition is due to its wide variety of practical applications, e.g., postal address recognition, automated word retrieval from scanned forms, and cheque amount recognition. Arabic HTR is divided into two distinct types of recognition; online and off-line text recognition. The applications listed earlier are good examples of off-line Arabic HTR, where we recognize texts after the writing is completed (e.g., scanned document). Whereas in online recognition, we recognize the texts in real-time, at the time of writing, character after another using an electronic device (e.g., tablet, smartphone) and a digital pen. Iman Yousif et al. [20] did a review on Arabic HTR which provides more details about this topic. We can further divide this task of Arabic HTR into segmentation-based and segmentation-free approaches. In segmentation-free approaches, we deal with each word to be recognized as a class of its own. Therefore, the number of classes in this type of recognition heavily depends on the number of words to be recognized. This type of recognition is often referred to as the holistic approach. In segmentation-based approaches [1, 2, 5, 7, 8, 12, 16], the words to be recognized are segmented into characters, and each character is recognized separately. In this type of recognition, the number of classes depends on the number of characters of the language. This type of segmentation can be furthermore divided into explicit and implicit segmentation approaches. During training, explicit segmentation approaches require character-level annotations, while implicit segmentation approaches only require word-level annotations. Some examples of explicit segmentation approaches are the work done by Jafaar Al Abodi et al. [2], Mohammad Tanvir Parvez et al. [16] and Moftah Elzobi et al. [8]. Khaoula Jayech et al. [12], Mustapha Amrouch et al. [5], Ramy El-Hajj et al. [7] and El Moubtahij Hicham et al. [1] used implicit segmentation strategies in their recognition systems. The segmentation-based approaches are often referred to as the analytic approach, and it generally works better than the former approach when we have a large input vocabulary. Very little research has been done on the recognition of Arabic HTR compared to other languages such as English or Chinese. The research on Arabic HTR has almost stagnated with classical approaches including hidden Markov models [1, 3–5, 7, 12] and handcrafted features [1–4, 7, 8, 12, 15, 16] to recognize the Arabic text. Whereat the same time we are witnessing a revolution in computer vision with the rise and development of deep learning. As an important research area in computer vision, HTR has been tremendously affected by the advances in deep learning, yet it seems that we are not seeing this affection in Arabic HTR. The reason could be the scarcity of data in Arabic compared to (e.g., English) where we have an abundance of data. For, we all know that deep learning needs huge amounts of data to bring the desired results. In this paper, we investigate the IFN/ENIT database developed by Mario Pechwitz et al. [17]. In the era of deep learning handcrafted features had been replaced with automatic features extraction using CNNs. All deep learning systems in HTR use CNNs as the features extraction module. Today, the main difference between the HTR deep learning systems is in the decoding module to decode the extracted features into texts. Two major techniques are used in the decoding step: The encoder-decoder framework (Ilya Sutskever et al.) [19] and the CTC technique (Alex Graves et al.) [9]. The encoder-decoder framework is originally proposed for machine translation. The encoder based on Recurrent Neural Network (RNN) reads an input sequence and passes its final hidden state to the decoder RNN, to which it will generate a new sequence as output in an auto-regressive way. The main advantage of this system is its ability to deal with variable length sequences, and only require word-level annotations, which, perfectly suits our case of study (text recognition). Usually, this system is combined with the attention mechanism to learn a better alignment between the input and the output sequences. One of the first good systems developed for text recognition using the encoder-decoder architecture is the system developed by lee et al. [14]. This system is end-to-end trainable and uses a recursive CNN as a features extraction module and passes these sequences of features to an RNN encoder and decoder module combined with a soft attention mechanism to perform implicit learning of character-level language model. The CTC method is originally proposed for speech recognition [9] and is end-to-end trainable needing only word-level annotations. Therefore, it applies an implicit segmentation strategy at its core. One of the famous systems that use CTC is the CRNN system developed by Baoguang Shi et al. [18]. This model is created by stacking RNNs on top of CNNs and using CTC for training and inference. This system can predict words of variable length. The word to be predicted by this system does not necessarily need to be present in the training data. Therefore, this system can work with either constrained or unconstrained environments. The CTC method usually needs less data than the encoder-decoder framework. In this paper, we investigate a recognition system that employs an implicit segmentation strategy based on the CRNN model, and we are going to adapt it to the Arabic language using the IFN/ENIT database to assess and test the performance of this system on off-line Handwritten Arabic texts.

2   The Proposed System

Our variation of CRNN consists of convolutional layers, a deep neural network of recurrent layers, and a CTC transcription layer as shown in Fig. 1.

At the bottom of the network, we have a CNN block that works as a feature extraction module. This CNN block extract features sequentially from left to right. The features extracted then passed to the next part of the network, the recurrent layers. The recurrent layers will take in the features sequentially, one after another, and make a prediction of sequence characters. The CTC layer, then will take the sequence predictions of the recurrent layers and convert them into actual label predictions. This network can be jointly trained end-to-end with one loss function.

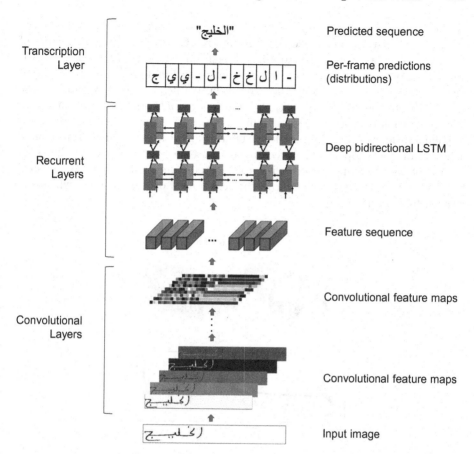

Fig. 1. The network architecture. The architecture consists of three parts: 1) convolutional layers, which extract a feature sequence from the input image; 2) recurrent layers, which produce contextual feature maps; 3) transcription layer, which translates the sequence of characters into a final label sequence.

2.1 Feature Extraction

In our CRNN model, the first part of the model (CNN block) expects a fixed input size of images. Therefore, the first thing we do is scaling the input images to the appropriate input size of the CNN block. In the conventional CNN architectures, they usually design their architectures as follow: convolutional layers followed by polling layer, followed by convolutional layers, repeated several times till finally they add fully connected layers to make actual predictions. In CRNN, we drop the fully connected layers from the CNN block, because, we are only interested in extracting feature maps using CNN, and not actually making predictions with it. Another difference between the conventional CNN architectures and our CNN architecture is that, our CNN architecture extracts a sequential feature maps vectors of the image where each vector represents a specific region in the original image, instead of extracting feature maps that represents the image in a holistic manner like the conventional CNN's. The feature vectors of our CNN block are generated from left to right and they are translation invariant. Therefore, each feature vector represents a rectangle region of the original image, and those regions are called the receptive fields, and they come in the same order of the vectors of features from left to right as illustrated in Fig. 2. The generation of feature vectors from left to right is due to the nature of the implementations of the convolutional operation in deep learning frameworks. The convolutional operation is reciprocal, consequently, it will yield the same vectors of feature maps even if it is implemented from right to left. As a result, this will not cause any worsening in the recognition of Arabic texts, which are written from right to left.

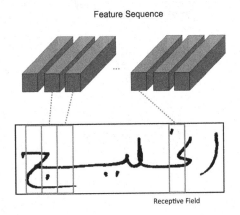

Fig. 2. The receptive field. Each receptive field represents a region in the original input image.

2.2 Label Prediction

We add Deep Bidirectional Recurrent Neural Network (DBRNN) layers on top of the CNN layers. The sequence of feature maps outputted by the CNN block is transformed to the appropriate input shape of the recurrent layers by a special layer in our architecture. Then, it will be fed to the recurrent layers as a sequence of feature vectors $X = x_1, ..., x_T$. Where each vector x_t from the sequence vectors is used to generate a new contextual feature vector y_t by the recurrent layers. Therefore, the generated sequence of contextual feature vectors is $Y = y_1, ..., y_T$. The recurrent layers are then followed by fully connected layers to make predictions of label distribution. These fully connected layers are required for the CTC part of the model. Labels in this case of study are the characters of the language we are trying to recognize; in our case, it is Arabic characters. Each contextual feature vector is used to predict one character. We used deep Bidirectional Long

Short-Term Memory (Bi-LSTM) recurrent layers to extract our contextual feature vectors. Some characters cover several neighbor regions in the original image as shown in Fig. 2. Therefore, non-contextual sequential features produced by the CNN layers are not the best option when we are trying to recognize characters that are spread out through several receptive fields. For that reason, we used RNN to extract contextual features, and we specifically used LSTM layers to capture long-term contextual information instead of the vanilla RNN, because it's known for its inefficiency in capturing long-term contextual information. We also used Bi-LSTM to capture past and future contextual information, following the nature of text recognition where contextual information from both directions is important. Furthermore, we stacked two Bi- LSTM layers on top of each other which results in a deep extraction of contextual information. An illustration of a deep Bi-LSTM network is shown in Fig. 3. We used the algorithm of backpropagation through time to optimize the weights of this part of the model.

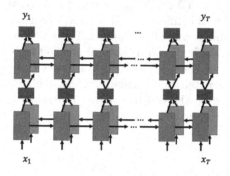

Fig. 3. The structure of deep bidirectional LSTM we use in our model. Using a forward (left to right) and a backward (right to left) LSTMs results in a bidirectional LSTM. Stacking several bidirectional LSTM on top of each other results in a deep bidirectional LSTM.

2.3   Transcription

Transcription is the operation of translating the sequence of characters predicted by the recurrent layers into a label sequence. We used the CTC method proposed by Graves et al. [9] to do the transcription. It's a powerful method to train and infer unsegmented sequence data with recurrent layers. There are two modes of transcription, lexicon-free mode, and lexicon-based mode. In lexicon-based mode, the transcription is made by taking the highest probable label sequence over a set of label sequences. In lexicon-free mode, the transcription is made without any lexicon. Lexicon-based transcriptions are computationally intensive. Therefore, we adopted the lexicon-free mode in our training and inference, because it's computationally more feasible, and we used BK-tree data structure [6] to search efficiently over the lexicon (Set of label sequences) after the transcription is made. We can also use the lexicon-free mode of transcription solely without the use of BK-tree data structure to make unconstrained general recognition or use BK-tree data structure over a lexicon for constrained task-specific recognition.

3   Experiments

We applied our method to the standard Arabic handwritten text database "IFN/ENIT database", which is a challenging database for Arabic text recognition. The dataset details are presented in Sect. 3.1. The training and implementation details are presented in Sect. 3.2. The experimental results are presented in Sect. 3.3.

3.1   Dataset Details

We experimented on IFN/ENIT database developed by Mario Pechwitz et al. [17]. It was written by 411 different writers from Tunisia. It consists of handwritten Tunisian Arabic town/village names. The first version of this database consisted of four sets; (a, b, c, and d) with about 26000 words containing more than 210000 characters. They further added 3 new sets; e, f, and s. Where set e and set f are also written by Tunisian writers, while set s is different. It's written by writers from UAE, which is more interesting and more challenging. Most researchers use the first four sets for training and the last three sets for validation and testing.

3.2   Training and Implementation Details

The network architecture of our model is described in detail in Table 1. The network is defined in the table from bottom to top. The first thing we do is to scale the input images to (64, 512), and apply normalization to those input images. For the first two layers, we used convolutional layers followed by the Rectified Linear Unit (relu) activation function. After those first layers, all convolutional layers are then followed by the BatchNormalization layer to speed up training. We followed a VGG like strategy where we used a 3 × 3 convolution kernels and we doubled the number of feature maps gradually from 64 feature maps to 512 feature maps. We also applied a 2 × 2 MaxPooling layers after the convolutional layers to reduce the size of the feature maps except for the last MaxPooling layer where we used a 2 × 1 MaxPooling window and a stride of 2 × 1. We applied this adjustment to produce wide feature maps so that we could cover all Arabic words when predicting. The final output of the CNN block is a sequence of 31 vectors of feature maps. 31 vectors are more than enough to predict all the words in the Arabic language. The network has a special layer: the "Map-To-Sequence" layer to convert the output of the CNN block to the appropriate shape of the input of the recurrent layers. We used two Bi-LSTM recurrent layers with a dropout of 0.2, and the last layer is a dense layer with a "softmax" activation function to predict the characters with L2 regularization.

Table 1. Network configuration summary. The first row is the top layer. 'k', 's', 'p' and 'd' stand for kernel size, strides, padding size and dropout respectively. "chars" is the number of characters in the language of the dataset.

Type	Configurations
Transcription	–
Dense	#units: chars + 1, activation: softmax, regularization: L2
Bi-LSTM	#hidden units: 128, d: 0.2
Bi-LSTM	#hidden units: 128, d: 0.2
Map-To-Sequence	–
Activation	relu
BatchNormalization	–
Convolution	#maps: 512, K: 2 × 2, S: 1, P: Valid
MaxPooling	Window: 2 × 1, S: 2 × 1
Activation	relu
BatchNormalization	–
Convolution	#maps: 512, K: 3 × 3, S: 1, P: Same
MaxPooling	Window: 2 × 2, S: 2
Activation	relu
BatchNormalization	–
Convolution	#maps: 512, K: 3 × 3, S: 1, P: Same
Activation	relu
BatchNormalization	–
Convolution	#maps: 512, K: 3 × 3, S: 1, P: Same
MaxPooling	Window: 2 × 2, S: 2
Activation	relu
BatchNormalization	–
Convolution	#maps: 256, K: 3 × 3, S: 1, P: Same
Activation	relu
BatchNormalization	–
Convolution	#maps: 256, K: 3 × 3, S: 1, P: Same
MaxPooling	Window: 2 × 2, S: 2
Activation	relu
BatchNormalization	–
Convolution	#maps: 128, K: 3 × 3, S: 1, P: Same
Activation	relu
BatchNormalization	–
Convolution	#maps: 128, K: 3 × 3, S: 1, P: Same
MaxPooling	Window: 2 × 2, S: 2
Activation	relu
Convolution	#maps: 64, K: 3 × 3, S: 1, P: Same
Activation	relu
Convolution	#maps: 64, K: 3 × 3, S: 1, P: Same
Input	64 × 512 black and white image

We implemented our network with the TensorFlow v2 framework, with custom implementations of the "Map-To-Sequence" layer, the transcription layer, and the BK-Tree data structure (in Python 3). We trained our model on pairs of training images and ground truth of label sequences with a batch size of 256 on Two types of GPUs: Tesla P100 of 16 GB of ram and Tesla T4 of 16 GB of ram. We activated the early stopping to stop training when the model reaches convergence and it stopped at epoch number 53. We used Stochastic Gradient Descent SGD algorithm for optimization with a momentum of 0.9 and we set the learning rate to 0.01. The training approximately took two hours with an average of 126 s for each epoch.

3.3   Experimental Results

We trained our model on the IFN/ENIT dataset on sets: A, B, C, and D, we validated on set E and we tested on set F. The results are shown in Table 2. Our model predicts a sequence of characters. For that, the calculation of the accuracy is tricky. Therefore, we calculate the accuracy by considering every label prediction that has no wrong prediction in any of its characters to be a true label prediction. We use the BK-Tree data structure to calculate the accuracy.

Table 2. Word recognition accuracy of training, validation and testing CRNN on IFN/ENIT database.

Database version: IFN/ENIT dataset v2.0p1e		
	Sets	Accuracy %
Training	A, B, C and D	99
Validation	E	80
Testing	F	79

In Fig. 4 we plot the training loss, and the validation loss of our model as a function of number of epochs. We can see that our model begins to converge at epoch 40.

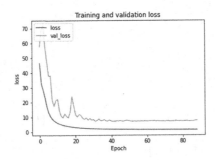

Fig. 4. Blue and orange graphs: training and validation losses as a function of number of epochs. (Color figure online)

In Table 3 we present the comparative results of previous systems. Several teams of researchers have conducted experiments on the IFN/ENIT database in many different training scenarios. Each team has used its own strategy of training, some of the teams trained on sets: a, b, c, and d using the cross-validation technique like [7,12], others trained on sets: a, b, c, d and tested on set e like [1,3,5,10,11,13,16], yet another team [15] trained on sets: a, b, c and tested on (d and e) and calculated the average accuracy of the two test sets. Our system has given promising results and has shown that deep learning systems can compete and gain the upper hand shortly just by focusing the research more on deep learning in the upcoming Arabic HTR research.

Table 3. Word recognition accuracy of several systems that have trained on Sets: a, b, c, d and tested on set E.

System	Training sets	Test set	Word Accuracy %
CNN-HMM (Mustapha Amrouch et al. 2018)	a, b, c, d	e	89.23
Mohammad Tanvir Parvez et al. (2013)	a, b, c, d	e	79.58
El Moubtahij Hicham et al. (2016)	a, b, c, d	e	78.95
Jawad H. Alkhateeb et al. (2011)	a, b, c, d	e	DBN (66.56) HMM (82.3)
Hamdani et al. (2009)	a, b, c, d	e	81.93
Kessentini et al. (2012)	a, b, c, d	e	79.6
DBN (Jayech et al. 2016)	a, b, c, d	e	78.5
CRNN our system	a, b, c, d	e	**80**

4   Conclusion

In this work, we have developed a new variation of CRNN, a deep learning text recognition system to recognize offline Arabic handwritten texts, and we have tested this system on IFN/ENIT database. We showed that this system gives promising results on IFN/ENIT compared with previous works. Therefore, it shows the significance of using deep learning systems instead of classical approaches and encourages more researchers to investigate deep learning models on offline Arabic HTR. Another reason to use CRNN is that it fully automates the training process (end-to-end training), and it opens doors for transfer learning and fine-tuning. Therefore, we eliminate the pain of hand-crafted features, and we can further use this trained model to other similar text recognition tasks with the use of transfer learning. For future work, we suggest using pre-trained Inception, ResNet, and MobileNet networks as CNN modules to extract robust features, and we also suggest integrating attention mechanisms into the RNN module to extract more intelligent contextual features.

Acknowledgments. We thank Dr. Volker Märgner for sharing the IFN/ENIT database with us.

References

1. Akram, H., Khalid, S., et al.: Using features of local densities, statistics and hmm toolkit (HTK) for offline Arabic handwriting text recognition. J. Electr. Syst. Inf. Technol. **4**(3), 387–396 (2017)

2. Al Abodi, J., Li, X.: An effective approach to offline Arabic handwriting recognition. Comput. Electr. Eng. **40**(6), 1883–1901 (2014)

3. AlKhateeb, J.H., Pauplin, O., Ren, J., Jiang, J.: Performance of hidden Markov model and dynamic Bayesian network classifiers on handwritten Arabic word recognition. Knowl.-Based Syst. **24**(5), 680–688 (2011)

4. AlKhateeb, J.H., Ren, J., Jiang, J., Al-Muhtaseb, H.: Offline handwritten Arabic cursive text recognition using hidden Markov models and re-ranking. Pattern Recogn. Lett. **32**(8), 1081–1088 (2011)

5. Amrouch, M., Rabi, M., Es-Saady, Y.: Convolutional feature learning and CNN based HMM for Arabic handwriting recognition. In: Mansouri, A., El Moataz, A., Nouboud, F., Mammass, D. (eds.) ICISP 2018. LNCS, vol. 10884, pp. 265–274. Springer, Cham (2018). https://doi.org/10.1007/978-3-319-94211-7_29

6. Burkhard, W.A., Keller, R.M.: Some approaches to best-match file searching. Commun. ACM **16**(4), 230–236 (1973)

7. El-Hajj, R., Likforman-Sulem, L., Mokbel, C.: Arabic handwriting recognition using baseline dependant features and hidden Markov modeling. In: Eighth International Conference on Document Analysis and Recognition (ICDAR 2005), pp. 893–897. IEEE (2005)

8. Elzobi, M., Al-Hamadi, A., Al Aghbari, Z., Dings, L., Saeed, A.: Gabor wavelet recognition approach for off-line handwritten Arabic using explicit segmentation. In: S. Choras, R. (eds) Image Processing and Communications Challenges 5. Advances in Intelligent Systems and Computing, vol. 233, pp. 245–254. Springer, Heidelberg (2014). https://doi.org/10.1007/978-3-319-01622-1_29

9. Graves, A., Fernández, S., Gomez, F., Schmidhuber, J.: Connectionist temporal classification: labelling unsegmented sequence data with recurrent neural networks. In: Proceedings of the 23rd International Conference on Machine Learning, pp. 369–376 (2006)

10. Hamdani, M., El Abed, H., Kherallah, M., Alimi, A.M.: Combining multiple HMMs using on-line and off-line features for off-line Arabic handwriting recognition. In: 2009 10th International Conference on Document Analysis and Recognition, pp. 201–205. IEEE (2009)

11. Jayech, K., Mahjoub, M.A., Amara, N.E.B.: Arabic handwritten word recognition based on dynamic Bayesian network. Int. Arab J. Inf. Technol. **13**(6B), 1024–1031 (2016). http://iajit.org/index.php?option=com_content&task=blogcategory&id=104&Itemid=387

12. Jayech, K., Mahjoub, M.A., Amara, N.E.B.: Synchronous multi-stream hidden Markov model for offline Arabic handwriting recognition without explicit segmentation. Neurocomputing **214**, 958–971 (2016)

13. Kessentini, Y., Paquet, T., Hamadou, A.B.: Off-line handwritten word recognition using multi-stream hidden Markov models. Pattern Recogn. Lett. **31**(1), 60–70 (2010)

14. Lee, C.Y., Osindero, S.: Recursive recurrent nets with attention modeling for OCR in the wild. In: Proceedings of the IEEE Conference on Computer Vision and Pattern Recognition, pp. 2231–2239 (2016)

15. Metwally, A.H., Khalil, M.I., Abbas, H.M.: Offline Arabic handwriting recognition using hidden Markov models and post-recognition lexicon matching. In: 2017 12th International Conference on Computer Engineering and Systems (ICCES), pp. 238–243. IEEE (2017)
16. Parvez, M.T., Mahmoud, S.A.: Arabic handwriting recognition using structural and syntactic pattern attributes. Pattern Recogn. 46(1), 141–154 (2013)
17. Pechwitz, M., Maddouri, S.S., Märgner, V., Ellouze, N., Amiri, H., et al.: IFN/ENIT-database of handwritten Arabic words. In: Proceedings of CIFED, vol. 2, pp. 127–136. Citeseer (2002)
18. Shi, B., Bai, X., Yao, C.: An end-to-end trainable neural network for image-based sequence recognition and its application to scene text recognition. IEEE Trans. Pattern Anal. Mach. Intell. 39(11), 2298–2304 (2016)
19. Sutskever, I., Vinyals, O., Le, Q.V.: Sequence to sequence learning with neural networks. In: NIPS (2014)
20. Yousif, I., Shaout, A.: Off-line handwriting arabic text recognition: a survey. Int. J. Adv. Res. Comput. Sci. Softw. Eng. 4(9) (2014)

DeBic: A Differential Evolution Biclustering Algorithm for Microarray Data Analysis

Younes Charfaoui[✉], Amina Houari, and Fatma Boufera

University Mustapha STAMBOULI of Mascara, Mascara, Algeria
{younes.charfaoui,amina.houari,fboufera}@univ-mascara.dz

Abstract. Biclustering is one of the interesting topics in bioinformatics and one of the crucial approaches to extracting meaningful information from data and performing high-dimensional analysis for gene expression data. However, since the colossal space complexity and the nature of the problem are proven to be NP-Hard, an approach to identifying valuable biclusters with a good quality measure is required in a reasonable amount of time. Moreover, metaheuristics and evolutionary computation algorithms have shown incredible success in this area. This paper offers a novel method of a Differential Evolution Based Biclustering algorithm to extract Biclusters called DeBic. The results of the experiments on the popular Yeast Cell-Cycle dataset indicate unique and interesting biclusters getting discovered with larger sizes.

Keywords: Differential evolution algorithms · Biclustering · Evolutionary algorithms · Microarray data · Knowledge discovery

1 Introduction

The analysis of genomic data and DNA microarray has gathered considerable interest from many researchers over the past few decades because this technology enables the study of several thousand genes under a variety of conditions (Temperature, Ph, tissue, etc.), which results in a large amount of data. Despite the data's high-dimensional size and massive search space occupation, we need to analyze it and use it to retrieve information and extract knowledge that will later serve essential tasks such as tumor classification [28], drug target identification [29], etc.

Data on gene expression is gathered and arranged in matrices, with rows representing genes and columns representing experimental conditions. The goal of analyzing these gene expression data is to find as many genes and conditions as possible that exhibit similar behavior under distinct conditions. This process is carried out using a specific method called **Biclustering** that consists of clustering rows and columns simultaneously. This method produces similar biclusters that represent a subset of genes which have a similar activity under a subset of conditions.

Evolutionary Algorithms (EA) are widely used to solve biclustering problems. The E.A.s can explore ample solution space by using the population and the appropriate

M. Salem et al. (Eds.): ICAITA 2022, CCIS 1769, pp. 288–302, 2023.
https://doi.org/10.1007/978-3-031-28540-0_23

operators. Furthermore, they have are able to control the search space and guide the search to amass a large number of solutions.

The Biclustering of gene expression data is a topic we address in this work using the **Differential Evolution** algorithm, which is known for excelling on complex problems and searching for reasonable solutions. Differential evolution improves a candidate solution based on an evolutionary process iteratively. These algorithms can quickly explore sizable solution spaces and nearly never make any assumptions about the fundamental optimization problem.

Differential evolution appears to be less effective than many other evolutionary methods at solving the biclustering problem. Specifically, Differential evolution is designed for continuous space solutions. For that, one of our contributions was also adapting custom operations suitable to the biclustering problem to use them with this kind of evolutionary algorithm.

This paper proposes a novel solution to the biclustering problem using the differential evolution algorithm. The proposed algorithm finds multiple biclusters at a time using a population-based approach. The structure of the paper is as follows: Sect. 2 provides an overview of existing biclustering algorithms; then Sect. 3 displays the main algorithm and its description; and Sect. 4 shows the experimental findings. Section 7 is devoted to conclusions.

2 Review on Biclustering

Biclustering is a data mining technique where the best biclusters of a given dataset are identified by sorting information into a matrix while simultaneously assigning the rows and columns of the matrix.

In genomic data, the DNA microarray dataset is represented by a matrix, where genes are represented by rows and conditions by columns, and cells represent the gene expression level of a gene under an experimental condition. The formal definition is as follows: a group of n gene indices $G = \{1, 2, ..., n\}$, a group of m conditions indices $C = \{1, 2, ..., m\}$ is, and the data matrix $M(I, J)$ related with I and J.

2.1 Biclustering (Definition)

A bicluster represents a group of co-expressed genes that behave similarly within a specific set of circumstances.

A bicluster consists of the pairing $(g; c)$, where g is a subgroup of G and c is a subgroup of C.

This study aims to detect and identify the maximum number of consistent biclusters where a subset of genes exhibits the same behavior under the same set of conditions.

2.2 Types of Biclusters

There are many different kinds of biclusters that have been described in the literature [4], some of which are listed below:

1. Bicluster with constant values

A biclustering algorithm will rearrange the matrix's rows and columns to group comparable genes and conditions, grouping biclusters with similar values ultimately in an attempt to search for a bicluster with a constant value. When the data is normalized, this technique is adequate.

2. Bicluster with constant values on rows or columns

The first step is to normalize the rows and columns to finish the identification of these types of biclusters. However, using different approaches, other algorithms can find biclusters with rows and columns without the normalization step.

3. Bicluster with coherent values

Biclusters with constant values on rows or columns should be given consideration as a general enhancement over the algorithms for biclusters with coherent values on rows and columns. That implies that a complex algorithm is required. This technique might use covariance between rows and columns to analyze variance between groups.

2.3 Biclustering as an Optimization Problem

Biclustering is the process of finding the largest possible biclusters that nevertheless comply to coherence requirements. It is possible to consider of the task of bicluster extraction from a data matrix as a combinatorial optimization issue since the biclustering technique's core is efficiency; it allows the computer to sift through and sort an enormous amount of data in a shorter amount of time compared to single clustering methods.

2.4 Metrics

Several evaluation metrics for biclusters have been proposed in the literature [5]. We will present a few notable metrics for assessing biclusters.

1. Variance (VAR)

The biclusters coherence is evaluated using variance measure, where it is intended to reduce the total bicluster variances.

$$Var(\mathrm{B}) = \sum_{i=1}^{|I|}\sum_{j=1}^{|I|}\left(b_{ij} - b_{IJ}\right)^2 \tag{1}$$

2. Mean Squared Residue (MSR)

The MSR is used to measure the coherence of the bicluster's conditions and genes containing J columns and I rows. It is the most common coherence measure used in

Biclustering algorithms [5]. MSR is defined as:

$$MSR(B) = \frac{1}{|I| \cdot |J|} \sum_{i=1}^{|I|} \sum_{j=1}^{|I|} (b_{ij} - b_{iJ} - b_{Ij} + b_{IJ})^2 \qquad (2)$$

3. Relevance Index (R.I.)

Compared to MSR, this metric is subtly different; the bicluster quality is calculated as the total of the columns' relevance indices. The definition of relevance index R_{Ij} for column $j \in J$ is defined as:

$$R_{Ij} = 1 - \frac{\sigma_{Ij}^2}{\sigma j^2} \qquad (3)$$

3 Related Work

An extensive search for the solution space is not practical due to the NP-Hard nature of the biclustering issue [3]. Instead of exhaustive search, heuristics are used by many biclustering techniques to identify biclusters. Heuristic search methods are therefore generally used to approximate the problem by discovering sub-optimal solutions. Generally, we can classify all biclustering methods into two categories: Systematic search algorithms (Heuristic) and stochastic search algorithms (Metaheuristic).

3.1 Systematic Biclustering Algorithms

Divide and Conquer Approach: The entire data matrix is used as the initial bicluster in this approach. Then, iteratively divide this bicluster into numerous biclusters that meet certain features until the termination criteria are verified, as shown in [8]. This method is fast, however it may exclude good biclusters by dividing them before evaluating them.

Greedy Iterative Search Approach: This method creates biclusters through an iterative process that aims to maximize or minimize certain functions. This procedure is based on adding or deleting rows or columns until no more additions or deletions are possible. This method is also distinguished by its speed. On the other hand, it can disregard good biclusters. OPSM [9] and BicFinder [10] are two algorithms that take this approach.

Bicluster Enumeration Approach: The data matrix is represented as a graph or search tree in this method. This method allows for the acquisition of the best solution. However, it necessitates a lengthy period of calculation and a significant amount of memory consumption. BiMine [11], BiMine+ [12], and DeBi [13] are three algorithms that use this approach.

3.2 Stochastic Biclustering Algorithms

Neighborhood Search: This method begins with a simple solution, which could either be a bicluster, or the entire data matrix. Then, in each iteration, it adds or deletes rows and columns to maximize or minimize specific functions in order to improve the current solution. This approach, unlike the Greedy Iterative Search Approach, permits the addition of previously deleted rows or columns. CC Algorithm [3, 14], PDNS [14], and LSM [15] are examples of algorithms that use this approach.

Evolutionary Algorithm (E.A.): The population is initialized as the first step in this approach, with each individual representing a potential solution for this population. Following that, it begins evaluating each solution based on an assessment measure in order to select the individuals who will be used to produce the next population using the crossover and mutation operators; this process is repeated until the stopping criterion are met. SEBI [16], SMOB [17], CBEB [18], and those proposed by Nepomuceno et al. in [19] and [20] are some of the algorithms that use this approach.

Hybrid Approaches: The evolutionary algorithm and the neighborhood search are combined in hybrid approaches. Both approaches are known for permitting the exploration of a large search space while compromising quality of the solution and runtime. However, the search could become entangled in a suboptimal solution. BiHEA [22] and SSB [21] are among the algorithms using this approach.

4 Review of Differential Evolution

In this section we will describe the DE algorithm:

4.1 Differential Evolution

Differential Evolution (D.E.) is a population-based, stochastic optimization algorithm for solving nonlinear and numerical optimization tasks using REAL number parameters. The D.E. algorithm was first introduced by Storn and Price in 1996 [23].

Unlike standard genetic algorithms, differential Evolution relies upon distance and directional information through unit vectors for reproduction. Its reproduction operator involves a mutation phase to produce a trial vector, It is subsequently used by the crossover operator to create one offspring. Weighted differences between individuals chosen at random are used to calculate mutation step sizes.

4.2 Algorithm of Differential Evolution

A potential solution among the population $x \in \mathbb{R}^D$, D represents the dimensionality of the optimized problem and the objective function that needs to be minimized is $f : \mathbb{R}^D \to \mathbb{R}$. The standard D.E. algorithm that follows the ***DE/rand/1/bin*** scheme is explained as follows [1]:

Data:

NP: population size, *F*: mutation factor, *CR*: crossover probability, *MAXFES*: maximum number of functions evaluations

INITIALIZATION $G = 0$; Initialize all *NP* individuals with random positions in the search space;

while *FES* < *MAXFES* **do**

> **for** $i \leftarrow 1$ **to** *NP* **do**

>> **GENERATE** three individuals x_{r1}, x_{r1}, x_{r1} from the current population randomly. These must be distinct from each other and also from individual x_i, i.e. $r_1 \neq r_2 \neq r_3 \neq i$

>> **MUTATION** Form the donor vector using the formula: $\mathbf{v}_i = \mathbf{x}_{r_1} + F(\mathbf{x}_{r_2} - \mathbf{x}_{r_3})$

>> **CROSSOVER** The trial vector $\mathbf{u}_i$ is developed either from the elements of the target vector $\mathbf{x}_i$ or the elements of the donor vector $\mathbf{v}_i$ as follows:

$$u_{i,j} = \begin{cases} v_{i,j}, & \text{if } r_{i,j} \leq CR \text{ or } j = j_{rand} \\ x_{i,j}, & \text{otherwise} \end{cases}$$

>> where $i = \{1, \ldots, NP\}$, $j = \{1, \ldots, D\}$, $r_{i,j} \sim \cup(0,1)$ is a uniformly distributed random number which is generated for each j and $j_{rand} \in \{1, \ldots, D\}$ is a random integer used to ensure that $\mathbf{u}_i \neq \mathbf{x}_i$ in all cases

>> **EVALUATE** If $f(\mathbf{u}_i) \leq f(\mathbf{x}_i)$ then replace the individual $\mathbf{x}_i$ in the population with the trial vector $\mathbf{u}_i$

>> *FES* = *FES* + *NP*

> **end**

> $G = G + 1$;

end

Algorithm 1 Differential Evolution with DE/rand/1/bin scheme

The complexity of this algorithm is shared in [32], where they measured the run time and averaging with a varied dimension size and applied it to different benchmarks. The results lead to an asymptotic bound computational complexity.

4.3 The Benefits of the D.E. Algorithm

The power of differential evolution is using directional information within the population to create offspring. The algorithm has been demonstrated to outperform the Genetic Algorithm (G.A.) in [25] and [30] alternatively, Particle Swarm Optimization (PSO) in [26] through numerical benchmarks and experiments because of its simplicity, efficiency, and local searching property, and speediness.

4.4 Mutation Operator

The mutation operator in Differential Evolution uses two components to calculate a difference and a target vector. The difference vector represents the differences between two or more parents, with the target vector representing the parent whose direction is prioritized when generating the unit vector, which is the result of the mutation process, it is then passed along to be crossed over with the current central parent, the parent of interest that was not included in the mutation procedure.

Here is an exact depiction below:

The following steps are taken to create a mutant vector for each target vector $x_i (i = 1, 2, \ldots, m)$:

$$h_i(i + 1) = x_{r1} + f(x_{r2} - x_{r3}) \tag{4}$$

By looking at the mutation operation formula above, we can easily conclude that it is capable of maintaining the closure only in real numbers field. So despite the simplicity and successfulness of the differential evolution algorithm in many branches of engineering, applying differential evolution to binary optimization issues with binary decision variables is still uncommon; and to fix this issue, We will employ a new mutation operator known as the semi-probability mutation operator, that incorporates the original mutation operator and a fresh probability-based defined operator [24].

5   DeBic Algorithm Description

We suggest a Differential Evolution Based Biclustering algorithm in this paper called *DeBic*. This new approach allows us to detect multiple biclusters simultaneously by using the evolutionary algorithm of differential evolution and adapting it to perform Biclustering on a binary search space with a different mutation operator.

Differential evolution is best known for its exceptional robustness in non-convex, multimodal, and nonlinear problem optimizations. It is also known for its quick convergence and straightforward programming.

The D.E. pseudo-code is provided as follows:

> *Generation of the Initial Population*
> **Repeat**
> | Mutation
> | Crossover
> | Selection
> **Until** (termination conditions aremet)

5.1   Algorithm Description

The proposed solution uses differential evolution to drive a population of biclusters solutions toward better solutions. First, it initializes the population by randomly selecting biclusters, like the most of current biclustering algorithms, the biclusters are represented by a binary sequence of predetermined length $(n + m)$. The binary sequence is usually made up of two parts: the first part represents the genes, and the second part represents the conditions.

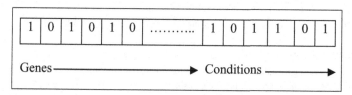

Fig. 1. Representation of binary string encoding for bicluster solution

After initializing the population, the algorithm employs the mutation operator to modify an existing individual; these modifications can be made at random or in accordance with a predetermined strategy.

As Discussed in Sect. 4.4, we adopted a new mutation operator that allows us to expand D.E. into the binary space search, proposed by Changshou et al. in [24], called the semi-probability Mutation operator, defined as the following:

$$h_i(t+1) = \begin{cases} x_{r1} + F(x_{r2} - x_{r3}), \, if \, h_i(t+1) = 0 \, or \, 1 \\ 0, pr < rand \\ 1, otherwise \end{cases} \quad (5)$$

In the next step, the DeBic algorithm uses the crossover strategy to create a new vector by mixing the information of the trial and the target vectors, and it is defined as the following:

$$u_i(t+1) = \begin{cases} h_i(t+1), rand \le CR \, or \, j = rand(i) \\ x_i(t), otherwise \end{cases} \quad (6)$$

where $i = 1, 2, \ldots, m, j = 1, 2, \ldots, n, \, rand(i) \in (1, 2, \ldots, n)$ is the randomly chosen index and $CR \in [0, 1]$ is crossover constant. In other words, certain elements of the mutant individual or at least one parameter chosen at random, as well as some of the target individual's other attributes, are present in the trial individual.

The effect of the mutation with the semi-probability operator alongside the inherited behavior of evolutionary algorithms such as crossover will help maintain diversity in the population since the different evolution will use those new individuals to create other improved individuals, which will get selected or discarded in the selection phase.

Then, we compare the trial individual to the corresponding to decide if it should be a member of the next generation. The selection process relies on the trial participant's fitness survival; if it is considerably better than the current solution, we replace it. After a specific number of iterations, the algorithm reaches stopping criteria and ends the process by returning the resulting population.

The fitness of an individual $X(I, J)$ used in this study is then given by the formula:

$$fitness(X) = \frac{MSR(X)}{\delta} + \frac{1}{size(X)} \quad (7)$$

δ is the threshold that represents the most dissimilarity that can exist within a bicluster., and the *size* function represents the size of the bicluster.

The different steps of DeBic are described in the following:

Algorithm 1: DeBic Algorithm Steps

Input : A Matrix of size $n \times m$, Population Size p, Crossover probability cp, Fitness function $fscore$

Output: p Bicluster solutions

1 $pop \leftarrow$ generating p random solutions for initial population solutions;
2 $popFitness \leftarrow$ calculate the fitness score using $fscore$ for all solutions in pop;

3 **while** *Termination Conditions Isn't met* **do**
4   **for** $i \leftarrow 0$ **to** p **do**
5        $a, b, c \leftarrow$ select 3 random different solutions from pop in different position than i;
6   /* Create mutant using semi probability mutation   */
7        $mutant \leftarrow semiProbabilityMutation(a, b, c)$
8        $trial \leftarrow$ Create Trial Vector using binomial crossover using the $mutant$ and the $current$ vectors with cp as crossover probability
9   /* Calculate the score of the new trial vector   */
10       $currentFitness \leftarrow fscore(trial)$;
11   **if** $currentFitness > popFitness[i]$ **then**
12          $popFitness[i] \leftarrow currentFitness$;
13          $pop[i] \leftarrow trial$;
14   **end**
15   **end**
16 **end**
17 return p Biclustera pop from last generation

18 **Function** *semiProbabilityMutation(a: vector, b: vector, c: vector) : vector* **is**
19

$$vector = \begin{cases} r = a + F(b - c), \text{if } r = 0 \text{ or } r = 1 \\ 0, pr < rand \\ 1, \text{otherwise} \end{cases} \qquad (1)$$

20   return $vector$;
21 **end**

Algorithm 2 Steps of DeBic method in pseudo-code format

6 Experimental Results

In order to assess the effectiveness of the DeBic Algorithm for bicluster detection in expression data, we conducted some experiments on the *Yeast Saccharomyces Cerevisiae Cell-Cycle* expression dataset. The yeast cell cycle is a particularly popular dataset for gene expression data experiments, as the genes functions are well-known. This dataset originated from [1], there are 17 experimental conditions and 2,884 genes in the associated expression matrix.

The parameter settings shown in Table 1 were used for all the conducted experiments, and all experiments were carried out on a PC equipped with a 4.30 GHz AMD Ryzen 7 3800XT 8-Core Processor and 64 GB of RAM running with Windows 10.

Table 1. Hyper parameter specification

Hyper parameter	Value
Population size	75
Generations	100
Crossover probability	0.75

14 out of 100 biclusters on the Yeast data set were discovered by the DeBic algorithm, as shown in Fig. 2:

The genes exhibit a similar response under a variety of circumstances, as can be shown by a visual examination of the additional biclusters suggested in Fig. 2.

The result of a selected set of genes from a bicluster with high biological relevance are shown below:

(YBR244W YCR102C YDL022W YEL024W YHR037W YHR104W YIL111W YIL124W YJR155W YKL150W YKR009C YLR395C YMR145C YMR152W YMR189W YNL037C YNL134C YOL059W YOR136W YDR185C YDR213W YDR240C YDR294C YDR299W YDR330W YDR466W YDR479C YEL005C YEL024W YEL046C YEL070W YER028C YER054C YER062C YER065C YER068W YER141W YER143W YGL013C YGL130W YGL156W YGL208W YGL219C YGR029W YGR100W YGR146C YHL039W YHL042W YHR037W YHR129C YIL055C YIL111W YIL124W YIL156W YJL003W YJL091C YJL101C YJL191W YJR078W YJR096W YJR155W YKL100C YKL150W YKR004C YKR013W YLL010C YNL336W YNR013C YNR038W YNR077C YOL017W YOL032W YOL059W YOL064C YOL079W YOL096C YOL108C YOL114C YOL154W YOL158C YOR094W YOR136W YOR177C YOR178C YOR181W YOR252W YOR283W YPL007C YPL057C YPL086C YPL094C YPL132W YPL165C YPL186C YPL204W YPL205C YPL208W YPL228W YPL236C YPL258C YPL268W YPL274W YPR017C YPR043W YPR068C YPR113W YPR168W YPR182W).

We also recorded the biological process from the previous bicluster with the corresponding P-values (which is a number describing how likely our data would have occurred by random chance) in Table 2.

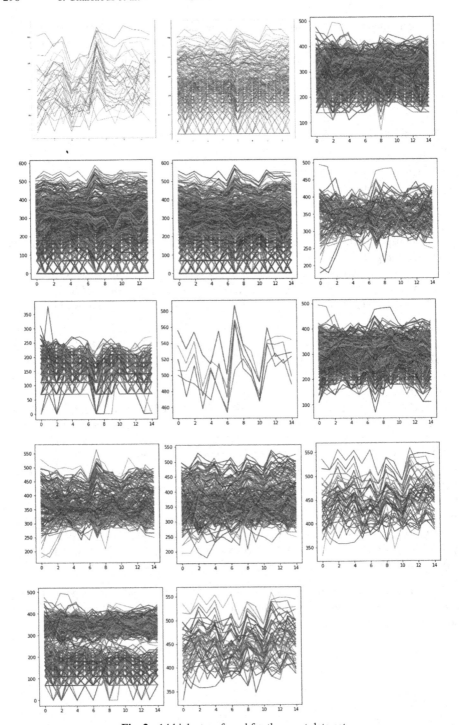

Fig. 2. 14 biclusters found for the yeast data set

Table 2. The biological process of biclusters

Biological process	P-value
Polyol metabolic process	5.07E-05
Oxidoreductase complex	1.63E-05
Oxidoreductase activity, acting on the CH-OH group of donors, NAD or NADP as acceptor	2.15E-07
Oxidoreductase activity, acting on the CH-OH group of donors	5.01E-07
Oxidoreductase activity	3.73E-10
Oxidation–reduction process	3.33E-10
Catalytic activity	1.32E-06

Table 3 compares DeBic's performance compared to Cheng and Church's method (hereafter CC) [3], Yang et al.'s algorithm FLOC [31], and MOEAB [27] for the average residue and average dimension of the discovered biclus-ters.

Table 3. Performance comparison

Algorithm	Avg bicluster size	Avg. residue	Average no. of genes	Average no. of condition
CC	1576.98	204.29	167	11
FLOC	1825.78	**187.54**	195	12.8
MOEAB	**10,301**	234.87	**1095**	9.29
DeBic	9889.58	272.13	628.04	**14.5**

The outcomes of this actual dataset demonstrate the ability of our suggested approach to find biclusters with significant biological relevance.

7 Conclusion

In this study, we proposed a novel approach for identifying biclusters from gene expression datasets using a new approach based on differential evolution. Using this technique, we can locate many biclusters simultaneously by specifying the algorithm's beginning population count. We discussed the problem's complexity, and we shifted the differential evolution continuous search aspect to fit the bi-clustering problem on binary search space with the right mutation operator.

The experimental results highlight the applicability of a differential evolutionary algorithm in binary search space as global optimization and allow us to capture interesting large biclusters with low residuals in a reasonable amount of time.

This work constitutes a first step on the path of study of differential evolution algorithms on bi-clustering. Thus, we will drive the subject to further studies in future work, particularly the evaluation or the metric function to instruct the algorithm toward even high-quality biclusters. It will be also interesting to see how biological evidence supports the results.

To further confirm the DeBic algorithm's effectiveness, we will test it on other datasets, perform a comparison study between the suggested technique and other biclustering algorithms, and employ additional evaluation metrics like MSR and Var in our future work.

References

1. Georgioudakis, M., Plevris, V.: A Comparative study of differential evolution variants in constrained structural optimization. Front. Built Environ. **6**, 102 (2020). https://doi.org/10. 3389/fbuil.2020.00102
2. Jose-Garcia, A., Jacques, J., Sobanski, V., Dhaenens, C.: Biclustering algorithms based on metaheuristics: a review. ArXiv:2203.16241 [Cs]. http://arxiv.org/abs/2203.16241 (2022)
3. Cheng, Y., Church, G.M.: Biclustering of expression data. In: International Conference on Intelligent Systems for Molecular Biology, vol. 8, pp. 93–103 (2000)
4. Madeira, S.C., Oliveira, A.L.: Biclustering algorithms for biological data analysis: a survey. IEEE/ACM Trans. Comput. Biology Bioinf. **1**(1), 24–45 (2004). https://doi.org/10.1109/ TCBB.2004.2
5. Noronha, M.D.M., Henriques, R., Madeira, S.C., Zárate, L.E.: Impact of metrics on biclustering solution and quality: a review. Pattern Recogn. **127**, 108612 (2022). https://doi.org/10. 1016/j.patcog.2022.108612
6. Mandal, K., Sarmah, R., Bhattacharyya, D.K.: POPBic: Pathway-Based Order Preserving Biclustering Algorithm Towards the Analysis of Gene Expression Data. In: IEEE/ACM Transactions on Computational Biology and Bioinformatics, vol. 18, no. 6, pp. 2659–2670, 1 Nov.–Dec. 2021. https://doi.org/10.1109/TCBB.2020.2980816
7. Maâtouk, O., Ayadi, W., Bouziri, H., Duval, B.: Evolutionary local search algorithm for the biclustering of gene expression data based on biological knowledge. Appl. Soft Comput. **104**, 107177 (2021). https://doi.org/10.1016/j.asoc.2021.107177
8. Prelic, A., et al.: A systematic comparison and evaluation of biclustering methods for gene expression data. Bioinformatics **22**(9), 1122–1129 (2006). https://doi.org/10.1093/bioinform atics/btl060
9. Ben-Dor, A., Chor, B., Karp, R., Yakhini, Z.: Discovering local structure in gene expression data: the order-preserving submatrix problem. J. Comput. Biol. **10**(3–4), 373–384 (Jun2003). https://doi.org/10.1089/10665270360688075
10. Ayadi, W., Elloumi, M., Hao, J.-K.: BicFinder: a biclustering algorithm for microarray data analysis. Knowl. Inf. Syst. **30**(2), 341–358 (2012). https://doi.org/10.1007/s10115-011-0383-7
11. Ayadi, W., Elloumi, M., Hao, J.K.: A biclustering algorithm based on a bicluster enumeration tree: application to DNA microarray data. BioData Min. **16**(2), 9 (2009). https://doi.org/10. 1186/1756-0381-2-9
12. Ayadi, W., Elloumi, M., Hao, J.K.: BiMine+: an efficient algorithm for discovering relevant biclusters of DNA microarray data. Knowl. Based Syst. **35**, 224–234 (2012). ISSN 0950 705. https://doi.org/10.1016/j.knosys.2012.04.017
13. Serin, A., Vingron, M.: DeBi: discovering differentially expressed biclusters using a frequent itemset approach. Algorithms Mol. Biol. **6**(1), 18 (2011). https://doi.org/10.1186/1748-7188-6-18

14. Ayadi, W., Elloumi, M., Hao, J. K.: Pattern-driven neighborhood search for Biclustering of microarray data. BMC Bioinform. **13**(7), 1–11 (2012). BioMed Central
15. Maâtouk, O., Ayadi, W., Bouziri, H., Duval, B.: Local search method based on biological knowledge for the Biclustering of gene expression data. Adv. Smart Syst. Res. **6**(2), 65 (2012)
16. Divina, F., Aguilar-Ruiz, J.S.: Biclustering of expression data with evolutionary computation. IEEE Trans. Knowl. Data Eng. **18**(5), 590602 (2006). https://doi.org/10.1109/TKDE.2006.74
17. Divina, F., Aguilar-Ruiz, J.S.: A multi-objective approach to discover biclusters in microarray data. In: Genetic and Evolutionary Computation Conference – GECCO '07, p. 385. ACM Press (2007). https://doi.org/10.1145/1276958.1277038
18. Huang, Q., Tao, D., Li, X., Liew, A.: Parallelized evolutionary learning for detection of biclusters in gene expression data. IEEE/ACM Trans. Comput. Biol. Bioinform. **9**(2), 560–570 (2012). https://doi.org/10.1109/TCBB.2011.53
19. Nepomuceno, J.A., Troncoso, A., Nepomuceno-Chamorro, I.A., Aguilar-Ruiz, J.S.: Integrating biological knowledge based on functional annotations for Biclustering of gene expression data. Comput. Meth. Prog. Biomed. **119**(3), 163–180 (2015). https://doi.org/10.1016/j.cmpb. 2015.02.010
20. Nepomuceno, J.A., Troncoso, A., Nepomuceno-Chamorro, I.A., Aguilar-Ruiz, J.S.: Pairwise gene GO-based measures for Biclustering of high-dimensional expression data. BioData Mining **11**(1), 4 (2018). https://doi.org/10.1186/s13040-018-0165-9
21. Nepomuceno, J.A., Troncoso, A., Aguilar-Ruiz, J.S.: Biclustering of gene expression data by correlation-based scatter search. BioData Mining **4**(1), 3 (2011). https://doi.org/10.1186/ 1756-0381-4-3
22. Gallo, C.A., Carballido, J.A., Ponzoni, I.: BiHEA: a hybrid evolutionary approach for microarray biclustering. In: Brazilian Symposium on Bioinformatics, pp. 36–47 (2009). https://doi. org/10.1007/978-3-642-03223-3_4
23. Storn, R., Price, K.: Differential evolution – a simple and efficient heuristic for global optimization over continuous spaces. J. Global Optim. **11**, 341–359 (1997). https://doi.org/10. 1023/A:1008202821328
24. Deng, C., Zhao, B., Yang, Y., Zhang, H.: Binary encoding differential evolution for combinatorial optimization problems. Int. J. Educ. Manage. Eng. **1**(3), 59–66 (2011). https://doi.org/ 10.5815/ijeme.2011.03.09
25. Hegerty, B., Hung, C.C., Kasprak, K.: A comparative study on differential evolution and genetic algorithms for some combinatorial problems. In: Proceedings of 8th Mexican international conference on artificial intelligence, vol. 9, p. 13 (2009)
26. Iwan, M., Akmeliawati, R., Faisal, T., Al-Assadi, T.M.A.A.: Performance comparison of differential evolution and particle swarm optimization in constrained optimization. Proc. Eng. **41**, 1323–1328 (2012). ISSN 1877-7058. https://doi.org/10.1016/j.proeng.2012.07.317
27. Mitra, S., Banka, H.: Multi-objective evolutionary biclustering of gene expression data. Pattern Recognit. **39**(12), 2464–2477 (2006). https://doi.org/10.1016/j.patcog.2006.03.003
28. Sahu, U., John, A., Alphonso, A., Kamath, A., Tripathy, A.: Cancer detection using biclustering. In: International Conference on Computer Communication and Informatics, 2013, pp. 1–5. https://doi.org/10.1109/ICCCI.2013.6466145
29. Xie, J., Ma, A., Fennell, A., Ma, Q., Zhao, J.: It is time to apply biclustering: a comprehensive review of biclustering applications in biological and biomedical data. Brief. Bioinform. **20**(4), 1450–1465 (2019). https://doi.org/10.1093/bib/bby014
30. Charalampakis, A.E., Tsiatas, G.C.: Critical evaluation of metaheuristic algorithms for weight minimization of truss structures. Front. Built Environ. (2019). https://doi.org/10.3389/fbuil. 2019.00113

31. Yang, J., Wang, H., Wang, W., Yu, P.: Enhanced biclustering on expression data. In: Third IEEE Symposium on Bioinformatics and Bioengineering, Proceedings 2003, pp. 321–327. https://doi.org/10.1109/BIBE.2003.1188969
32. Tasgetiren, M., Liang, Y.-C., Gencyilmaz, G., Eker, I.: A differential evolution algorithm for continuous function optimization (2022)

Author Index

Printed in the United States
by Baker & Taylor Publisher Services